GEOMORPHOLOGY TEXTS

General Editor: K. M. CLAYTON, University of East Anglia

3

SLOPES

Most of the land surface of the earth is formed by valley slopes, so it is not surprising that the study of slopes and surface processes has become one of the foremost branches of geomorphology. Apart from brief accounts in textbooks of general geomorphology, there has been no systematic treatment of this important subject, and Dr Young has brought together the considerable amount of literature which is scattered throughout numerous sources, making this volume one of the first adequate accounts of slopes beyond single chapters in lower level textbooks.

In addition to summarizing existing knowledge of the subject, the author sets out the major problems that still exist, and the research methods by which they can be approached. Particular attention is given to questions of scientific method in the study of the natural environment, the types of evidence available, and the relations between the processes that bring about change, the existing form of the land and the past evolution of form.

The text reviews the historical development of the geomorphology of slopes, the main theories of slope evolution that have been proposed, and surface processes, including soil creep, solifluction, surface wash and landslides. Theoretical approaches, including the construction of process-response models, are critically examined, and each of the main aspects of slope form is discussed. The final chapter outlines some practical applications of slope studies in engineering, agriculture and soil science. Emphasis is placed on the role of the cover of weathered material, regolith or soil, as a dynamic agent in slope evolution: it is particularly in this respect that the text is relevant to soil science and other branches of the environmental sciences.

Dr Young is a Reader in Environmental Sciences at the University of East Anglia.

OTHER TITLES IN THIS SERIES

1 *Geomorphology* F. MACHATSCHEK 1969
2 *Weathering* CLIFF OLLIER 1969

ANTHONY YOUNG
Reader in Environmental Sciences
University of East Anglia

SLOPES

Edited by K. M. Clayton

OLIVER & BOYD · EDINBURGH

Dedicated to the memory of
W. VAUGHAN LEWIS

OLIVER AND BOYD
Tweeddale Court
14 High Street
Edinburgh EH1 1YL
A Division of Longman Group Limited

ISBN 0 05 002448 5

First published 1972

Printed in Great Britain by T. and A. Constable Ltd
Hopetoun Street, Edinburgh

PREFACE

When I began to study slopes, nearly 20 years ago, I resolved to read everything that had been published on the subject. This was then a manageable task, involving no more than 200 papers. To someone starting at the present day, when the total number of publications is approaching 1000, such a task would be well-nigh impossible.

I have therefore tried to summarize the existing material on the geomorphology of slopes and surface processes: to indicate the main branches of the subject, the nature of the problems that arise, and the present state of knowledge about them.

Some explanation of the frequency with which statements are supported by the citation of references is called for. It would be unrealistic to write a definitive account of slopes, along the lines of, say, a textbook of chemistry. The reason for this is partly the imperfect state of present knowledge. It arises also from the nature of studies of the physical environment, and may be illustrated by an example. In 1960 Fourneau published the results of a study of slope form in relation to lithology, based on surveys in part of Belgium; he found, *inter alia*, that on sandstones the convexity occupied the greater part of the slope profile. Now it would certainly not be justifiable to infer from this the general proposition that on all sandstones under humid temperature climates the convexity is dominant. The position is similar with respect to field studies of surface processes, for example the relative importance of soil creep and surface wash. Until numerous convergent studies indicate that a general relation has been established, all that can safely be said is that in a given region and environment a particular result was obtained. A citation of the reference is intended as a shorthand means of stating that this result is the product not only of the local environmental conditions but also of the techniques and sampling methods employed, for details of which reference must be made to the original.

I confess to a weakness for citing early works. Some of them are deficient, by modern standards, in scientific and statistical rigour, and might be ignored if the results obtained were the only consideration; but in the best there is a freedom of observation and hypothesis (or speculation) that has often nowadays been lost. It is a refreshing and rewarding exercise to turn from modern studies of hillslopes and pediments and to read Bryan's monograph of 1925 on the Papago Country, Arizona.

There are two aspects of the geomorphology of slopes that have become large and specialized subjects, and in which I am aware of my incomplete knowledge of the literature and imperfections of understanding. These are soil mechanics in relation to landslides, and periglacial geomorphology, including the study of solifluction. That other books have been written on both subjects is some justification for the brevity of treatment here.

Two important collections of papers on slopes appeared whilst this book was in the press. These are 'New contributions to slope evolution', the Sixth, and final, Report of the Commission on Slope Evolution of the International Geographical Union

(*Z. Geomorph. Suppl.*, **9**, 1970), and the British Geomorphological Research Group compilation 'Slopes: form and process' (*Inst. Br. Geogr. Spec. Pubn*, **3**, 1971). Except for a few cases where the papers were available to me in manuscript form, results from the 26 papers contained in these have not been incorporated.

This book is dedicated to W. Vaughan Lewis, whose teaching revealed to me the fascination and challenge to be found in the study of landscapes, and whose lecturing could not be more appositely described than as inspired.

I owe a deep debt of gratitude to David L. Linton, under whose guidance I first studied slopes, and from whom I gained an insight into the true nature of scholarship.

I should like to thank Ronald A. G. Savigear for many helpful discussions; and to acknowledge help from my present colleagues, many of whom have read sections or, in some cases unwittingly, contributed ideas.

I am grateful to Barbara Satchell, who was responsible for the drawings, to Peter Scott, for photographic reproductions, and to Lyn Wilson-North and others who typed the manuscript.

Lastly may I record my gratitude to my wife, for assistance with fieldwork, for reading the entire manuscript twice and making numerous suggestions for improving the expression, and for constant help and encouragement.

ANTHONY YOUNG

ACKNOWLEDGEMENTS

FIGS 4, 8, 10, 11, 12, 13, 21, 22, 23, 25, 26, 28, 32, 34, 36, 37, 38, 40, 41, 42, 43, 47, 49, 55, 61, 62, 66, 67, 69, 70, 80, 86 and 94 are reproduced or modified from the publications cited in the captions, and listed in full in the bibliography. Grateful acknowledgement is made of permission to do this which has been granted by authors and publishers. Figs 24, 64, 79 and 88 are printed by kind permission of the Executive Secretary of the Royal Society. All of the photographs are by the author.

CONTENTS

I SLOPES AND GEOMORPHOLOGY

MOST of the land surface of the earth is formed by valley slopes. At a given point on the ground surface it is normally possible to follow the line of maximum slope downwards until a drainage channel is reached; moreover a causal relation between the slope and the channel is apparent. This feature has been little remarked upon, possibly because it is so widespread as to seem part of the natural order. It is nevertheless a fundamental and remarkable feature of the earth's surface form.

FIG. 1. Steeply-dissected landscape consisting almost entirely of valley slopes. Himalayan foothills north of Katmandu, Nepal; monsoon rainforest climate.

The main exceptions to this generalization are landforms of depositional origin. It is possible to drive for hours across the alluvial plains of the Indus and Ganges and see nothing but level ground stretching to the entire horizon. The central regions of the continental structural basins, such as the Amazon, Lake Eyre and those of the African continent, are similarly dominated by depositional topography. Areas of wind-blown sands may include abundant depositional slopes not organized into valley systems; the same is true of certain regions of glacial deposition, for example drumlin fields.

If regions of erosional relief are considered at a local scale, valley slopes occupy most of the area. Flood-plains, river terraces and other local depositional landforms are of variable but relatively small extent. Erosion surface remnants are another exception, to which much prominence has been given in studies of denudation chronology; yet the maps accompanying these studies show that the remnants rarely occupy more than 10% of the total area. The remainder of the landscape consists of erosional slopes. In certain special cases, such as escarpments and sea cliffs, these are not directly related to lines of concentrated drainage; but most do show such a relation, and are termed *valley slopes*.

The areal predominance of valley slopes is found not only amid well-dissected relief but also on plains. In the Great Plains of the American Mid-West, truly level plains, not dissected into valleys, occupy only 7% of the total area (Lewis, 1962). The vast erosional plains of tropical Africa, when viewed from the summits of inselbergs rising

FIG. 2. Landscape formed by moderate slopes, with limited level areas on crests of the main interfluves. Nyika Plateau, northern Malawi, 2500 m, granite; montane savanna climate.

above them, appear to be flat, and were misleadingly described as such in some early accounts. But in travelling across them, their organization into a succession of interfluve crests, valley sides and valley floors is striking; it is only the skyline, a projected profile of crests, that remains level. Slope profiles on the Lilongwe Plain of central Malawi show that, even in areas with a local relief of less than 20 m and maximum slopes below 2°, truly level land, with indeterminate direction of slope, is of small extent (Fig. 3; cf. Fig. 86, p. 234). In a part of the southern Mato Grosso, Brazil, which on a macro-scale is an exceptionally well-developed erosion surface, over 95% of the ground is formed by valley slopes (Young, 1970b).

These facts indicate that slopes provide a basis on which, by successive combination into larger units, a general geomorphology could be constructed. There have been only two attempts to do this. *Les formes du terrain*, the remarkable work of de la Noë and de Margerie (1888), was written at a time when it was still necessary to start a geomorphology text by showing that landforms are of erosional and not diastrophic origin. They follow this with a discussion of surface processes. The profile form of slopes under a variety of rock structures is analyzed as a function of process and relative rock resistance, and the effects on the slope of river erosion at its foot are then added. Slopes are next combined, in a consideration of valleys, and it is argued that since the slopes between adjacent streams intersect, 'la considération du modèle des versants suffit pour expliquer l'ensemble des formes topographiques' (p. 97). *Les formes du terrain* has never received the acknowledgement that is its due. The second attempt to construct a general geomorphology on the basis of slopes was *Die morphologische Analyse*, by W. Penck (1924); this suffered comparative neglect by the mainstream of geomorphology for nearly 30 years.

Fig. 3. Gently-undulating plain with level interfluve crests, formed largely of valley-side slopes. Lilongwe Plain, central Malawi. Basement Complex rocks; savanna climate.

Two methods that approach geomorphology by working upwards in scale from the individual component to the total landscape have more recently been developed. The most widely applied quantitative technique is morphometry, introduced by Horton (1945) and subsequently developed by Strahler (1958) and others. Morphometric analysis takes the river basin as the fundamental unit; this is a consequence of its origin in hydrological studies, in particular in the attempt to predict flood discharge from the landform characteristics of river catchments. As the basin is the larger unit, slope characteristics occur in morphometric data in generalized forms, e.g. mean angle of slope. The second technique employs a unit smaller than the slope. This is the system of morphological mapping evolved by Waters (1958) and Savigear (1965). The basic unit is the facet, a portion of the ground surface with uniform angle of slope. This method has found later application in the land systems method of natural resource survey (Brink *et al.*, 1966). In building up the facets into higher-order classes, however, the individual slope, from interfluve crest to valley floor, is not taken as a stage in generalization.

A means by which slopes can be employed as a unit basis for landform analysis is illustrated in Fig. 4. The first stage is to identify rivers and streams, together with all

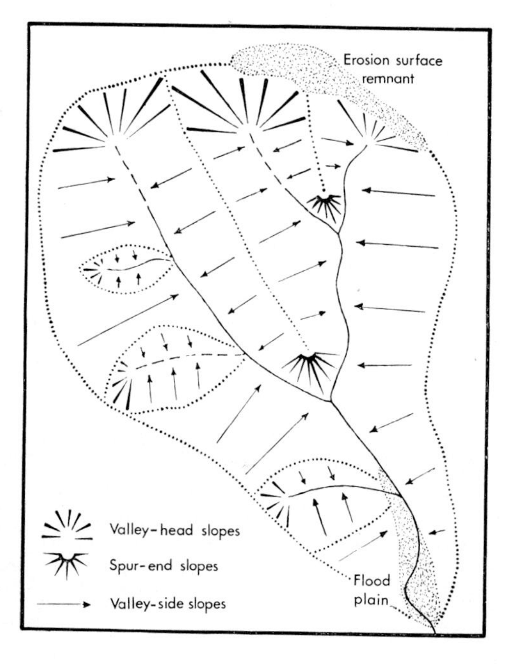

FIG. 4. Division of a landscape into *component slopes*. After Young (1964b).

other lines of concentrated drainage; areas which do not consist of valley slopes are delimited. This can be done from air photographs, or contoured maps at scales of the order of 1:25 000. The next stage is to mark the watersheds, including those between main valley sides and minor gullies. The slopes are then divided into three classes: *valley-head slopes*, which are concave in plan form; *spur-end slopes*, which are convex in plan; and *valley-side slopes*. The latter are usually relatively straight in plan, although in exceptional cases, for example above incised meanders, there may be substantial plan curvature. Each class of slopes may be divided according to the stream order with which they are associated; in Fig. 4 there are twelve 1st-order, four 2nd-order and two 3rd-order valley-side slopes. Although sharing with morphometry two types of boundary, drainage lines and watersheds, this system focuses attention on the slope, not the stream or catchment basin. The opportunity of relating slope form to stream characteristics is retained. One application of the system is in siting slope profiles, either by purposive or stratified random sampling (see p. 145). It is not known whether a system of quantitative empirical relations comparable to that of morphometry can be developed from this basis.

SLOPES, GEOMORPHOLOGY AND ENVIRONMENT

Since the Second World War, the study of slopes and surface processes has come to be recognized as an integral part of other branches of geomorphology. The various systems of climatic morphology, developed mainly in France and Germany, give much attention to surface processes and their effects upon slopes, either as the central point of the system, or by comparing the efficiency of linear river erosion and areal slope denudation (Louis, 1961). In the study of tropical landforms, both arid and humid, the origin of pediments and inselbergs have long been central problems; both are primarily questions of slope evolution. Periglacial geomorphology has provided much evidence of alterations to slope form resulting from past changes in climate. In the field of applied geomorphology, studies of landslides and of accelerated erosion are concerned with slopes and surface processes, and involve links with the techniques of soil mechanics, hydrology and soil science.

In the relations between geomorphology and other factors of the environment it is the form of the ground surface, not its origin, that is principally involved. Landforms are usually shown on maps by contours, which directly represent altitude; yet this factor is of little direct significance for soils, vegetation and animal life. Altitude acts indirectly, through its effects on temperature and, in a more complex manner, rainfall. The direct influence of landforms on soils is primarily through the effects of slope; steep slopes tend to carry lithosols, and concave slopes are liable to gleying. There are also indirect effects of slope, as in the way aspect affects microclimate, which in turn influences vegetation. In practice, contour maps are often interpreted indirectly, as maps of slope.

The widespread use, since the Second World War, of air photographs in the environmental sciences has meant that absolute altitude has received less attention. The greatest stimulus to geomorphology in this century has been its application, especially through air photograph interpretation, to soil survey and other branches of natural resource appraisal. The salient point is that only two factors of the environment can be directly seen on air photographs: landforms and, less completely, vegetation. Geological structure, soils and hydrology must be indirectly inferred. It is the form of the ground surface, described as combinations of relative relief, slope form and drainage lines, that is seen on photographs. In one of the principal methods of natural resource survey, the land systems approach, descriptions such as 'gently undulating', 'steeply dissected' are the most frequent means of distinguishing areas considered to be distinct units of the total environment.

In respect of its influence on the activities of man, it is arguable whether geomorphology is the most important environmental factor; a case could equally be made for climate or soils. There is no doubt, however, that slope angle is the main property of landforms that affects man, whether in agriculture, transport or urban activities. For example, in the land classification schemes used in agricultural development planning, slope angle appears equally with soil characteristics as a limiting factor to particular uses of land. Slope substantially affects road alignment and constructional costs, and is significant in the detailed siting of industrial plant.

The study of slopes was developed for the intellectual challenge that it offered. It originated, like geomorphology in general, from curiosity about the natural world. It has come to have some part to play in one of the greatest challenges of our time: planning of the use and conservation of the earth's resources for its present and future population. This need not mean that the more theoretical aspects will receive less attention than otherwise would have been the case. An increase is necessary in the proportion of geomorphological research effort devoted to aspects of practical application; but this is likely to be more than countered by the increased growth of the subject as a whole. As in other sciences, a balance between pure and applied research should be maintained.

TERMINOLOGY

Certain terms used throughout this book may here be defined, although definitions of more specialized terms are to be found in the relevant chapters. Exceptions to the general use of the following terms are made in discussing specifically the work of individual authors, when the original terminology is generally retained.

Erosion is used to refer to the destructive action of rivers upon their banks and bed, and in the sense of marine erosion; *denudation* refers to the weathering of bedrock into regolith, the transport of regolith across and away from slopes, and the consequent lowering or retreat of the ground surface. Denudation acts areally, whereas erosion, on the scale of the landscape as a whole, may be considered to act linearly. To conform with normal usage the term *accelerated erosion* (soil erosion) is retained for the transport and removal of soil as a result of human interference. The agencies which act areally to cause denudation are *surface processes*, being distinct from the processes of river erosion. *Surface transport* is the movement of regolith from one part of a slope to another, by processes taking place both within the regolith and on its surface. *Direct removal* is the removal of regolith from a slope in ways other than by transport across its surface, for example in solution in ground water.

Surface processes include *weathering* and the processes of *surface transport*. The main processes of surface transport are *soil creep*, *surface wash*, *solifluction*, and *rapid mass movements*. The term surface wash is preferred to rain wash, slope wash and rill wash, the last being a specific type of surface wash. Solifluction is excluded from the class of rapid mass movements. Although the word landslide is often loosely used to refer to all rapid mass movements, this leaves no general term for movements that are initiated by failure along a defined slip plane. Therefore rapid mass movements will be used to refer to all movements involving either rapid flow or slip; following Sharpe (1938), the subgroup in which movement is mainly by slip are termed *landslides*.

Ground loss is the lowering or retreat of the slope surface, as a result of the net removal of regolith. The converse, an advance of the ground surface upwards or outwards as a result of net accumulation, is *ground gain*. Ground loss and gain are assumed to be measured perpendicularly to the ground surface unless otherwise specified.

Profile refers to a slope profile; it is not used in the sense of a soil profile. The highest and lowest points on a profile are referred to as the *crest* and *base*. In relation to ground surface form the epithets upper, lower, above and below can be used without ambiguity; but when considering the regolith, it is necessary to distinguish first between the surface and deeper layers, and secondly between one part of the surface and another; therefore the terms *upslope* and *downslope* are used for the latter where there is the possibility of confusion.

Basal removal is the removal of regolith material from the base of a slope. Basal removal may be *unimpeded*, *partly impeded* or *impeded*, according to whether all, a part, or none of the material is removed.*

The symbol θ (theta) is used throughout the book to refer to the slope angle at a point, measured in degrees and decimals of a degree.

Structure is commonly employed in two ways: in the broader sense, as used by Davis, it refers to all geological properties of the environment, including lithology, while in the narrower sense it refers only to the attitude of strata. Except where otherwise stated, it is used here in the wider sense.

Regolith refers to the cover of unconsolidated material overlying solid rock. Where the bedrock is unconsolidated material, such as sand or clay, this is not applicable, and the lower boundary of the regolith may then be conveniently taken as the lower limit of weathering. Regolith may be *in situ*, i.e. in the same position on the slope as the rock from which it was derived, or it may have undergone transport. For the textural composition of the regolith, the terms used in the English translation of Penck (1924) are adopted; *reduction* is the breaking up of the regolith material into finer particles, and the *degree of reduction* of the regolith is the extent to which it is broken up into fine particles. Reduction is brought about by the processes of weathering.

Features of existing landforms which originated under climatic conditions substantially different from the present are termed *relict features*. The term *periglacial* was introduced by von Łoziński (1909), who used it to refer to the climatic conditions in the zone marginal to the ice-sheets during periods of Pleistocene glaciation; although currently used also to describe contemporary cold climate environments, for convenience it is only employed here to refer to relict conditions or features. The corresponding present day climates are described as *polar* or *montane*.

ARRANGEMENT OF THE BOOK

This book is concerned primarily with sub-aerial slopes of erosional origin. Submarine slopes and slopes formed by glacial erosion are not considered. Screes are included, together with slopes that are of possible depositional origin, such as concave footslopes, but sand dunes, alluvial fans and other depositional slopes are not discussed.

The following chapter surveys historical developments in the study of the geomor-

* These terms are not identical to the use of 'unimpeded removal' etc. by Savigear (1952).

phology of slopes. Chapter III is concerned with some basic concepts. The problem of slope evolution is central to the subject, and in Chapter IV the main systems of evolution that have been proposed are reviewed. Chapters V-VIII cover surface processes and Chapters IX and X theoretical approaches. Chapters XI-XVI are concerned primarily with slope form. Chapters XVII-XIX cover process, form and evolution under differing environmental conditions; synthesis of the individual aspects is attempted in the sections on structure and climate. The final chapter is a brief treatment of some applied aspects.

II THE DEVELOPMENT OF IDEAS

THE NINETEENTH CENTURY

THE earliest scientific observations on slopes and surface processes are found in the classic foundation works on geology. The great concept set forth by Hutton (1788, 1795) and Playfair (1802), that the rocks of the earth pass in a continuous cycle between land mass and sea floor, involves the recognition of rock weathering and the transference of soil into rivers as stages in the total scheme. Hutton notes that 'A soil is nothing but the materials collected from the destruction of the solid land' (1788, p. 214). Playfair specifically comments on slope form; he notes that rugged mountains are found, on examination, to be formed of rocks of 'unequal destructibility. . . . Where, on the other hand, the rock wastes uniformly the mountains are similar to one another; their swells and slopes are gentle and they are bounded by a waving continuous surface' (1802, p. 112). With respect to process he is mainly concerned with river erosion, but notes also that the surface wash of rain moves particles, which in turn may cause corrasion: 'The parts [of the earth] loosened and disengaged by the chemical agents, are carried down by the rains, and, in the descent, rub and grind the superficies of other bodies' (p. 99). Perhaps the most remarkable insight for its time is that Playfair envisaged the soil mantle as of permanent existence yet ever-changing composition: 'The soil, there, is augmented from other causes, just as much, at an average, as it is diminished by that now mentioned; and this augmentation evidently can proceed from nothing but the constant and slow disintegration of the rocks' (p. 106)—an enunciation of the neutral balance of denudation as put forward 152 years later. Lyell (1841) gave an instance of the changing form of slopes with time; having correctly interpreted raised beaches and the cliffs which back them, he asks why such cliffs are not more widely found, if, as he believed, marine erosion was responsible for much of what are now subaerial landforms; it is because 'the cliffs crumble down . . . they are soon reduced to a gentle slope' (p. 161). He correctly interprets the origin of earth pillars, as caused by 'the effects of the denuding action of rain . . . as distinct from those of running water' (1867, p. 335).

A major topic of geological concern in the mid-nineteenth century was whether landforms originated by marine or subaerial erosion; attention was directed mainly towards rivers, as the subaerial agency primarily responsible for the cutting of valleys. Until this matter was settled, problems of the transference of weathered rock into river channels, and the consequences upon valley form, remained largely in abeyance. Nevertheless it was out of this controversy that the earliest discussions of slopes arose. The first paper in which observed slope form is applied to the elucidation of the origin of landforms is the study by Sorby (1850) of the origin of the striking, steep-sided valleys of the North Yorkshire Moors. He came down in support of marine currents, but his field observations are quoted; the four cross-profiles given of the valley of Yedmandale are somewhat diagrammatic, but the text mentions valley sides of 30° and 14°, the latter at least suggesting field measurement.

There followed what may, by the standards of the time, be regarded as a burst of publications on slopes; five papers appeared in the *Geological Magazine* of 1866 and 1867, of which four were intended as contributions to the subaerialist cause. Scrope (1866) clearly envisaged slope denudation as distinct from river erosion: the 'direct fall of rain' removes particles from the ground surface and carries them away to the lowest accessible levels, thus 'the general surface is more or less lowered'. Maw (1866) gives a diagram of four stages in the development of valley cross-profiles; he was the first to reason that near the watershed the ground is affected only by rain falling directly upon it, but downslope 'a progressive concentration of water and consequent power of excavation takes place'. Wynne (1867) described how the fragments on mountain sides weather, are reduced, and carried to lower parts; the gentle slopes of plains, on the other hand, were probably formed by the sea, for 'rain seems to act vertically, its tendency always being to produce steep ground where it is not accumulating materials', a comment which might be held to anticipate pedimentation. Whitaker (1867), arguing for the subaerial origin of escarpments, analyzed processes and then applied a deductive argument; a cross-section of an escarpment, if it were a degraded cliff, would show an accumulation of talus at the base; this is contradicted by observation of quarry sections. Also at this period, although not part of the same controversy, appeared the deductive mathematical treatment by Fisher (1866) of the weathering and consequent evolution of a cliff face, demonstrating that the solid rock core progressively buried beneath the talus would have the form of a semi-parabola; it is doubtful if the numerous later papers on this topic have added much that is of significance.

Contributions to slope studies during the last three decades of the nineteenth century were not numerous, but they are noteworthy for the freedom of field observation and conclusions, uninhibited by any existing body of received theory. Tylor (1875), in a paper far in advance of its time, gave a valley side profile surveyed in detail; he plotted binomial curves that showed a close fit with observed slope form, and he supposed that slopes are eroded into this curve because it is the 'form of greatest stability. . . . It is the form which gives the nearest possible approach to uniform motion of water on its surface.' Soil creep was first described by Thomson (1877): 'a number of causes tend to make the whole soil cap . . . creep down even the least slope'. The causes cited are expansion and contraction on wetting and drying, the decay of vegetable matter in the soil, and its weight. He observed and correctly interpreted outcrop curvature in slates. Kerr (1881), in a pioneer work on both processes and periglacial phenomena, was struck by the same feature, 'the gradual drawing out, . . . attenuation of the coloured bands [hornblenditic and chloritic strata], as parts of them in succession were moved down the slope'. This movement was produced by 'the alternate freezing and thawing of the saturated mass of decayed rocks', producing a movement not only of the whole mass but of the particles *inter se*—the feature now recognized to be distinctive of soil creep as contrasted with other forms of mass movement. Kerr considered that much, but not all, of this movement took place in the glacial period. Moseley (1869) and Davison (1888) had carried out experiments recording quantitatively the downslope movement of blocks resting on an inclined plane and subjected to temperature changes. On reading Kerr's paper, Davison (1889) measured movements of the ground surface during and after frosts; expansion is perpendicular to the ground surface and contraction vertical, hence there is a net downslope movement; since varying intensities of frost reach different depths in the soil, the rate of creep decreases with depth. By making certain explicit, quantitative assumptions, Davison deduced that the section sketched by Kerr could have been produced by contemporary processes in 531 years. This paper is exemplary for its combination of hypothesis, field testing and deduction, as well as for its conciseness.

The earliest quantitative measurement of the rate of surface processes under natural conditions was made by Darwin (1881), for movement on a 9·5° slope caused by worm casting. It is a tribute to the care exercised in this measurement that the value obtained, equivalent to 0·52 $cm^3/cm/yr$, is fully consonant with modern estimates for rates of soil movement.

G. K. Gilbert's *Geology of the Henry Mountains* is the greatest single contribution to geomorphology ever published. Written with a logic, fertility of ideas, and clarity and conciseness of expression never surpassed, it is one of the classics of scientific literature. More than half is geological in content, elucidating the structure of a laccolite. The 50 pages devoted to land sculpture contain, besides the basis of fluvial geomorphology, an astonishing range of the major concepts in slope studies. He notes that the accumulation of disintegrating rock retards weathering by preventing frost action. Weathering is also slow where there is complete soil removal, because rain immediately runs off; this anticipates one modern view of the development of inselbergs. A limitation is that Gilbert did not here recognize soil creep, and his deductions concerning surface processes are in terms of rain wash. Transportation is favoured by rocks that weather to fine-textured debris. Vegetation retards rain wash. 'In regions of small rainfall, surface degradation is usually limited by the slow rate of disintegration; while in regions of great rainfall it is limited by the rate of transportation' (p. 99). This concept was only re-stated nearly a century later, by Savigear (1960). 'Steep slopes are worn more rapidly than gentle' (p. 109), tending toward uniformity of slope. Gilbert's 'Law of structure' is an explicit application of an equilibrium concept to slopes: 'When the ratio of erosive action as dependent on declivities becomes equal to the ratio of resistances as dependent on rock character, there is equality of action' (p. 110). 'Every slope is a member of a series, receiving the water and waste of the slope above it, and discharging its own water and waste upon the slope below' (p. 118); this concept is now the basis for a class of process-response models.

The most often quoted sections of the *Henry Mountains* are those on the 'Law of divides' and its 'exception'. Rivers become steeper upstream because their volume is less. 'The same law applies . . . to the slopes over which the freshly fallen rain flows in a sheet before it is gathered into rills' (p. 110). Hence slopes have a generally concave curvature, as exemplified in an area of badlands, where the rock is homogeneous. There is, however, an exception to this law; near the watershed the curvature changes from concave and becomes convex. 'Evidently some factor has been overlooked'

(p. 117). This short, forthright and intellectually humble sentence was to give rise to a sequence of papers which furthered the concepts of relating process to form. W. M. Davis (1892), in his first publication on slopes, pointed out briefly that the missing factor is soil creep, which can account for the divide convexity. Hicks (1893), again citing Gilbert's 'exception', termed convexity 'the weather curve', produced by weathering penetrating inwards from an initially angular intersection of summit plane and rectilinear slope; concavity, in contrast, was 'the water curve', produced by transportation. Following from Davis, Gilbert (1909) then detailed the causes of soil creep, including all those now recognized as important. He attempted to show that a convexity necessarily results from the manner of action of soil creep. Finally, he put forward one of the most often discussed, and as yet unresolved, hypotheses on slope formation: that the convex-concave form of slopes is due to the respective actions of creep on the upper part of the profile and water flow on the lower.

Working inductively from observations of steep cliffs and canyon walls in the American West, Dutton (1880-1, 1882) set out clearly the development of successive slope profiles by recession. One of his diagrams (reproduced in Chorley *et al.*, 1964, Fig. 123) shows a scheme of slope retreat identical with that to be re-formulated by Fair in 1947. In attempting to explain slope profile form on rocks of varying hardness Dutton put forward two other basic ideas. First, that talus acts as a control on slope form: 'The recession of the hard beds is accelerated by undermining, while the recession of the soft beds is retarded by the protection of the talus. The result is the final establishment of a definite profile, which thereafter remains very nearly constant as the cliff continues to recede. Thus the talus is the regulator of the cliff profile' (1880-1, pp. 163-4). Secondly, he proposed that a debris cover protects the slope from weathering, and the basal concavity is caused by the progressive downslope increase in amount of protection.

Towards the close of the nineteenth century appeared *Les formes du terrain*, noted above, and the beginning of the long sequence of publications by W. M. Davis, which will be discussed in the following chapter.

1900-1918

For geomorphology as a whole the formulation in 1899 of Davis's geographical cycle may be taken as the division between two eras. No such break is apparent in slope studies; they remained few in number but with a high standard of field observations and a fertility of ideas. Marr (1901), writing two years after the publication of Davis's cycle, and in the same journal, is uninfluenced by it. He explains Gilbert's 'exception' as due to the action of vegetation. This aids weathering through organic acids, but checks transport by running water. In arid lands, where there is little vegetation, opposing concavities meet at the summits. In Britain and, especially, in the forested tropics, vegetation checks the formation of the concave curve of water denudation, hence the convex curve of weathering predominates. This rarely cited paper could have been a precursor of climatic morphology over 40 years before it was in fact developed.

There were further studies of processes in the period 1900-1918. Andersson (1906) proposed the term solifluction, described its mechanism, and considered that, of all surface processes, it represented the 'optimum of destructive action'. Fenneman (1908) made the first detailed study of unconcentrated wash, describing how it flows as rills, with impermanent courses and, in his view, 'overloaded'. He suggested that the downslope increase in the volume of wash may, through a corresponding increase in erosion, produce a convexity. Hogbom (1914) studied frost action in soil. Lawson (1915) discussed slope evolution in deserts, where sheetwash from occasional storms was the main denudational process. He postulated that alluviation, commencing on the lower part of the slopes, will progressively extend towards the mountains; this, combined with scarp retreat, will give rise to a suballuvial bench of solid rock, deduced as having the form of a hyperbola tangent to the surface of the alluvial fan. This is an extension of the deductions on cliff retreat made by Fisher (1866).

Two further works of this period are outstanding for their careful field observation, those of Götzinger (1907) and Lehmann (1918). Although not widely read outside Germany they are important in having influenced Walther Penck. Götzinger studied slopes in the Flysch zone of the Wiener Wald. He gave particular attention to the regolith cover, noting variations in its thickness and giving descriptions of outcrop curvature. He observed that the regolith increased in thickness downslope, giving more protection from weathering to the lower than to the upper parts; from this he concluded that rectilinear slopes will decline in angle. Lehmann's study is still more detailed, combining observations of the surface form, regolith thickness and denudational processes to explain the origin of the valley sides of small ephemeral streams. Penck (1924) cites Götzinger eight times, and remarks of Lehmann's paper: 'This is an investigation unequalled for delicacy of observation'.

By 1918 some 60 publications concerned wholly or in part with slopes had appeared, of which perhaps a third can be read with profit at this day. Soil creep, surface wash, solifluction and other types of mass movement had been recognized; there had been studies of slope form, and the concept of slope evolution was accepted; the approach of relating the action of processes to the resulting form and evolution had been attempted. These studies rested on a sound basis of field observation, to which the amount of theoretical reasoning was kept in reasonable proportion. Moreover in *Les formes du terrain* there existed a comprehensive study linking the various aspects.

In relation to the modest resources available to geomorphology as a whole, here was a sound and substantial foundation for slope studies. Why, then, was this not to develop as a recognized aspect of the subject? Chorley (1964b) gives four difficulties associated with the study of slopes: their complexity of form, the multivariate nature of processes, the doctrinaire attitude of most researchers, and the feedback nature of the problem. The stricture of doctrinaire attitudes, however, does not apply to pre-1918 work, whilst the complexities of both form and process had on occasion been appreciated. What appears to have been lacking was simply an appreciation that here was a related group of studies, the results of the various branches bearing upon each other and forming a coherent whole.

1919-1939

The inter-war period saw an increase in the number of publications but not, with few exceptions, an improvement in their quality. The period should have been dominated by developments arising out of Penck's *Die morphologische Analyse*, appearing in 1924. Yet, apart from a few comments in German-language periodicals (Henkel, 1926; Morawetz, 1932; Louis, 1935), this massive challenge to the Davisian scheme did not receive the attention it deserved. The reasons for this have often been given: the obscurity of expression in the original, and the lack of a French or English translation. Among standard English language textbooks only Von Engeln (1942) set out some of the views of Penck. Davis (1932) attempted a comparison of Penck's system of slope evolution with his own; but, as Simons (1962) has shown, he misrepresented Penck's view, and Davis's erroneous version was temporarily accepted. Penck received further attention at a symposium of the Association of American Geographers (Von Engeln *et al.*, 1940), but the discussions at this symposium were still handicapped by not having access to a full English text. It was only the translation by Czech and Boswell in 1953, together with a masterly summary, which brought Penck's work fully into the field of international discussion.

The mainstream of geomorphological thought between the wars moved in a different direction. Having in the nineteenth century concentrated upon rivers, it turned towards denudation chronology, in particular the study of erosion surfaces. These latter, of proper significance in their context, came to exert a fascination for their own sake, such that 90% of the land surface was ignored by geomorphologists 'bemused by long, though mild intoxication on the limpid prose of Davis' remarkable essays' (Bryan, 1940).

Hence the substantive contributors to slope geomorphology during this period were still in the position of pioneers. Bryan's monograph on the Papago Country, Arizona (1925) is a classic. In discussing boulder-controlled slopes he introduced the concept now known as limiting angles, and he gives an extended treatment of pediment formation, stating the hypothesis that the pediment is a transportation slope. Lawson (1932) continued the discussion of the relative importance of soil creep and rain-wash, arguing for the latter, but in this paper one sees the start of an over-dependence on abstract reasoning which was to afflict slope studies for some 20 years. The work of the United States Department of Agriculture on accelerated soil erosion, begun during the 1930s, was later to contribute to geomorphological process studies. In Germany Lehmann (1931) analyzed the energy and work-performance of falling rain and surface run-off, and subsequently (1933, 1934) contributed to the theory of cliff recession. Morawetz (1932) made quantitative measurements of the movement of stones on Alpine screes. He also raised an important question that has since been largely neglected: the causes for the dissection of a single, smooth slope by tributary gullies (Morawetz, 1937). In Britain, the isolated achievement of Lake (1928) in making a form analysis of surveyed slope profiles deserves mention.

1940-1960

The modern period has seen an unprecedented growth in the volume of literature on slopes (Fig. 5). The rapid increase between the 1920s and 1930s was checked in the following period by the Second World War; otherwise, from a base of ten in the 1910s, there has been a doubling in the number of papers every decade. This geometric growth is similar to that for geography and for science as a whole.

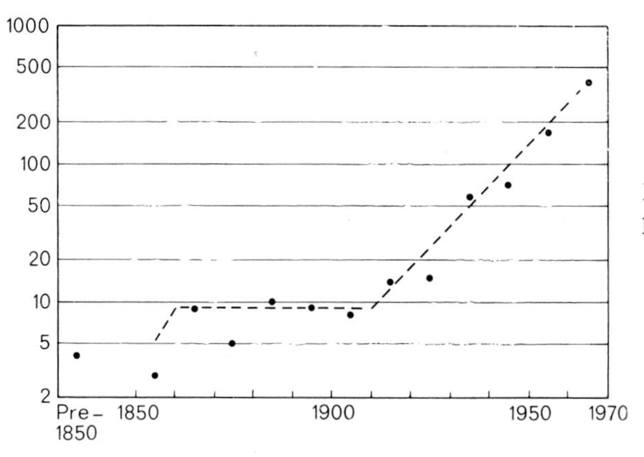

FIG. 5. Growth in the number of publications on slopes, by decade.

A necessary condition for such geometric growth was the greatly increased man-power and financial resources devoted to research in universities, but this is not a total explanation, for geomorphological research might have continued to concentrate on the topics prominent between the wars. It is of interest to try to identify the seminal works out of which grew modern slope studies. With the increase in the range of the subject, no single modern work is able to equal the scope and breadth of vision of the *Henry Mountains* and *Morphologische Analyse*. But there are certain papers which, whilst not necessarily of themselves major contributions, directed attention to central problems, the investigation of which gave rise to the identification of slope studies as a major branch of geomorphology.

In seeking a minimum list of such seminal works, one may cite Bryan (1940), Baulig (1940), Fair (1947, 1948, 1948b), Savigear (1952) and Czech and Boswell's 1953 translation of Penck. Bryan defined the general problem of the retreat of slopes; whilst responsible for some misconceptions in presenting Davis *v.* Penck as a simple opposition, he re-discovered *Les formes du terrain* and many of the classic papers, showing how each formed part of a unified branch of knowledge. Baulig also identified and critically discussed the significant previous works, and achieved a characteristically French synthesis. It is unfortunate that this wholly theoretical treatment, forced upon Baulig by the circumstances of wartime France, was to be copied so frequently by others. Fair's surveys laid the foundations for field studies of slope form, demonstrating the value of the technique of profiling; he showed how a comparison of the form of different present-day slopes could be used as an inductive basis for ascertaining past slope evolution. His studies in Natal (1947, 1948), in which slope forms under different structural conditions are compared, are discussed below (pp. 36, 215). The short account of hill-slopes and pediments in semi-arid Karroo (1948b) is a magnificent work of field observation, and remains a rich source of information on this class of landforms. Fair's work might have been neglected had not Savigear (1952) independently applied similar techniques and inductive reasoning to a coastal succession of former marine cliffs, which had been progressively protected by the extension of a sand bar along their base.

It remains to identify the modern foundation works of some of the separate branches that have become established. In the field of process studies Horton (1945) brought to geomorphology the results of work on hydrology and on soil erosion; this placed knowledge of surface wash on a firm experimental foundation, as well as giving rise to the science of morphometry. The engineering techniques of soil mechanics were first applied to landforms by Ward (1945) and Skempton (1953). The now widespread application of statistical methods to slopes was first made by Strahler (1950). Measurements of the rate of processes were first extensively made in polar environments, notably by P. J. Williams (1957 *et seq.*) and Rapp (1960). Among field studies that relate measurements of processes to their effects on slope evolution, those of Schumm (1956 *et seq.*) are noteworthy. Young (1956 *et seq.*) drew attention to the dynamics of regolith movement in relation to slope retreat, and first measured soil creep. The principles of slope profile analysis were established by Savigear (1956 *et seq.*) and subsequently developed by Young. 1956 saw the establishment of the Slope Study Commission of the International Geographical Union; the six Reports of this Commission, containing 142 papers, include much that is of ephemeral value only, but amount in sum to a contribution of substance.

The number of publications on slopes during the 1960s exceeds that of all previous time. An historical perspective of these is not yet possible; summaries are given in the individual chapters below.

III CONCEPTS AND APPROACHES

FORM, PROCESS AND EVOLUTION

In the study of any aspect of the physical environment there is a distinction between form and process. *Form* applies to what is there, the morphology at a given moment in time; *process* to what is happening, the agents active in causing form to change. In soil science, form refers to the existing characteristics of soils, the soil types present in an area, their distribution, profile morphology, mineralogical composition, and also non-visible properties such as reaction. Process applies to such phenomena as hydrolysis, oxidation and reduction, mechanical eluviation, and the leaching of exchangeable cations, agents that have given the soil its form and are currently modifying it. In plant ecology the vegetation communities present, their floristic composition and physiognomic structure, constitute form; photosynthesis, transpiration, and the various mechanisms of growth and reproduction are among the processes.

In the case of slopes, as in geomorphology in general, the distinction between form and process is particularly clear. Form is most obviously represented by the shape of the ground surface; the techniques of contour survey, slope profiling, morphometry, and some types of morphological mapping are concerned with this aspect. Form *sensu lato* is not confined to the surface, but includes the thickness and composition of the regolith. Process refers to agents such as soil creep, surface wash, and the processes of weathering. The environmental conditions of slopes, such as the geology, climate and vegetation cover, are elements of form.

Both form and process have existed in the past. The succession of past forms, leading to that of the present, constitutes the *evolution* of a slope. For the past with respect to process, no recognized comprehensive term exists; the study of the influence of climatic change upon surface processes, and therefore upon slope evolution, falls into this category.

DESCRIPTIVE AND GENETIC APPROACHES

A distinction which cuts across that of form and process is between the descriptive and genetic approaches to the environment. Form may be studied purely from the descriptive aspect; a contour map, an isopleth map of slope angle or a graph of the frequency distribution of slope angle in a specified region are statements about slope form. Studies of process, if they are concerned with ascertaining which processes are currently active, how they operate, or at what rate, may also be essentially descriptive.

In many academic studies, information on slope form is treated primarily as a source of evidence for the manner of evolution; but in the practical application of data on slope form, for example to road engineering or soil conservation, it is the form characteristics that are relevant and not their origin. This dichotomy is undesirable. The need for a foundation of descriptive work, of considerable mass, is recognized in all of the environmental sciences. Slope studies have until recently suffered from a preponderance of theoretical discussion, without an adequate basis of detailed field observation. The scientific description of slope form *per se* is a legitimate and necessary part of research.

The second approach is that of genetic studies, directed towards the origin of landforms. Such studies are concerned primarily with slope evolution, and involve consideration of past as well as present form. Frequently, also, evidence of present processes and inference about changes of process in the past is employed. The general problem of slope evolution is therefore central to the subject as a whole, combining evidence from each of its branches.

Studies of process or process-form relations may be empirical or rational. Empirical results are obtained by inductive studies, and are generalizations about observed phenomena; it might, for example, be found in a certain area that maximum slope angle was some function of relative relief. No reason for the observed relation is intrinsically implied. Rational studies seek to explain observed phenomena in terms of the fundamental laws of behaviour of matter and energy. Both empirical and rational equations can be used for predictive purposes. Strahler (1952) notes that as research progresses, the empirical and rational models should converge. The complexity of phenomena on slopes is such that a high degree of explanation in purely physical terms is unlikely to be attained; the empirical approach, particularly in conjunction with statistical methods, will retain its value.

In order to combine the evidence of process and form an intermediate stage is logically necessary. To illustrate this, consider a hypothetical situation in which it is assumed that processes in the past have been identical to those of the present. A further simplifying assumption is that there is no change in variables external to the slope, such as river erosion at the foot. The first stage of an investigation is the description of existing slope form. Secondly, the existing processes are studied. The information required is not only which processes are operative, but more particularly the manner in which these processes act in causing change of form; it might be found, for example, that soil creep was the active process, and that the rate at which soil was moved downslope by creep was proportional to the sine of the slope angle. This permits the construction of a process-response model (Whitten, 1964), showing the succession of form-changes through which the slope passes consequent upon processes acting in a specified manner. If the further assumption is made that knowledge of present form and process is complete, then the only remaining unknown factor is the initial conditions, that is, the form of the slope at the start of the current phase of evolution. These conditions may therefore be inferred. Process-response models are discussed further in Chapter X.

The assumptions in the hypothetical case outlined above are unrealistic. The form of present-day slopes has evolved mainly over the past million years or less. The climatic changes over this period have, in most parts of the world, been on a scale sufficient to have a considerable effect on surface processes. Moreover, the world climatic pattern was substantially different as recently as 10 000 years ago, so that even slopes presumed to be of quite recent origin will have been affected by this factor. Apart from actual glaciation the greatest changes to surface processes have been in two zones: the temperate latitudes bordering areas glaciated in the Pleistocene, and the sub-tropical desert-margin belts. Changes have probably been least in the equatorial humid tropics. It may be that rainforest lands are geomorphologically unique in offering the opportunity to study landforms in conjunction with the processes that have produced them.

The problem of the effects of climatic change on slopes can be approached through the standard geological technique of the study of deposits. The regolith cover on or near the foot of slopes may provide evidence of the type of processes that produced them. The most widespread use of this type of evidence has been the interpretation of periglacial deposits, the product of frost-shattering, solifluction, and related processes. Various techniques are applied to interpret such deposits, for example stone orientation, pollen analysis and the study of land mollusca. More recently, under tropical conditions, mineralogical analysis has been employed to distinguish successive strata within the regolith, each representing a climatic phase of slope evolution (e.g. Mabbutt and Scott, 1966). A substantial advance in knowledge would become possible if means were developed for distinguishing regolith that has been transported by soil creep and surface wash respectively.

The general problem of relating form to process, at the present and in the past, is as follows. Present form can be surveyed. Present processes can also be observed, although this is technically more difficult because of their extreme slowness. Evidence of past processes is derived mainly from deposits, but the interpretation of such evidence is frequently uncertain. Past form cannot be observed; to reconstruct it is the primary aim in the study of slope evolution. Past and present process and form can be related by process-response models, a necessary prerequisite being knowledge not only of the types of processes but also their manner of action. The complexity of processes, the slowness of form change, the spatial diversity of the environmental conditions and their variation in time, are among the difficulties in the way of achieving an understanding of the relations between process, form and slope evolution.

DEDUCTION AND INDUCTION

Both deductive and inductive reasoning are employed in geomorphology (Johnson, 1939-40). The deductive method may be illustrated with reference to slope evolution. A hypothesis is first made, and is expressed as, or translated into, a set of assumptions; the assumptions might consist of the initial conditions of a slope, for example a rectilinear slope of 35°, the conditions at the slope foot, and the manner of action of surface processes. The consequences to be expected from these assumptions are then deduced, by mathematical or other reasoning. The results could refer not only to the form of the ground surface but also to the thickness and other properties of the regolith. Finally the deduced consequences are compared with observed form. If the deduced forms do not match, then either the deductions are at fault or else at least one of the assumptions is incorrect. If the forms agree, then all that can strictly be said is that the assumptions have not been disproved. They have not positively been proved, for in theory an infinite number of different sets of assumptions can lead to the same consequence. This has been termed the principle of equifinality (Von Bertalanffy, 1950).

Using the inductive method a quantity of observed data is gathered, which then forms the basis for further inference. An example of inductive reasoning is as follows: at the foot of a slope a deposit containing sharp-edged, fractured stones is found; observation in polar regions, and laboratory experiment, shows that stones with these characteristics are produced by frost-shattering; therefore the slope has been affected by this process. For a less simple example, assume that a random sample of slope angles has been taken for a given area. When the frequency distribution of these is plotted graphically, it is found that there are peaks of 25° and 35°, with fewer slopes at intermediate angles. An assumption is made that the landforms of the present include slopes at various stages of evolution. The inductive inference is that slopes evolve relatively rapidly from 35° to 25°, but remain at 25° for a longer period (steepening to 35° is also logically a solution).

It may be argued that a purely inductive approach is impossible, since an element of hypothesis must have been present to guide the selection of observations. There is some truth in this, but to apply it rigidly ignores the way in which scientific research is actually conducted. Observations are often made, not with the aim of proving a specific hypothesis, but from an intuitive feeling that by applying a particular technique, or collecting a particular group of data, useful results will emerge. Some of the most interesting discoveries occur when a research worker finds that his results fail to prove what he had hoped they would. This does not become widely known, due to the convention in scientific publication of presenting results as if there had been an undeviating course from hypothesis via critical experiment to verification. In the study of slopes, as in many of the environmental sciences, it is rarely possible to obtain controlled conditions in the manner of a laboratory experiment. Under these circumstances it is usually desirable to apply to problems a combination of deductive and inductive reasoning. The distinction between them remains useful as a means for logical analysis of the steps by which a result is obtained.

SYSTEMS THEORY AND TIME-INDEPENDENCE OF FORM

General systems theory has been proposed as a framework for geomorphology (Chorley, 1962). Insofar as it can be shown to be either valid or useful for landforms in general, it is likely to be equally so in respect of slopes. The boundaries of an open system that contains a valley-side slope are shown in Fig. 6. Vertical planes through the watershed and the river bank form two boundaries; two perpendicular vertical planes are arbitrarily placed to mark off a portion of the valley side. The lower boundary is conveniently taken as base level, the zero level for energy deriving from gravity. The upper limit is a horizontal plane at the presumed initial surface or at an arbitrary distance above the highest point on the watershed. Some flows of energy and materials into and out of

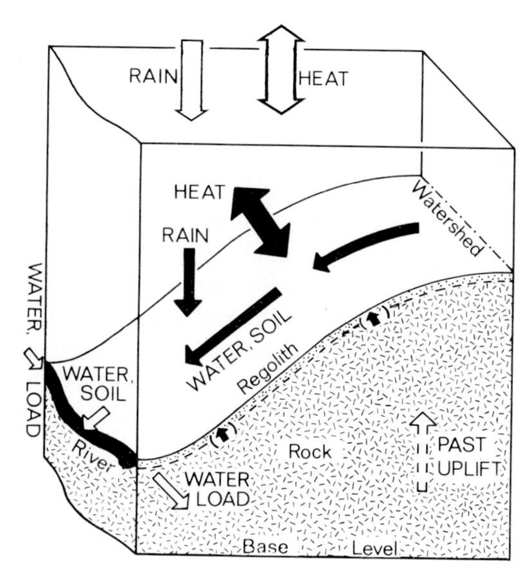

FIG. 6. A physical system containing a valley slope, with arrows showing flows of energy and materials. Solid arrows = intra-system flows; open arrows = flows across boundaries of the system.

the system are indicated. Energy continuously enters the system from the atmosphere, as heat and as the kinetic and potential energy of rainfall. In addition, large inputs of chemically-bonded energy occurred when the rock was formed, and of potential energy when it was raised above base-level. The flows for such a system have not been computed. It is possible, however, that the total energy balance for the system will not prove to be geomorphologically useful. The amount of energy entering and leaving the system as heat flux is probably of a considerably higher order than any other component of internal or external flow. All flows that are of interest for slope evolution, for example the thermodynamics of rock weathering, would then be completely compensated for in the system as a whole by a fractional raising or lowering of the temperature of the surrounding atmosphere. It has yet to be ascertained whether a drainage basin or other landform can be regarded as an open system in a steady state.

A related but not identical concept is that of time-independent form (Hack, 1960). Fig. 7 shows two sets of hypothetical stages in slope evolution (the fact that slope decline is represented is immaterial to the present argument). In Fig. 7*A*, form is time-dependent throughout the evolution shown; the form of a particular slope depends on

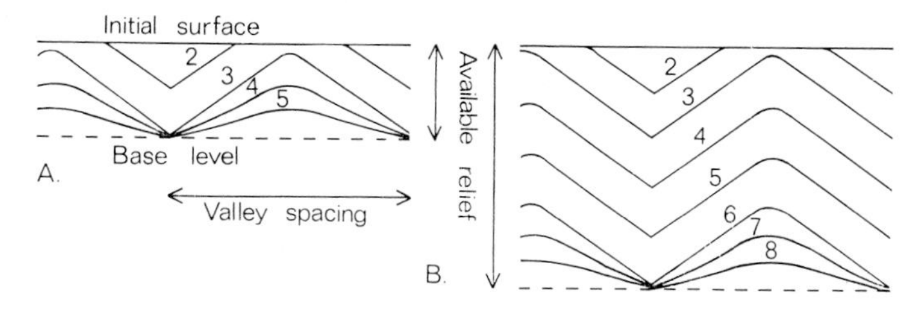

FIG. 7. Slope evolution in relation to available relief and valley spacing.

the time that has elapsed since the inception of evolution. In Fig. 7*B* the internal form of the valley-side slopes remains unaltered from stages 3 to 6, although the external altitude-relation to base level changes; between these stages the shape of the slope is independent of the time elapsed. But a slope in such a time-independent condition is not an open system in a steady state. The two concepts are not equivalent. The existence of an open system in a steady state necessarily implies the absence of any change in form, including that of uniform slope lowering; this supposition is geomorphologically unrealistic. The difference in evolution between *A* and *B* in Fig. 7 is a function of the relative magnitudes of available relief and valley spacing, where available relief is defined as the altitude difference between base level and the initial surface following uplift (Glock, 1932; Johnson, 1933). It is possible for the form of a slope, *except* for its altitude, to remain unchanged for a large part of its evolution; this restricted meaning of time-independence of form is geomorphologically the more useful concept.

UNIFORMITARIANISM AND THE TIME SCALE

There has been some discussion of the meaning of the concept of uniformitarianism (Davis, 1896; Gould, 1965). The term originated from Hutton's phrase 'the present is the key to the past' (1788). The original meaning was that the study of present-day processes is an aid to interpreting the consequences of processes in the past, as exhibited in present form. Subsequently, uniformitarianism has been variously taken to mean that all processes act slowly but continuously; that environmental conditions, including geomorphological processes, have been the same during the past as at the present; and, at the opposite extreme, as no more than a statement of the temporal constancy of physical laws. None of the latter interpretations is necessary and the original meaning holds good. The principle calls, for example, for the study of landforms and processes under polar climates of the present in order to interpret relict periglacial deposits.

Of the assumptions made in the hypothetical case of a process-response model discussed above, that of an accurate knowledge of present form is attainable. Such knowledge of process is not attainable. Geomorphological processes in general act either extremely slowly or, in the case of catastrophic processes, at long intervals. In either case the change of form produced over a century may be imperceptible. The problem is greater with processes on slopes as compared with river erosion. A calculation illustrates the difficulty. To detect a movement or other change of form of the order of 0·1 mm/yr, not under laboratory conditions but amid the natural environment, presents considerable problems of instrumentation; yet such a small annual change represents 100 m in a million years, sufficient for the entire formation of a valley of moderate size. A consequence is that extrapolation of the rates of present processes is, on statistical grounds alone, subject to a high degree of uncertainty. The difference in time-scale between a human lifetime and geological time is considerable; for the onset and termination of Pleistocene glaciation, the difference is of the order of 10 000 and 100 respectively. The slowness of process, and the consequent time-scale involved in change of form, is one of the major problems in the study of slopes.

It is not usually possible to refer simply to the 'age' of a slope, or other landform. All slopes have been modified, to a greater or lesser extent, by processes acting in the immediate past. Conversely, most contain some features of form which owe their origin to events and processes dating further back. By definition, slopes with time-independent form retain no record of events preceding attainment of that state; this is the case for most slopes on badlands, and may apply also to other areas of well-dissected relief. The historical approach, as for example the interpretation of the upper, more gently sloping, part of a V-in-V valley as originating in an earlier cycle, or the attribution of periglacial features to a slope, implies that form is time-dependent. All such features, however, have necessarily had their form modified in some degree by processes subsequent to their origin. An account of the age of a slope should aim to specify the nature and amount of form change in different periods of time.

GENERALIZATION AND PARTICULARIZATION

In the environmental sciences, progress in knowledge may either be towards generalization, the formulation of basic principles which underlie a wide variety of phenomena, or towards particularization, the explanation of individual natural phenomena in terms of these principles. Generalization here has a deeper meaning than merely the grouping together of similar but non-identical statements, an example in slope studies being the progression from empirically-derived equations for soil erosion by surface wash to the attempted explanation of these equations in terms of the laws of hydraulics. Such a progression leads ultimately, in all sciences, to the physics of fundamental particles. Particularization involves the application of general principles to specific landforms, for which purpose it may be necessary to assume that the principles are established, even when this is not entirely true. Thus if the evidence of polycyclic valley forms is used in a study of denudation chronology, it is usually assumed that slopes at a late stage of evolution are gentler than at an early stage.

Features of the natural world rarely result from a few simple causes but are greatly influenced by what, in respect of fundamental principles, appear as extraneous accidents. In practice an interchange of information between generalization and particularization is necessary. General principles in geomorphology are built up partly from case studies of individual landforms. Conversely, explanations of particular forms are dependent on progress in understanding the functioning of processes at a more fundamental level.

CONTROL OF SLOPE FORM

Certain concepts concerning the manner in which slope form and evolution are determined are fundamental to discussions of their origin. A *denudation slope* is one on which ground loss is occurring. A *transportation slope* undergoes neither ground loss nor ground gain, because at each point the material brought in from upslope is equal to that carried away downslope. A slope on which ground gain is occurring is an *accumulation slope*. The forms of denudation and accumulation slopes are determined by their form at some previous time, minus or plus the amount of ground loss or gain, respectively, undergone by each point on the surface subsequent to that time. The rate of ground loss will not normally, therefore, be constant over different parts of a denudation slope, nor the rate of ground gain over an accumulation slope. A slope with *uniform ground loss* represents a limiting case for denudation slopes, in which the rate of ground loss is everywhere the same.

In the case of denudation slopes, there is an important distinction between control by weathering and control by removal (cf. Gilbert, 1877, p. 99; Savigear, 1960). On a slope subject to *control by weathering*, the potential rate at which weathered material can be removed from the slope exceeds the rate of weathering; hence weathered material is removed shortly after it is formed, and the rate at which ground loss occurs is controlled by the rate of weathering. Removal may either be direct (e.g. by wind or in a solution in groundwater) or by surface transport. Under these conditions, if a slope is composed of two or more rock types, it is their relative resistance to weathering that determines the slope form. *Control by removal* occurs where the potential rate at which regolith can be produced by weathering exceeds the rate at which it can be removed; ground loss is then dependent on the rate of removal (direct or downslope). Under control by removal the form of a slope of varied lithology is influenced not by rock resistance to weathering but by the properties of the regolith derived from each rock as they affect the processes of removal.

The form of accumulation slopes is subject to *control by accumulation*. A transportation slope is subject to control by removal, the rate of net removal being everywhere nil. It is thus a limiting case separating denudation slopes with uniform ground loss from accumulation slopes with uniform ground gain, in which rates of gain and loss become zero.

These terms are strictly applicable only to a point on a slope; slopes need not be subject to the same type of control throughout their length. Control is likely to vary with rock type and climate. The types of control as here stated are hypothetical, or model, concepts; it is for field and experimental investigation to determine what are the actual conditions on a given slope. If there is no regolith the slope must be subject to control by weathering (rejecting the logical alternative, that the rate of weathering is nil). This applies to most marine cliffs in consolidated rocks, mountains in arid climates, hills of bare limestone and domed rock forms of the humid tropics. The converse is not necessarily true. The existence of a thick regolith suggests control by removal; but a thin regolith may nevertheless be present on a slope subject to control by weathering, if the rate of weathering decreases rapidly with increase in soil depth. A thick and highly-altered cover of weathered rock *in situ* suggests control by removal. If regolith showing indications of downslope transport adjoins relatively unaltered rock, control by removal may be present. Many previous discussions of slope evolution have failed to make clear the distinction between control by weathering and control by removal.

IV SLOPE EVOLUTION: THE MAIN THEORIES

EVOLUTION concerns the change of slope form with time. One of the principal questions is whether a slope, under given circumstances, becomes steeper, less steep, or retains the same form after it has been acted upon by processes for a period of time. Among studies of the different aspects of the geomorphology of slopes, evolution occupies a central position. Knowledge of processes is a basis for the construction of models representing evolution. The observed characteristics of slope form can be brought to bear upon the problem, using either deductive or inductive methods, and in investigating the effects of environment, for example rock type or climate, one of the chief aims is to discover how environmental factors affect evolution. In this chapter, the principal systems of slope evolution that have been proposed are set out, as an introduction to the nature of the problems involved.

There are three main theories of slope evolution, which will be termed slope decline, slope replacement and parallel retreat. These theories are associated respectively with the names of W. M. Davis, W. Penck and L. C. King. There is sometimes confusion between two types of question, which may be represented by the paradigm 'Was Davis right?' and 'Do slopes decline?' The two are not identical, the first being primarily a matter of the history of science, the second concerning the natural world. It is a mistake to discuss the views of an individual as if value-judgements were involved. Discussion of such views in a scientific context is justified insofar as they are a convenient label for a particular set of theories or type of approach; but it is the theory that matters, and not the person.

SLOPE EVOLUTION IN THE DAVISIAN CYCLE

Of the numerous published works of W. M. Davis, those specifically on slopes comprise two short notes of 1892 and 1898, a discussion of Penck's system (1932), and two later articles concerned in part with arid conditions (1930, 1938). His views on slopes under humid climates are set forth mainly in general accounts of the geographical cycle (1899, 1899b, 1902, 1909); the most complete statement is in the textbook of 1912. Davis's account of slope evolution must be taken in the context of his cycle of erosion; he saw the movement of rock waste down slopes as one stage, separating those of weathering and of river transportation, in a continuous process of transfer of the material of rocks into the sea, the whole resulting in a time-dependent change of form as the cycle advanced.

Davis holds scientific priority regarding the hypothesis that the convex upper parts of slopes are produced by soil creep. He described qualitatively the mechanism of movement in creep, by alternate dilatation and contraction of the soil under the influence of gravity, and correctly interpreted outcrop curvature. Reasoning on *a priori* grounds that surface wash increases in volume downslope, he supposed that, near drainage divides, the ratio of creep to wash is large. Creep produces 'rounded contours', and is responsible for the convex profile of divides. Being impressed by the smoothly-rounded appearance of many convexities in humid-climate regions, he saw this as a reason to advocate the overall importance of creep as compared with wash in such climates.

Davis next put forward the concept of graded waste sheets and graded valley sides. This was presented by analogy with the condition of grade that he supposed was attained in rivers, arguing that both rivers and waste sheets are moving mixtures of water and rock waste, albeit in different proportions. 'A graded waste sheet . . . is one in which the ability of the transporting forces to do work is equal to the work that they have to do.' In the youthful stage of the cycle, 'the rocky cliffs and ledges that often surmount graded slopes are not yet graded; waste is removed from them faster than it is supplied by local weathering and by creeping from still higher slopes, and hence the cliffs and ledges are left almost bare'. The graded condition first becomes established at the base of the slope, and works upwards. 'As late maturity passes into old age, even the ledges on ridge-crests and spur-fronts disappear, all being concealed in a universal sheet of slowly creeping waste. From any point on such a surface a graded slope leads the waste down to the streams. At any point the agencies of removal are just able to cope with the waste that is there weathered *plus* that which comes from further uphill' (1899, pp. 495-6). It is clear from these quotations that Davis does not separate two concepts, one about form and the other about process; he equates a graded slope, defined as one having a continuous soil cover, with a graded waste sheet, one on which there is equality between the supply and removal of debris (cf. p. 100).

This leads to Davis's view of slope decline, in which the method of argument by analogy with rivers is continued. 'Just as graded rivers slowly degrade their courses after the period of maximum load is past, so graded waste sheets adopt gentler and gentler slopes. . . . When the graded slopes are first developed they are steep, and the waste that covers them is coarse and of moderate thickness. . . . In a more advanced stage of the cycle the graded slopes are moderate, and the waste that covers them is of finer texture and greater depth than before; here the weakened agencies of removal are favoured by the slower weathering of the rocks beneath the thickened waste cover, and by the greater refinement (reduction to finer texture) of the loose waste during its slow journey' (1899, p. 497). 'The retreat of a valley side is usually accompanied by a decrease in the steepness of its slope as well as by the development of a convex profile at its top and a concave profile at its base' (1932, p. 408). As the cycle advances, the convexity and concavity become longer and acquire a larger radius of curvature. Davis's diagrams of the evolution of valley form are reproduced as Fig. 8.

With regard to the reasons for slope decline, Davis in several places makes statements similar to the following: 'The soil-cloaked, tree-covered slopes of the mountains, mounts and hills will be worn to gentler and gentler declivities, because the downwash of soil from their convex upper slopes will be faster than its removal from the base of the slopes' (1930, p. 141). Other reasons are that the waste sheet protects the underlying rock from weathering, the degree of protection increasing downslope; and that the coarse detritus on the upper part of the slope 'requires' a steeper angle for its transport than the finer debris at the slope foot. Here there is some confusion as to whether the change in slope form is controlled by the rate of weathering or by removal.

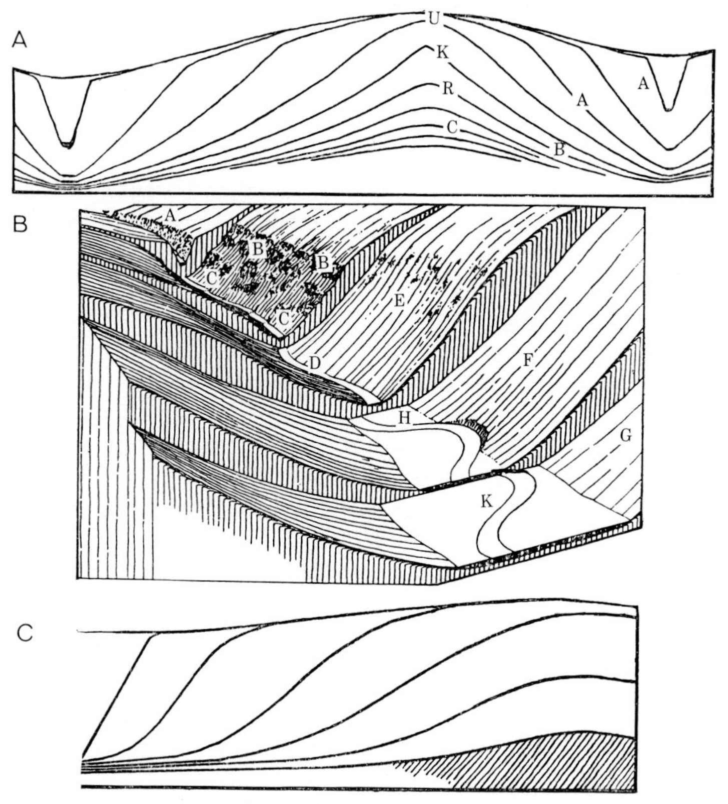

FIG. 8. W. M. Davis's diagrams of slope evolution. *A* and *B* after Davis (1912), *C* after Davis (1932).

Davis considered that in arid climates, rill-wash is a more important process than creep. Under these conditions slopes retreat with a constant angle, developing a pediment below. In 1930 he compared the evolution of landforms under humid and arid conditions, seeking similarities. This paper contains an equivocal passage: he writes of, and illustrates by block-diagrams, 'lateral valley-floor strips', which King (1953) took to be a belated recognition that pedimentation occurs in humid climates. This interpretation is contradicted by other passages from the same paper, which state unambiguously that slope angle remains constant under arid conditions but decreases under humid, and that the respective late-cycle residuals are very different in form. This position is maintained in his last published work, in which it is stated that, even in semi-arid conditions, slopes decline with time (1938, pp. 1361-2).

Davis's faults of scientific methodology have been well rehearsed (Lewis, 1959; Chorley, 1965). The least venial is his use, for descriptive purposes, of terminology with genetic implications; to do so begs the question, if such descriptions are used to *investigate* the origin of landforms. Davis did not, of course, intend them to be: he was convinced he was right. The Davisian model has, however, great flexibility combined with internal logical consistency, which renders it unassailable by critical attacks with its own weapons, those of abstract qualitative reasoning.

The hypotheses on slopes are in varying degrees open to testing. The relative importance, under specified conditions, of soil creep and surface wash is answerable by direct measurement; *a priori* reasoning is a poor substitute. The argument that creep produces convexities and the presence of convexities demonstrates the existence of creep is circular. The statement that slopes decline because more material is removed from the upper than from the lower parts is tautologous. If a graded slope is defined as one possessing a continuous soil cover, this is a simple but potentially useful term, since whether a given part of a slope is graded can be determined by observation, and this opens the way to investigation of the spatial relations of graded and non-graded slopes. Using the Davisian definitions, the existence of a graded slope is neither a necessary nor a sufficient condition for the presence of a graded waste sheet. It is in principle possible to test whether the latter condition is present by measurements of rates of weathering and soil movement, although this would require more refined instrumentation than has yet been developed. The hypothesis of slope decline is more difficult to test. Because of the slowness of geomorphological change, slope evolution can be observed directly in the field only under abnormal conditions, such as badland topography. Davis's theory is a clear statement of a type of slope evolution, and is capable of being translated into quantitative terms. Consonant with its relation to the cycle of erosion, the theory implies that slope form is time-dependent; possibly this is the feature that offers an opportunity of indirect testing by field observation.

WALTHER PENCK'S SYSTEM OF MORPHOLOGICAL ANALYSIS

The method by which W. M. Davis and Walther Penck treated questions of slope development were in some respects similar. Both had made visual observations of slopes in many parts of the world; they explained the forms of these mainly by deductive considerations of the effects of processes, but in neither case were the results of the deductions compared in detail with the forms of specific slopes. There the similarity ends. Davis wrote with lucid expression, but his reasoning does not always bear close analysis; whereas Penck's mode of expression was difficult to the extent that in some cases his meaning remains obscure, but his arguments were constructed with the firmly-based logic of the German academic tradition.

The initial neglect, subsequent misinterpretation and belated recognition of

Penck's work has been noted above. Historically, its importance depends not only upon the intrinsic value of its content but also upon the challenge that it presented to orthodox Davisian ideas. The broader aim of Penck was to relate exogenetic processes and landforms, both of which can be observed, to endogenetic processes, and so discover the nature of the latter by a process of deduction. This approach has subsequently rarely been attempted, although it has not been entirely discredited (Bryan and Wilson, 1931; cf. Simons, 1962). Penck's examination of the processes of mass-movement was in advance of its time, but has been largely superseded. The main parts of his system that are of continuing value are the discussion of the manner of action of processes, and the deductive model of slope evolution.

Penck's analysis of exogenetic processes rests on his discussion of the properties of the regolith cover on slopes, and the mechanisms of weathering and surface processes, among which he considered primarily mass movement. His particular contribution was to analyze the manner of action of processes, that is, the ways in which their operation is influenced by properties of form and in turn cause changes in these properties (1924, Chapter III). Seven properties of form and three rates of action of processes are considered. The following summary gives the definitions of each factor, together with the properties upon which each is dependent. The analysis is essentially Penck's, the distinctions between form and process, between independent and dependent variables, and the comments in brackets having been added here. The translation of terminology follows the English edition of 1953, except that regolith has been preferred to soil; subsequent page references are to this edition.

Properties of form

The *degree of reduction* (*Aufbereitung*) of the regolith is the degree to which it is broken up into fine particles. A fine-textured soil, or one with few stones, has a higher degree of reduction than a coarse-textured, or stony, soil. The degree of reduction depends on the rate of weathering (direct relation) and the rate of denudation (inverse relation).

The *mobility* of the regolith is the ease with which it may be moved by denudational processes. The greater the mobility, the more readily will the regolith be affected by denudation. For each angle of slope a critical degree of mobility is required before the regolith can be moved. Mobility is dependent on the character of the rock (non-parametric relation) and the degree of reduction (direct relation).

The *regolith thickness* is self-defining. It depends on the rate of reduction (direct relation) and the rate of denudation (inverse relation).

The *exposure* of the slope surface is the degree to which the rock beneath the regolith (and in some respects the regolith itself) is exposed to the process of reduction. Exposure depends upon regolith thickness.

The *character of the rock* (*Gesteinsverhältnisse*) refers to all properties of the rock that influence both the rate of reduction and the mobility of the regolith. It is an independent variable. (Rock character is referred to in non-parametric terms, hence its influence cannot be stated as direct or inverse.)

Climate is taken into account insofar as it affects the rates of reduction and of denudation. It is an independent, non-parametric variable.

The *angle of slope* is in the initial stage an independent variable. Subsequently it becomes partly dependent on the rate of denudation (differential relation: a downslope increase in the rate of denudation increases the slope angle).

Properties of process

Reduction is the breaking up of the regolith into finer particles, by the processes of weathering. The rate of reduction is dependent on rock character and climate (non-parametric relations) and on exposure (direct relation).

Denudation (*Abtragung*) is the removal of regolith material from the slope surface (note that this term is used in a special sense, not in its normal meaning). The rate of denudation depends on the climate (non-parametric relation), the mobility (direct relation) and the angle of slope (direct relation).

Renewal of exposure is the exposure of the rock surface beneath the regolith (and in some respects the regolith itself) to the process of reduction. The rate of renewal of exposure depends on the rate of denudation (direct relation). (Renewal of exposure is hardly a process in itself, but is treated as such.)

The apparently complex relations between factors may be rationalized. The first step is to distinguish between form and process, and between independent and dependent variables, as has been done above. The second is to note that there are three closely-related pairs of properties: degree of reduction and mobility, regolith thickness and exposure, and the processes of denudation and renewal of exposure. This is not going beyond what is implicit in the original. What is doubtful is whether Penck appreciated that the relations between properties could differ according to the time-scale involved (cf. Schumm and Lichty, 1965). Influences upon the rates of action of processes are operational over a period of a year or less; these will be termed short-term effects. Factors affecting the regolith are medium-term effects, and are functional over periods of time sufficient to cause substantial change in the properties of the regolith, of the order of 1000-10 000 years. Factors influencing slope form act over the long term, possibly requiring periods of the order of 10 000-100 000 years to cause substantial change.

Penck's system of interactions between properties, modified as described above, is illustrated in Fig. 9. It would now be termed a process-response model. If quantification were to be attempted, modifications to two variables would be necessary. Rock character would require separation into independent variables, each with numerical values, one representing the properties of rocks as they affect the rate of reduction, and the other their effect on the mobility of the regolith. Similarly climate would need to be split into two independent variables, with values representing separately the climatic influences upon the rates of reduction and denudation.

One error, which has been noted in most discussions of this system (Morawetz, 1932; Louis, 1935; Birot, 1949; Beckett, 1968), concerns the relation between the rates of denudation and of renewal of exposure, the latter being equivalent to slope retreat.

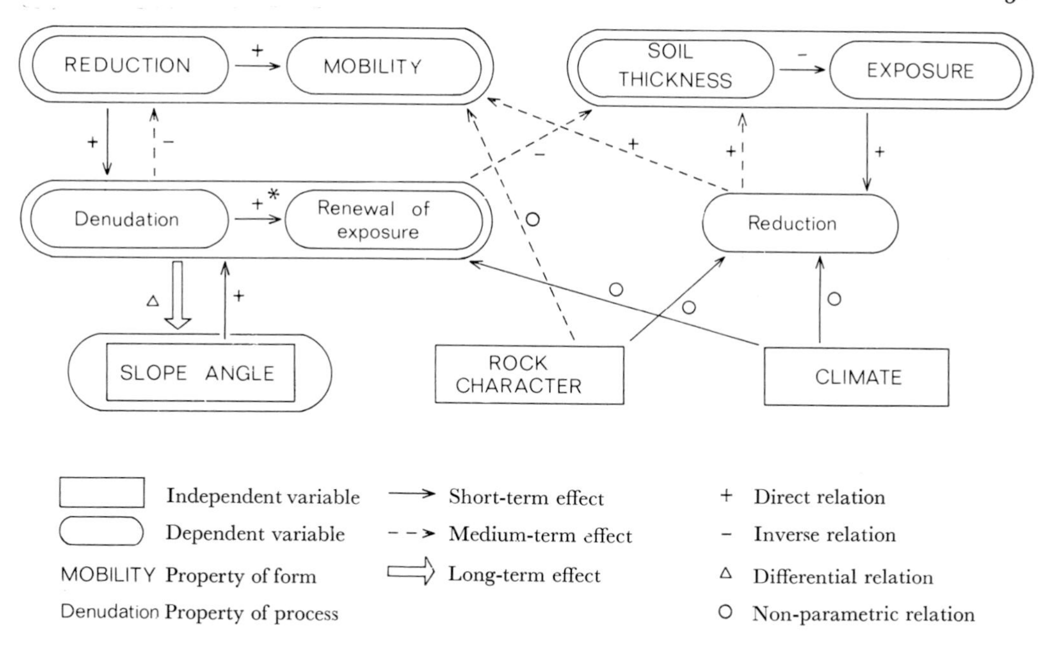

FIG. 9. Representation of the system of interaction between processes and form properties described by W. Penck (1924).

Penck assumed that as soon as rock waste resting on a slope of given angle attains the mobility necessary to be affected by denudation it is instantaneously removed from the slope. He failed to appreciate that, under the very processes of mass-movement that he discussed, the insoluble portion of rock waste derived from any point must be transported across all parts of the slope lying below it. This debris in transit would, to use Penck's own terminology, cause a progressive downslope decrease in the degree of exposure. Birot expresses it by saying that Penck assumed the rate of transportation to be infinite, once sufficient mobility was attained; this is effectively true on rock cliffs, but not on soil-covered slopes. The error is one of matching the model to reality, not in construction of the model itself. The German term for denudation, *Abtragung*, literally means 'carrying away', which might be contributory to this error. In Fig. 9 it means that the relation marked * is not direct, as Penck assumed, but of a differential type. If 'denudation' is re-defined as downslope soil movement, the rate of renewal of exposure depends on a function of the rate of change of denudation with respect to distance from the slope crest.

Penck applied this system to a deductive model of slope evolution (1924, Chapter VI). He took as initial conditions a steep rock cliff of homogeneous composition (Fig. 10*A*). The slope is surmounted by a level surface. A river at the slope foot removes all debris supplied to it, but does not actively erode. 'In unit time a superficial layer of rock, of a definite thickness the same everywhere, is loosened and removed. The method of removal is that loosened particles of rock crumble away and fall down.

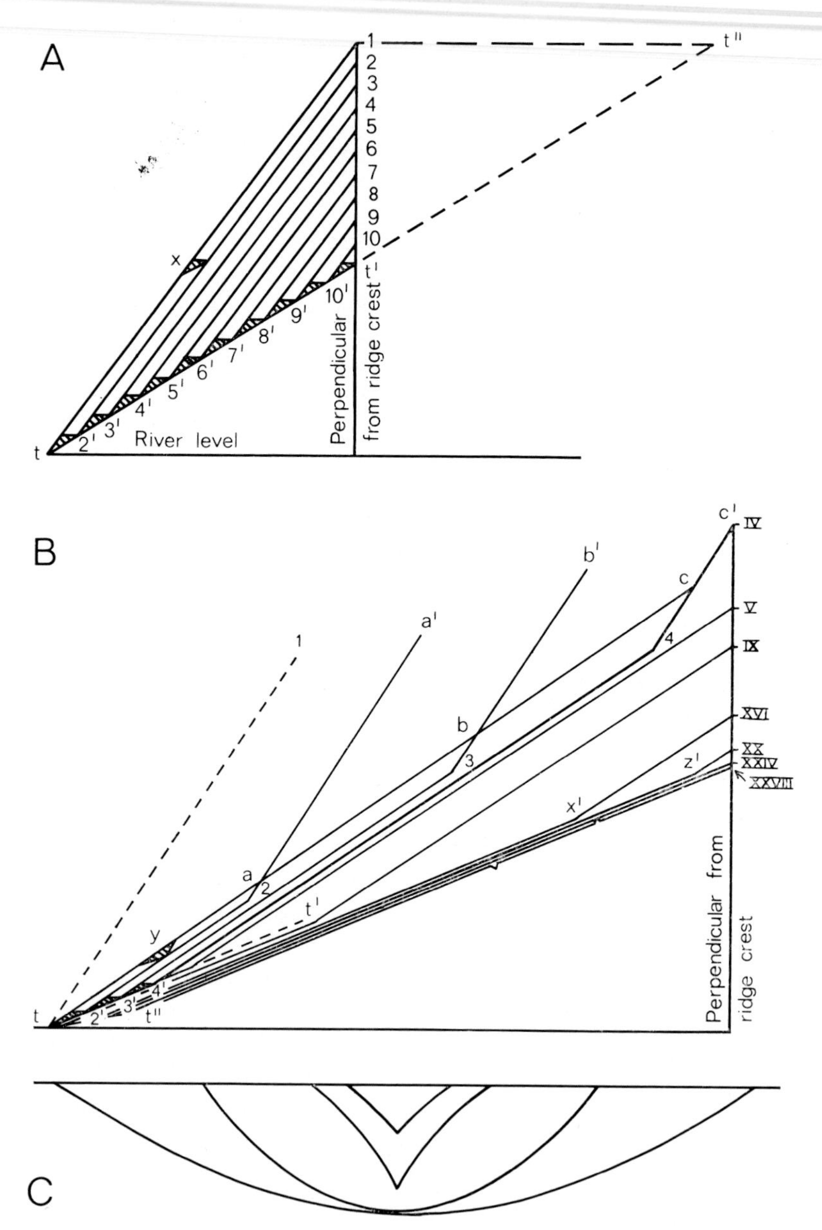

FIG. 10. W. Penck's diagrams of slope evolution. *A* and *B* after Penck (1924). *C* after Simons (1962), based on Penck (1925).

For this to happen the gradient must be too great to allow the little pieces of rock, just loosened by weathering but not further comminuted and reduced, to remain at rest. This gradient is available for each unit of the rock face except the lowest' (p. 134).

Thus the whole slope retreats parallel to itself except for the fragment *t*-2′, which does not rest on a slope angle sufficient to permit its removal at that degree of mobility. The same sequence is repeated during a second time interval, except that now 'the lowest particle (2′-3′) is without the same gradient at its disposal, i.e. it has not in the interval acquired the mobility essential for movement on the much smaller gradient. The rock face moves into the position 3′-3. In the third unit of time it retreats to 4′-4, and so on. If sufficiently small units are taken, we come very near to the actual process, and the exceedingly small ledges *t*-2′, 2′-3′, 3′-4′, and so on, combine to give a continuous slope of uniform gradient (*t*-*t*′). This is the *basal slope*. . . . The following statement may therefore be made: *a steep rock face left to itself, moves back upslope, maintaining its original gradient; and a basal slope of lesser gradient develops at its expense*' (p. 135).

This model is a correct deduction from Penck's assumptions. Moreover, as applied to the transition from rock cliff to the slope which first replaces it there is broad correspondence to reality. Penck, however, applied identical reasoning to the debris-covered slope (Fig. 10*B*). Reduction proceeds on the basal slope *t*-*c* until a stage is reached when the whole of this slope has been changed into a sufficiently mobile form for movement to take place on it; the required degree of mobility is higher than that for the rock cliff, since the angle is gentler. 'All [rock particles] move down except the lowest, that adjoining the general base level of denudation (*t*), since it is the only one not provided with the requisite gradient for movement. Still maintaining the same inclination, the slope now moves from the position *t*-*a* to 2′-2' (pp. 136-7). This is repeated until below the basal slope there is formed a new slope of lesser gradient, *t*-*t*′. The latter is in turn replaced by a still gentler slope, the commencement of which is shown as *t*-*t*″. 'It is now quite clear that the process obeys a law: *Flattening of slopes always takes place from below upwards*' (p. 138).

There are two errors here. It has already been noted that the transition from the case of cliffs, with reasoning based on the process of rock fall, to the case of regolith-covered slopes, without modifying the assumption of quasi-instantaneous removal of regolith, is unrealistic. The second error is to assume that the whole of the basal slope, or its later replacement slopes, is equally exposed to weathering; it is apparent from Fig. 10*B* that the period of exposure will decrease from a maximum at the slope foot to nil at the intersection with the slope above. Again this is a case where what is a correct assumption for a cliff becomes incorrect for other slopes. Penck appreciated that down-slope soil movement could cause rounding of an initially sharp summit intersection into a convexity, but argued that there was a distinct downward limit to this 'flattening from above' (pp. 141-3).

Penck's application of his system of analysis to cases with river erosion at the slope foot will not be discussed in detail. The deductions follow an identical method. They lead to the conclusion that river erosion at a progressively increasing rate will generate a convex slope*; erosion at a constant rate will be indicated by a rectilinear slope rising from the river, its angle being proportional to the rate of erosion. It follows that if the

* Penck does not say, as some commentators have assumed he does, that a smoothly-accelerating rate of erosion will produce an angular break of slope.

rate of erosion abruptly increases, a convex break of gradient will be produced; the steeper slope below it will work progressively upslope, replacing the gentler unit above.

One further aspect of his model has acquired significance in the light of more recent slope profile studies. This is the question of whether he considered that slope evolution actually took place by discrete stages, or whether this was just a device adopted for explanatory purposes, the stages being regarded as infinitely small. It affects whether a slope evolving according to his model would be expected to consist of a series of intersecting rectilinear segments or a smooth concavity. Textual passages can be found in support of either interpretation of his meaning; on balance the latter alternative seems indicated, and is supported by the diagram of slope evolution in his later paper on the Black Forest (Fig. 10*C*).

Penck's system of morphological analysis is not only of historical significance. It stands in its own right as a powerful deductive model, applicable to cases where the manner of action of denudational processes simulates instantaneous removal of material, and where there is unimpeded basal removal. Rockfall is one such process; removal in solution, insofar as it takes place within and immediately below the regolith and not in the body of the rock, may be another. The assumptions of the model become invalid, however, for slopes on which regolith removal takes place in part by downslope transport; it is also questionable whether Penck's assumption of parallel retreat of a weathering cliff is correct (Mortensen, 1960). Given these qualifications, Penck's model presents a hypothesis of slope evolution in which gentle slopes succeed steeper, but in a manner very different from the Davisian scheme. To refer to Penck's hypothesis as one of parallel retreat was an error (attributable to Davis, 1932) which died hard. In Penck's system parallel retreat only takes place on an initial rectilinear slope; such a slope is transformed into a concave form relatively early in its evolutionary history, and development then proceeds by successive replacement from below. If this replacement is not in discrete stages but continuous, then no part of the concave slope retreats parallel to itself. This system of slope evolution is therefore more appropriately referred to as the hypothesis of slope replacement.

THE HILLSLOPE CYCLE OF L. C. KING

L. C. King's theories on slopes are part of a wider scheme of landscape evolution that he proposed. This scheme is presented as having three parts: the river cycle, the hillslope cycle and the landscape cycle (King, 1951, 1953, 1962). The river cycle follows accepted, in part Davisian, theory. The hillslope cycle sets out a system of evolution differing fundamentally from those of both Davis and Penck (the most detailed account is King, 1957). It is an eclectic scheme, derived mainly from the work of Wood (1942) and Fair (1947, 1948), with some points taken from Penck and Horton (1945). King saw that Wood's concepts were applicable to conditions in southern Africa, in particular to some of the results obtained from field studies in Natal by Fair. He considered that sub-tropical semi-humid climates had a better claim to be regarded as 'normal' for landform evolution than the humid temperate lands on which much geomorphological thinking, including that of Davis, had been based; such lands are now known to contain abundant features relict from periglacial conditions and, King argued, do not therefore allow processes to be studied in conjunction with the resulting landforms. King's landscape cycle involves the formation of erosion surfaces by a process termed pediplanation, comprising scarp retreat and the production of extensive plains by the growth and coalescence of pediments. It is the work on surfaces that is King's main contribution to geomorphology; this lies beyond the scope of this book. The ideas on slopes are not original, but deserve consideration both as a comprehensive (and vigorously proselytized) synthesis, and because the landscape cycle is dependent upon their validity. The precursors of the scheme should first be examined.

Wood's discussion of the development of hillside slopes (1942) was a product of its time, a wholly theoretical paper in which the deductions are not tested against quantitative field observations. He started by considering the retreat of a rock cliff, terming a rock slope uncovered by soil a *free face*. As the free face retreats, talus will accumulate at its base; the talus initially stands at a uniform angle, and is termed the *constant slope*. Subsequently the weathered scree is affected by rain-wash, which carries the finer particles further than the coarser, producing a concavity, the *waning slope*. At the crest of the cliff weathering attacks the initial angular intersection from two faces to produce a convexity, the *waxing slope*. The scheme of evolution for a valley slope below which the river has ceased erosion is shown in Fig. 11. Wood emphasized that the manner of

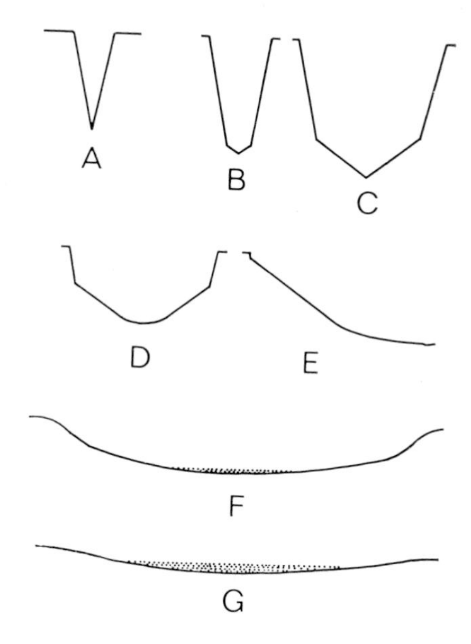

FIG. 11. Wood's diagram of the slope cycle. A = free face only; B and C = constant slope forms; D and E = waning slope develops; F = waxing slope forms, waning slope rises up side of constant slope, alluvial filling represented by dots; G = constant slope has been consumed, alluvial fill deepens, slopes gradually flatten and approach a peneplain. After Wood (1942).

evolution was not the same for all slopes, but depended on structure, climate, and conditions at the slope base. Under particular conditions one or more of the elements may be absent. The form and relative properties of the four slope elements do not remain unchanged during their evolution. Wood's contribution was to define these elements, and to suggest that they might be identified on slopes of apparently very dissimilar form.

The study of Natal by Fair (1947, 1948) has a claim to be considered the finest single contribution ever made to the study of slopes. It is based on field surveys of slope form, by profiles, under different climatic and structural conditions. The slope elements identified were the waxing slope, free face, talus or detrital slope, and pediment. The detrital slope is the equivalent of Wood's constant slope even where, in humid regions, the boulders which fall onto it from the free face become buried by soil and vegetation. These observations were the basis for deriving information on slope evolution, by the method of induction. Prima facie evidence existed that the observed slopes represented various stages in a time-sequence of evolution: some were close to river courses, others separated from them by broad pediments. Putting together the presumed stages, Fair found that the manner of evolution differed according to structure, climate and conditions of debris removal at the slope foot. Among the findings was that in certain cases slope form remained almost identical for long periods of evolution; the characteristics and relative proportions of the waxing slope, free face and detrital slope remained unchanged, whilst the pediment increased in length. This implies that the first three elements retreat parallel to themselves. This type of evolution only occurred when two conditions were present: a relatively resistant cap rock (dolerite or sandstone) overlying weaker strata, and adequate basal removal.

L. C. King adopted and redefined the hillslope elements of Wood. His definitions are given in Fig. 12.

In a 'standard' hillslope all four elements are present. King argues that they are the 'natural' product of slope evolution, whether by flowing water, mass movement, or both. Full development of the elements depends on local conditions, chief among which are a strong bedrock and adequate relief. If either of these conditions are lacking the scarp may be absent, the debris slope is necessarily also missing, and 'a decadent convexo-concave hillslope then results'.

During slope evolution the scarp element wears back parallel to itself, and controls the evolution of the slope as a whole. By observation the debris slope does not grow upwards to cover, and eliminate, the scarp, therefore there must be a balance between the supply of debris and its removal. A balance of a similar type keeps the form and extent of the crest constant. Consequent upon the retreat of the upper elements, the pediment extends in length. Since the concave curve which the pediment acquires early in its growth would, if extended upslope, progressively eliminate the upper elements, it must be that the pediment is regraded; that is, as it extends in length its angle is slightly but continuously reduced.

Where the scarp face is not present, because of weak rocks or insufficient local relief, slope retreat may follow a different pattern, with a decline in the maximum angle. King emphasizes, however, that slopes having a scarp are 'normal', and if it is absent the 'active' element in slope retreat is gone. This applies particularly to humid temperate regions where slopes are largely relict, owing their form to evolution under periglacial conditions. 'The long concavo-convex slopes of certain northern lands must be regarded as degenerate, evolving no longer by scarp retreat but more or less atrophied'

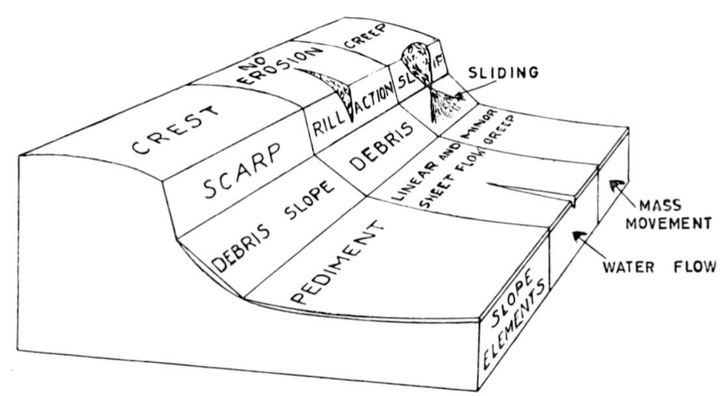

FIG. 12. The standard hillslope of L. C. King (1962).

Crest (waxing slope)—the summit area of a hill or slope; the profile is usually convex. Weathering and soil creep are the main processes forming this convexity.

Scarp (free face)—the bedrock outcrop on the steepest part of the slope. 'It is the most active element in backwearing of the slope as a whole.' Backwearing is caused by rillwash and landslides.

Debris slope (constant slope)—formed by detritus fallen from the scarp above. Its angle is determined by the angle of repose of the coarser material. Weathering reduces it to finer particles, which are then removed by wash, flowing as rills or turbulent sheet-flow.

Pediment (waning slope)—is the broad concavity extending from the base of the other elements to the stream or alluvial plain. Although frequently veneered with detritus, it is essentially a rock-cut feature. It is produced by surface wash, and its profile may approximate to an hydraulic curve.

(1953, p. 734). He disagrees with the approach based on climatic geomorphology, in which slope evolution is considered to differ in basic respects in different climates. 'Our thesis will be that the basic physical controls of landscape remain the same in all climatic environments short of the frigid or extremely arid' (1957, p. 82).

King's opinions lack a basis of detailed field observations; no measurements of either process or form appear in his published work. Although this is a severe indictment of any study of the natural environment, it could be maintained that he was concerned to present a broad view, based upon reconnaissance observations on a world scale previously unparalleled in geomorphology, and that this is no bar to subsequent detailed investigations of particular aspects. A more serious fault is the doctrinaire manner in which the views are presented. Repeated assertion is employed in place of evidence as a means of establishing a hypothesis. Evidence from the work of others is selectively mustered in support of preconceived views. An example is the argument put forward involving the hydrological distinction between laminar and turbulent flow of water; it is asserted that turbulent flow on the upper slope elements gives place to laminar flow on the pediment. This has never been experimentally verified. Where there exists clear field evidence contradicting a view, the circumstances are dismissed as not 'normal'. King rightly censures Davis for calling the humid temperate cycle of erosion 'normal', yet himself applies this term to the sub-tropical landscapes of southern Africa, and to slopes that possess a scarp (free face). Whether or not a landform is 'normal' in this implied sense is an unproductive subject for debate. If information on a world scale about slopes with and without a free face were desired, a scientific approach would be in terms of frequency. By means of sampling, an estimate could be obtained of the proportion of slopes on the earth's surface having a free face, together with confidence limits for this estimate. The resulting figures would stand as impartial factual evidence, without the special pleading implied by calling such slopes normal. His writing lacks objectivity. The defence is that he was advocating an approach that ran counter to received opinion, and revolutionaries cannot be moderates. The value of King's work is in setting up hypotheses for testing; they are not proven.

THREE HYPOTHESES OF SLOPE EVOLUTION

Before summarizing theories of the evolution of entire slopes, consideration of change in form on a part of a slope is necessary. Parts *A* to *D* in Fig. 13 each show two stages

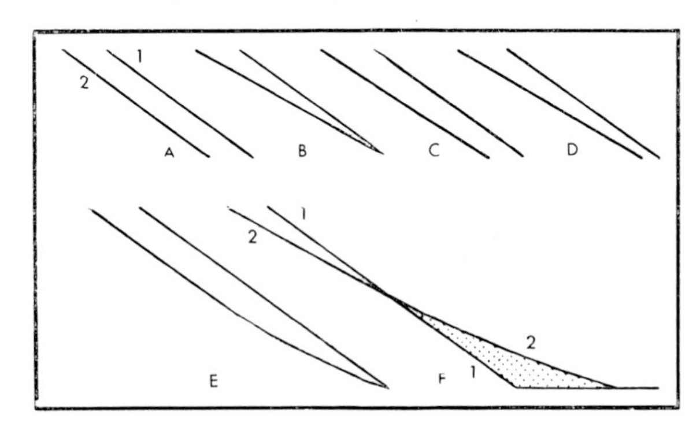

FIG. 13. Parallel retreat and slope decline in parts of a slope. For explanation of lettering, see text. After Young (1963).

in the evolution of a short rectilinear segment of a slope. In *A*, all parts of the segment have retreated by the same amount, resulting in *parallel retreat*. In *B*, the lower end of the segment has not moved, whilst the upper end has retreated; this is the ideal case of *slope decline*. In both *C* and *D*, all parts of the segment have retreated, but the upper parts by a greater amount. This type of evolution consists partly of parallel retreat and partly of slope decline, the former being relatively more important in *C* than in *D*. If the base of the slope remains unchanged, *B* shows that slope decline alone cannot give rise to the formation of a concavity. A proportion of parallel retreat must be present in the segment above, as in *E*, for a concavity to form. Slope decline alone can only be associated with a concavity if the base of the slope advances through accumulation, for example on a flood-plain, as in *F*; in the latter case the concavity will be distinguishable in the field as an accumulation form.

The three types of slope evolution illustrated in Fig. 14 are termed the hypotheses

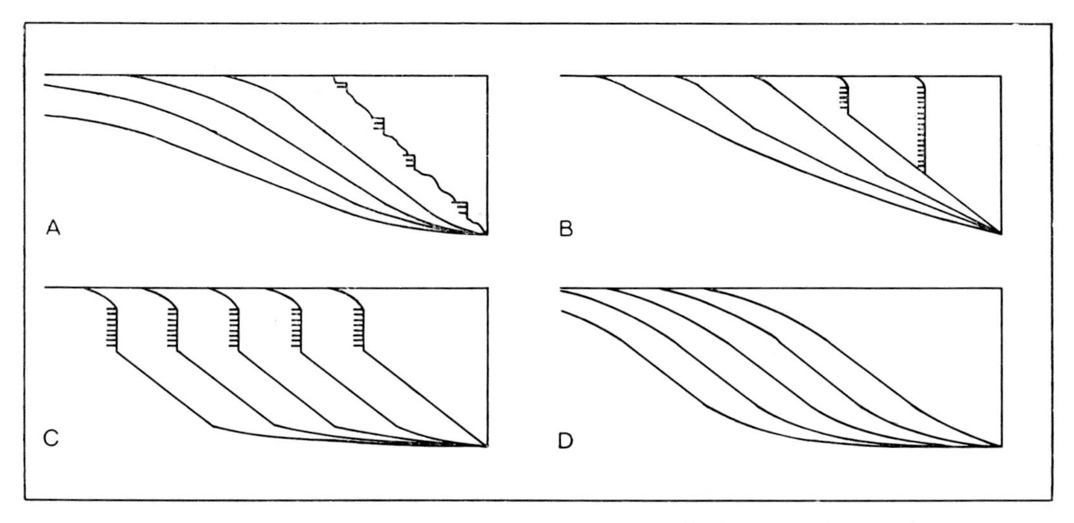

FIG. 14. Three hypotheses of slope evolution. *A* = slope decline; *B* = slope replacement; *C* and *D* = parallel retreat, in *C* with a free face, and in *D* without a free face.

of *slope decline*, *slope replacement* and *parallel retreat*. The assumed initial conditions are homogeneous rock, a vertical or steep cliff above which is a level surface, and conditions at the slope base of unimpeded basal removal, but no erosion. There is no association with any particular climate, vegetation cover or surface processes, nor is any time-equivalence between the stages illustrated implied. The types of evolution are defined as follows:

Slope decline—the steepest part of the slope progressively decreases in angle, accompanied by the development of a convexity and a concavity.

Slope replacement—the maximum angle decreases through replacement from below by gentler slopes, causing the greater part of the profile to become occupied by the concavity. The concavity may be either smoothly curved or segmented.

Parallel retreat—the maximum angle remains constant, the absolute lengths of all parts of the slope except the concavity remain constant, and the concavity increases in length; the angle at any (vertically constant) point on the concavity decreases. This type of evolution may be subdivided into parallel retreat *with a free face* and *without a free face*.

In slope decline, the convexity normally occupies at least 20% and the concavity at least 10% of the total length of the profile. In slope replacement, the concavity normally occupies at least 60% and the convexity not more than 5% of the total length. These values do not apply in the early stages of evolution.

Models of evolution intermediate between these basic hypotheses may be devised, but to be of value, they would require quantitative specification. Such models do not exhaust the possible modes of evolution; for example, they do not provide for the possibility that at any vertically constant point on the slope the angle may progressively steepen, although such behaviour has been observed (Wurm, 1936; Swan, 1970b). Nevertheless, the three basic hypotheses represent critical models, or end-points in a series of modes of evolution. It is a fundamental aim of slope studies to discover to which of these models actual slope evolution most closely approximates, under different structural and climatic conditions.

V SURFACE PROCESSES: GENERAL CONSIDERATIONS, WEATHERING AND SOLUTION LOSS

THE cycle undergone by rock material that comes close to the land surface comprises six stages: uplift, above sea-level, by endogenetic processes; weathering, converting the rock to regolith; surface transport, the removal of regolith from slopes; transport by rivers; deposition, on the sea-bed or other sites; and consolidation together with other processes that convert the deposited sediment into rock. Glacial and arid conditions cause modifications to these stages. In the study of slopes it is convenient to consider both endogenetic processes and river erosion as external variables. The stages directly responsible for slope formation are therefore weathering and surface transport.

Two distinct aspects of denudational processes are their mechanism and manner of action. The *mechanism* of a process is how it operates: the forces and agents that are involved, and their system of interrelation. The *manner of action* is the way in which the process affects form. For example, part of the mechanism of surface wash is that runoff occurs when rainfall intensity exceeds the infiltration capacity of the soil; whereas it is a feature of the manner of action of surface wash that the rate of transport increases with distance from the slope crest. Manner of action is dependent upon mechanism. An empirical equation relating to process is an equation which describes the manner of action without explaining it in terms of mechanism. It is the manner of action of processes that is the basis for the construction of process-response models.

The methods for investigation of processes may be classified as follows:

Field Methods

(*i*) Measurement of the present rate of action of processes. Such observations usually relate to the total, or net, form-change produced by processes. Field measurement of soil creep is an example. The results of work of this type are normally of an empirical nature.

(*ii*) Field investigation of particular aspects of processes. Recording the shear strength of regolith *in situ*, or the periodic moisture changes taking place in the regolith, fall into this category. Such work is normally directed towards understanding the mechanism of processes.

(*iii*) Indirect investigation of processes, through study of their consequences in the form of deposits. This method has been most widely applied to deposits interpreted as being of periglacial origin. It is a standard method in geology, and is frequently resorted to owing to the extreme slowness of many processes compared to the human time-scale. It is not, however, the study of processes *per se*, and in work of this type great care must be taken to avoid circular reasoning.

Laboratory methods

(*iv*) Laboratory testing of the properties of materials. These may either be artificial materials, e.g. glass spheres, utilized to obtain controlled conditions to the sacrifice of field verisimilitude; or rock and regolith sampled in the field. Examples are the laboratory testing of shear strength, rock fracture on heating and cooling, and the rates of percolation of water through soil.

(*v*) The use of scale models, for example simulated rain on sand models of landforms. The main reason for using such models is to speed up the time scale. Interpretation of the results requires careful consideration of the effects of scale change upon physical constants, for example density, viscosity and strength.

Theoretical methods

(*vi*) Theoretical considerations, based on the application of physical laws to the circumstances of the regolith on slopes.

All these methods can contribute to knowledge of processes, but they are most valuable when applied in conjunction. Empirical field investigations can normally only suggest, but not prove, basic mechanisms. On the other hand, laboratory and theoretical investigations are severely prone to become divorced from the reality of field conditions, as is abundantly exemplified in the literature of the natural sciences. In general, the most profitable means of approach is to commence with field observations, leading, by inductive reasoning, to a hypothesis; and to devise and carry out a critical experiment to test this hypothesis under controlled laboratory conditions, or partly-controlled field circumstances. Given a degree of control over external variables, it may then be possible to relate the experimental results to fundamental physical laws, thence carrying through this explanation in part to the results of the original fieldwork. The study of talus weathering and scarp recession in the Colorado Plateaus by Schumm and Chorley (1966) exemplifies this approach.

The standard account of surface processes is that by Sharpe (1938); it is a matter for regret that surface wash was not included in this classic work. Other accounts are contained in textbooks on general geomorphology, amongst which those of Cotton (1941, Chapters 2, 3) and Thornbury (1954, Chapters 3, 4) may be noted.

STRESS, STRAIN AND MATERIALS

One means of analyzing processes is in terms of type of stress, types of strain, and the physical properties of materials. The following account is based, with modifications, on that of Strahler (1952; cf. also Leopold *et al.*, 1964; Farmer, 1968). Stress is the force applied to a body; strain is the resulting movement, of the body as a whole or its parts. The stresses that occur in surface processes are of gravitational, molecular and biological origin. *Gravitational stress* (g) acts vertically on the particles of the regolith. Since downward movement is normally constrained by the bedrock, it is the component of gravitational force acting parallel to the ground surface that is effective in surface transport; this component is proportional to the sine of the slope angle (Fig. 15).

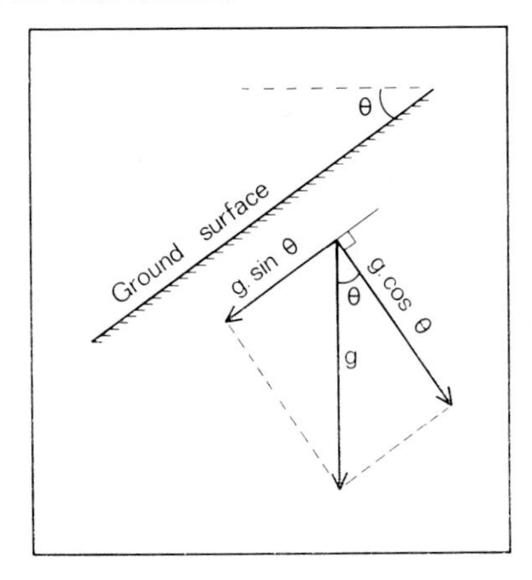

FIG. 15. Components of gravitational stress acting parallel and perpendicular to the ground surface on a slope.

Gravitational stress is superimposed on all other forces to which the regolith particles are subjected, tending to give a downslope direction to the total, or net, stress. *Molecular stress* originates from phenomena such as the swelling and shrinkage of colloids on wetting and drying, thermal expansion and contraction, and the growth of ice crystals. *Biological stress* is caused by the growth of plant roots and movement of soil animals. In relation to certain kinds of creep it is convenient to distinguish a further class of *inter-particle stress*, the forces acting between particles of greater than molecular and colloidal dimensions; particles may be defined as solid fragments with a diameter exceeding 2μ, the conventional upper size limit for clays. Inter-particle stresses are ultimately derived from gravitational, molecular and biological stresses.

Strain may take place by fracture, laminar flow, turbulent flow, or inter-particle movements (Fig. 16). In *fracture*, strain takes place entirely along a plane, the material

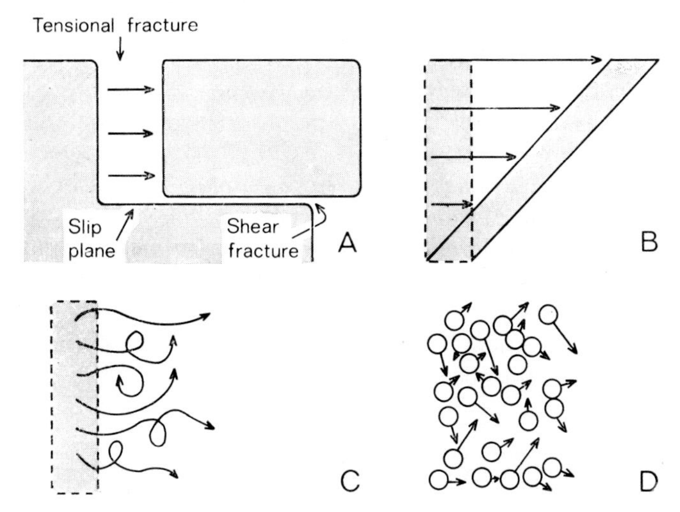

FIG. 16. Types of strain. A = fracture; B = laminar flow; C = turbulent flow; D = net shear. A and B based on Sharpe (1938).

on either side of the plane remaining undeformed. It may be divided into *shear fracture*, occurring along a slip plane, and *tensional fracture*, in which the material on either side of the plane becomes separated. In *laminar flow*, strain is distributed throughout the material, as if by an infinite number of parallel slip planes; the deformation need not be linearly distributed. *Turbulent flow* involves irregularly distributed deformation of the entire material, with mixing and some movements in all directions, including the direction opposite to that of the net flow. The deformation of a material consisting of a large number of solid particles, moving by shear and tensional fracture along numerous irregularly distributed planes, is not satisfactorily described by any of the preceding types of strain; such strain is therefore termed *net shear* (Kirkby, 1967).

According to the relation between stress and strain, materials may behave as rigid solids, elastic solids, plastic solids, fluids, or particulate matter (Fig. 17). In a completely *rigid solid* there is no strain until the *strength* of the material is exceeded, upon

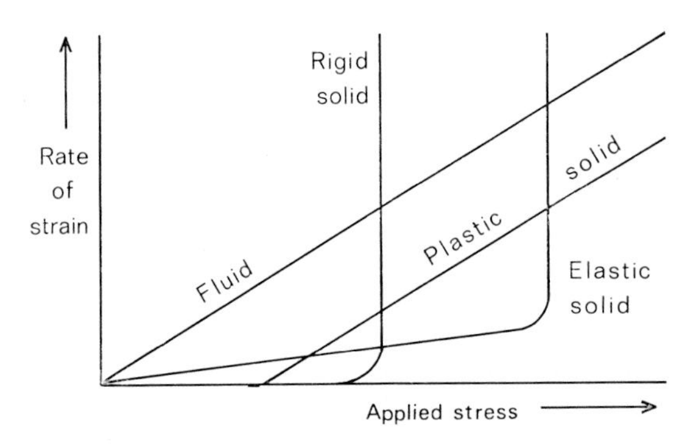

FIG. 17. Stress/strain relations in ideal materials.

which fracture occurs. In *elastic solids*, slow continuous deformation, proportional to the applied stress, occurs prior to fracture. *Plastic solids* possess a finite strength, termed the *yield limit*; the rate of deformation is proportional to the amount by which applied stress exceeds the yield limit. *Fluids* possess no strength, and the rate of deformation is directly proportional to the applied stress and inversely proportional to the *viscosity*. These are idealized conditions, and many natural materials possess more complex stress-strain relations. Consolidated rocks approximate to quasi-elastic solids. The regolith consists of a mixture of solid particles, colloidal clay and water. If it becomes saturated it may behave temporarily as a fluid. In its normal moist or dry state, however, relatively little is known of its behaviour, particularly when undergoing net shear, occurring slowly and over a long period. Previous attempts to treat the regolith as a plastic solid, and movement within it as laminar flow, are an oversimplification. It is therefore provisionally termed *particulate matter*, the properties of which remain to be investigated.

WEATHERING

Weathering and surface transport are the two main groups of processes responsible for slope formation. The mechanism of weathering, together with its relation to landforms, is treated by Ollier (1969); details of chemical processes are discussed by Reiche (1950) and Loughnan (1969). Attention is here confined to aspects of weathering that are of particular significance for slopes.

Some critical questions concerning the manner of action of weathering in relation to slope evolution are as follows:

(*i*) What is the volume change undergone by a given rock on alteration to regolith?

(*ii*) What proportion of the rock mass is ultimately susceptible to chemical weathering?

(*iii*) What proportio 1 of the rock mass is ultimately lost, or is susceptible to loss, in solution?

(*iv*) How does the rate of weathering vary with rock type?

(*v*) What are the order-values for actual rates of weathering under given conditions?

(*vi*) How does the rate of weathering vary with slope angle, curvature and position on the slope?

(*vii*) How does the rate of weathering vary with thickness of regolith?

As regards volume change, both Reiche (1950) and Ollier (1969) stress that most forms of chemical weathering are either constant-volume or involve an increase in volume. Weathering alone, without removal of the altered products, will therefore not cause ground loss. Questions (*ii*) and (*iii*) are not equivalent. The constituents of a rock may be divided into material almost unaffected by weathering, material altered by weathering into insoluble forms, such as silicate clay minerals and some iron compounds, which must then be removed from a slope by surface transport, and material lost in solution. The latter includes material removed in solution from non-calcareous rocks, e.g. as silicic acid and dissolved iron oxides. The weathering properties of rocks may be approached in terms of their constituent minerals; for example, quartz is almost unaffected by weathering, feldspars and ferromagnesian minerals lose part of their matter in solution whilst part is converted to clay minerals, and calcium carbonate is largely lost in solution. Thus quartzite and pure limestone lie at opposite ends of a scale; the former can lose little of its volume by solution, hence slope retreat is dependent on surface transport, whereas slopes on the latter may result largely from direct removal in solution. Non-calcareous shales and sandstones have less potential for solution loss than most igneous rocks; igneous and metamorphic rocks of acid (siliceous) composition have less than basic rocks. The rate of weathering for different rock types does not necessarily vary directly as the proportion of the rock susceptible to loss in solution. Properties of the rock other than its composition must be taken into account. For further discussion of weathering in relation to rock type reference may be made to Ollier (1969, Chapter 7).

Absolute rates of weathering vary widely, and order-of-magnitude values are not available. Questions (*vi*) and (*vii*) above involve variations in the rate of weathering on an individual slope, and assumptions concerning them are necessary for the construction of process-response models. Chemical weathering is more rapid the moister the regolith; therefore it might be assumed on prima facie grounds that the rate of chemical weathering is independent of slope angle, but is faster on concavities than on convexities, and faster on the lower than on the upper parts of a slope. These assumptions have not been confirmed experimentally. Of still greater importance is the variation in rate of weathering with thickness of regolith. Physical weathering becomes much less effective beneath even a few centimetres of soil. Chemical weathering is probably slower on bare rock surfaces than below soil. It has been widely assumed that the rate of chemical weathering decreases with increasing depth of regolith, but there is no direct evidence available on this point. It is desirable that research organizations should establish long-term experiments, involving the burying of weighed rock particles at different depths in the regolith, recovering them after a period of years, and determining the variation in loss of weight with depth of burial.

SOLUTION LOSS

It is remarkable that so little attention has been given to solution loss as a denudational process. For the special case of limestones the mechanism is well understood and its importance recognized: calcium carbonate is converted by carbonation into the relatively soluble form of calcium bicarbonate, which is subsequently removed in solution in groundwater. In the more common case of siliceous rocks the main weathering process is hydrolysis. The products of hydrolysis are secondary clay minerals, which remain in the soil; silicic acid, of which part remains as amorphous silica (sometimes called allophane) and part is lost from the soil; and metallic cations, which are leached from the soil unless adsorbed by the clay-humus complex. Iron released on weathering may remain or be removed, according to the nature of the soil-forming processes operative. In most igneous, metamorphic and sedimentary rocks there is a loss in solution and colloidal suspension of 10-50% by weight of the original rock. For example, in the regolith on granitic gneiss under rainforest in Ghana, Nye (1955) found that rock still clearly *in situ* has a specific gravity of 1·53, showing that 'at least 41% of the original material in the fresh rock has been lost by solution in the ground water'. By volume the loss is smaller or may be nil, since most weathering products are of lower density (Ollier, 1969). Analyses of river water show that large quantities of silica are removed annually in solution from river basins; part of this, possibly a high proportion, is derived from the regolith on valley slopes. In a quantitative comparison of surface processes in a polar environment Rapp (1960) found that the transport of dissolved salts caused a net movement of materials equal to that of all other processes combined. Rougerie (1967) studied the variation of dissolved substances in drainage waters; the amounts are of the same order of magnitude in areas of igneous and metamorphic rocks and sandstones as in limestone areas, showing that the latter are not unique in susceptibility to loss by solution. With respect to climate, the highest values were found in the tropical arid zone. On Flysch sandstones in the Carpathians, Gerlach (1967) found a mean ground lowering by solution of 26 mm/1000 years.

The manner of action of solution loss differs from that of most other surface processes. The material lost from a point is not necessarily transported across all lower parts of the slope; it may pass directly into the bedrock. The manner of action is therefore one of direct removal (cf. *innere Abtragung*, Rohdenberg and Meyer, 1963). When applied to process-response models, direct loss acts differently from processes of downslope transport; and an assumed low rate of direct loss has an equivalent effect on change in slope form to that of high assumed rates of downslope transport (p. 115).

Solution loss is a major process by which rock material is removed from slopes. The work to be done by processes of surface transport is correspondingly reduced to that of removing unweathered primary minerals, principally quartz, together with secondary minerals and other products of weathering. In quantitative studies of surface processes, solution loss should be included. Estimates of its importance *vis-à-vis* soil creep and surface wash, under given rock and climatic conditions, are needed.

VI CREEP AND SOLIFLUCTION

SOIL CREEP

Soil creep is the slow downslope movement of the regolith as a result of the net effect of movements of its individual particles. This definition corresponds to 'seasonal creep' as described by Terzaghi (1950). It is more restrictive than the definitions of Sharpe (1938) and Parizek and Woodruff (1957b), who applied the term creep to all downslope movements taking place at a rate imperceptible except to observations of long duration. Essential features of soil creep are its dependence on inter-particle stresses, net downslope movement being activated by gravity and occurring through net shear (p. 44), and the fact that no continuous external stress is applied to the regolith. Sharpe included soil creep in the category of slow flowage, but this is misleading, as the nature of movement differs from that of both laminar and turbulent flow. The common engineering use of the term creep refers to slow but continuously-distributed strain, resulting from stress that is either external to the material affected or is caused by the weight of a considerable thickness of such material; this essentially different type of movement is termed *continuous creep* by Terzaghi (1950), and is discussed in a later section (p. 57).

The existence and nature of soil creep, the field evidence for it, and its principal causes were first recognized by Thomson (1877: outcrop curvature, plant roots, wetting and drying), Keeping (1878: outcrop curvature, freeze-thaw, roots), Coppinger (1881: wetting and drying), Kerr (1881: freeze-thaw) and Davison (1889: freeze-thaw, a quantitative study). All of the main causes as now recognized are included in accounts by Davis and Snyder (1898) and Gilbert (1909). An outstanding early field study of outcrop curvature and the mechanism of creep is that of Götzinger (1907). The term outcrop curvature is due to Cotton (1926). The most comprehensive qualitative survey of causes is by Sharpe (1938, Chapter 3).

The cause of soil creep is twofold. Regolith particles, normally at rest, undergo disturbance by some agent, and the force of gravity, acting upon the weight of the particles, then adds a vertical component to the stress. Vertical movement is prevented by the underlying rock, so the effect of gravity is to add a downslope component to all individual particle movements, and therefore to the net movement of the regolith. Where a void is present, a disturbed particle can move freely into it. Frequently this will not be the case, and the disturbance of one particle will apply stresses to, and cause movements of, adjacent particles; stresses are thereby transmitted through part of the regolith to such a distance as enables all movements to be taken up by voids. Gravitational force is virtually uniform over the earth's surface, therefore the rate of soil creep may be expected to vary with the sine of the slope angle and the magnitude and frequency of disturbances to individual particles. Agents of disturbance that cause volume change generally affect large numbers of particles; the result is an expansion and contraction of the soil as coherent blocks, which is similarly affected by gravity. Movement in this manner may be termed the *heave mechanism.*

The main agents of disturbance are:

(*i*) Expansion and contraction due to temperature changes.
(*ii*) Expansion and contraction due to wetting and drying.
(*iii*) Freezing and thawing of soil moisture.
(*iv*) Plant roots: growth, decay, and forces transmitted from swaying of vegetation in the wind.
(*v*) Soil fauna: worms, moles and other burrowing animals, termites, and microfauna.
(*vi*) Volume changes on weathering, including loss of material in solution.
(*vii*) Temporary increases in load on the soil surface: precipitation, animals.

Agents (*i*)-(*iii*) operate through the heave mechanism; agents (*iv*)-(*vii*) produce more irregularly distributed stresses.

The effect of temperature changes may be shown by an approximate calculation. The diurnal temperature change in the upper few centimetres of soil in temperate climates is of the order of 3°C, giving a total annual change of about 1000°C; the coefficient of cubic expansion of silica is 0·0000013; therefore the annual volume change is of the order of 0·1%. Extreme values for arid climates may be ten times greater. This agent is therefore only a minor cause of creep, except possibly in deserts. Creep caused by temperature changes is possibly one surface process operative on the moon.

Volume changes on wetting and drying are caused by the expansion and contraction of silicate clay minerals. The amount of volume change is dependent on the magnitude of variations in soil moisture content, the proportion of clay, and the types of clay minerals present. Montmorillonitic clays swell considerably more on wetting than illitic and kaolinitic clays. The coefficient of linear shrinkage of a soil is the change in linear dimensions for a 1% change in moisture content (moisture as a percentage by weight of dry soil). It may be determined by filling a greased tin with saturated soil, drying, and measuring the change in dimensions. During the initial stages of drying the soil consists entirely of solids and water, and the loss in volume is equal to the loss of water; below a certain moisture content air enters, and shrinkage continues at a slower rate. The latter is named residual shrinkage. The annual cycle of soil moisture content may be determined by repeated sampling or other measurement. For the upper 5 cm of a sandy loam in Northern England, values found were a coefficient of linear shrinkage of 0·06-0·08%, a moisture change of 35% per annum, giving a change in linear dimensions of 2-3% per annum (Young, 1960). A comparison of other available data on shrinkage coefficients (e.g. Haines, 1923) and seasonal moisture changes suggests that on soils other than sands, the value for temperate climates may commonly exceed 5%.

The volume changes resulting from freezing and thawing are proportional to the moisture content, and the number of times the soil temperature crosses freezing point.

The latter is related to the atmospheric frost-change frequency (Troll, 1943), and falls off rapidly with increase in soil depth. Below 50 cm no more than one annual crossing is probable. In fine-grained soils with a high moisture content, the frost heave may be augmented by drawing up of water from beneath into the frozen layer; in extreme cases ice-lenses may form (Taber, 1929, 1930; Everett, 1966). The volume change of water on freezing is 9%; thus a soil containing, for example, 50% moisture by volume will expand by 4·5%. *Frost creep* is therefore a major cause of disturbance in polar and temperate latitudes. An extensive study of frost creep was made by Schmid (1955). Freezing can also cause movement of fragments on a bare ground surface through the formation of needle-ice (Roberts, 1903; Sharpe, 1938, p. 27; Everett, 1966).

Few evaluations of the order of particle movements caused by plants and animals have yet been made. Consideration of the worm population of temperate soils (other than strongly acid soil types) and the observed frequency of termite channels in soils of the tropics suggests that faunal disturbance may be substantial. Darwin's observations on worm casts have been noted above (p. 11). Evans (1948) lists observed rates for worm casting on the ground surface as 0·025-0·6 g/cm^2/yr. Under savanna conditions, M. A. J. Williams (1968) estimated that termites carry to the ground surface the equivalent of a 5 cm thick layer of soil per 1000 years. If an average downslope movement of one metre during this operation is assumed, the rate of creep from this cause alone would be 0·5 cm^3/cm/yr (cf. p. 56). For rainforest conditions in Ghana, Nye (1955) estimated that termites bring to the surface 0·0126 g/cm^2/yr, which is equivalent to a layer 8 cm thick in 1000 years. Kirkby (1967) found disturbance by the growth and decay of grass roots to be very small. Weathering is not a repeated, cyclic change, and therefore causes relatively little disturbance. The effects of temporary increases in load have not been measured, but are probably also small.

Kirkby (1967) attempted to evaluate the relative magnitude of movements caused by different agents, under a humid temperate climate. In descending order of importance, the agents were wet-dry, freeze-thaw, worms and other burrowing animals, temperature and plant roots. The order of the two first-named agents agrees with the conclusion of Young (1958b). Owens (1969) found frost creep to be the main cause. In the humid tropics, faunal agents possibly rank second to moisture changes.

In cases of movement by the heave mechanism, if the volume change within the regolith over a period of time is estimated, there remains the problem of converting it into downslope movement. As regolith thickness is normally small in relation to slope length, in considering conditions at a point the slope may be regarded as infinite. A possible assumption is that during the expansion phase of the cycle, movement parallel to the ground surface being constrained, all expansion takes place perpendicular to the ground surface. During contraction there is no lateral constraint, and movement is vertical. Net downslope movement would then be equal to the amount of expansion times the tangent of the slope angle. Perpendicular-vertical movement is approximated to in the special case of needle-ice. In soils, contraction takes place at an angle between vertical and perpendicular, probably due to cohesion. Results from wetting and drying experimental soil blocks (Young, 1960; Kirkby, 1967), and from field measurements of frost heave (Schmid, 1955) agree in showing that the angle is approximately intermediate; net downslope movement is therefore half that which would occur in a cohesionless soil (Fig. 18). Kirkby (1967) attempted an analysis of the forces involved in movement by net shear; the resulting equation shows that the rate of movement varies with the sine of the slope angle, and decreases exponentially with depth below the surface.

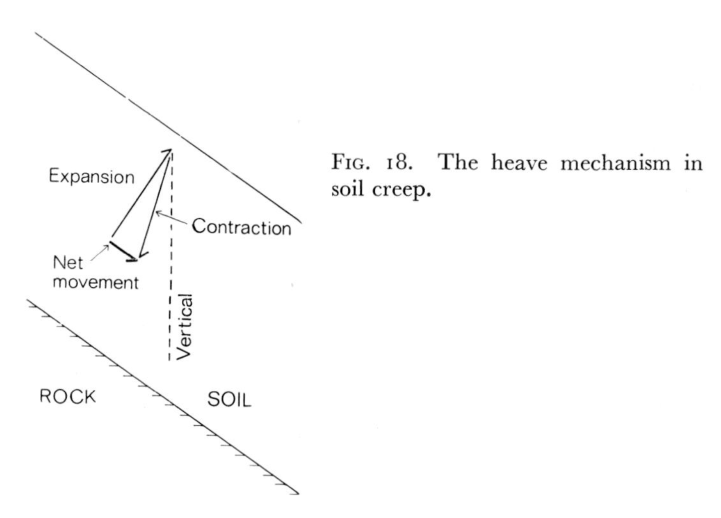

FIG. 18. The heave mechanism in soil creep.

The manner of action of soil creep, as determined by its mechanism, is concerned with the variation in its rate with regolith and slope parameters. Movement from both wet-dry and freeze-thaw mechanisms is greater on fine than on coarse textured soils. As regards regolith thickness, the magnitude of nearly all the main agents is greatest at the ground surface and decreases rapidly with depth. The rate of soil movement may therefore be expected to fall off exponentially with depth below the surface. This is confirmed by most available field studies, which show that movement almost ceases, or becomes very much smaller, below 20 cm (Schmid, 1955; Young, 1960; Leopold *et al.*, 1966; Kirkby, 1967). The total rate of creep will thus vary little with regolith thickness, provided that this exceeds 20 cm.

Variation in the rate of creep with parameters of slope depends partly on the associated variations in regolith properties. Where the soil becomes finer-textured on the concavity, creep will be faster. Young (1960) found the change in dimensions from moisture changes on a 7° concavity to be approximately double that on steeper slopes above. The mean moisture content, and therefore the heave on freezing, will also normally become greater toward the base of a slope. Thus the rate of creep is not, as it is sometimes assumed to be, independent of position on the slope. Leaving aside this complication, it has been shown above (p. 42) that the rate should vary with either the sine or the tangent of the slope angle; in relation to field accuracy of measurement the difference between these is unimportant. This argument is largely on prima facie grounds; it has been confirmed experimentally only for the case of rock creep (p. 58). Schumm's data (1964) for exceptionally rapid soil creep under semi-arid conditions was interpreted by Kirkby (in discussion) as proportional to sin θ, but the markers used in this work were subject to considerable disturbance by livestock. For slopes in humid climates, attempts to correlate recorded creep with slope angle have been unsuccessful; the evidence is insufficient to decide if this failure is due to the large magnitude of experimental errors. If theoretical considerations are ignored, and only the results of field measurements taken, the hypothesis that in humid climates soil creep varies with slope angle is not proven.

Attempts have been made to apply to soil creep methods of mathematics and mechanics originally devised for other purposes. Souchez (1963, 1964, 1966) applied the theory of plasticity to slow mass-movements. This assumes that movement is by laminar flow, which conflicts with evidence from laboratory experiments on soil blocks. Culling (1963, 1965) suggested on theoretical grounds that creep is a process similar to molecular diffusion. Some particles will, after moving, return to the void they have vacated; such 'vibration', even if it constitutes a high proportion of all particle movements, may be ignored. The remaining movements are then of a statistically random character, analogous to movements of particles suspended in a still liquid. This approach is of potential interest, but has no experimental confirmation. It could be investigated by placing in a glass-sided trough soil containing a proportion of visually identifiable sand grains, subjecting these to disturbance, and taking high-magnification photographs of the movements of the grains. Most experimental work on creep, however, suggests that expansion and contraction of the soil in the form of coherent masses, as in wetting and drying, is more important than isolated movements of single particles.

Types of field evidence that have been held to demonstrate the existence of soil creep include outcrop curvature (Fig. 19), stone lines, tree curvature, tilting of structures, and soil accumulations upslope of retaining structures. Outcrop curvature is unambiguous evidence that relative movement has occurred within the regolith; some indication of the relative rate of movement compared with the rate of weathering can be deduced, but there is no indication of absolute rate. *Stone lines* are layers of stones forming a soil horizon (Parizek and Woodruff, 1957; Ruhe, 1959; Alexandre, 1967; M. A. J. Williams, 1968). Their origin is not settled, and there is consequently the danger of circular argument in citing them as evidence of soil creep. Striking evidence of creep is provided, however, where a section reveals a quartz vein successively *in situ*, affected by outcrop curvature, and 'feeding' a stone line (Fig. 20). The curved trunks sometimes seen on trees growing on steep slopes has been attributed to the attempt of the growing plant to regain vertical growth whilst the base of the trunk is tilted downslope. Schmid (1955) found that such curvature develops during the first 10 years of growth, after which the tree is strong enough to resist the pressure of superficial earth accumulating above the trunk. The relative importance of phototropism and geotropism in this phenomenon is not known; a long-term experiment would be possible, growing trees on a steep slope and artificially varying the shade conditions. The tilting of posts, gravestones, etc. may be due to the weight of the structures themselves, and not to movement generated within the soil. Accumulation upslope of field boundaries, as is common in the long-settled landscapes of western Europe, takes place under agricultural

FIG. 19. Outcrop curvature in shales. Central Wales.

use of land, conditions which geomorphologically are highly unnatural; the accumulation may result also from surface wash (Kittler, 1955; Gerlach, 1963).

Laboratory measurement of creep may be made by placing blocks of soil in glass-sided troughs, inserting pegs in the sides, and recording movements of the projecting ends of the pegs. The soil may be bare or covered with turf; the ideal is a grass cover sown from seed and of several years standing. The trough may be exposed to natural weather, or to simulated temperature and rainfall. In small troughs, basal slipping and severe end effects occur (Young, 1960; Kirkby, 1967). To avoid such, a minimum trough length of 3 m and soil depth of 50 cm is recommended. By comparing movements of pegs in the surface and sides of the trough, this method can aid the interpretation of field measurements using surface pegs.

FIG. 20. Quartz vein feeding into a stone line. Northern Malawi; savanna climate.

The slowness of soil creep renders it one of the most difficult of geomorphological processes to measure in the field. Only a summary of methods is given here; for further descriptions and a comprehensive bibliography reference should be made to Selby (1966). A compendium of descriptions is also included in *Revue de Géomorphologie dynamique* (1967). Methods that have been used are:

(*i*) Markers placed on the ground surface, e.g. painted stones; positions before and after movement determined by survey.

(*ii*) Pegs inserted in the ground surface; positions determined by survey (Fig. 21*A*).

(*iii*) Tilt-bars (T-bars) inserted in the ground surface; downslope or upslope tilting determined by clinometer.

(*iv*) Plastic tubes, buried perpendicularly to the surface; deformation determined by inclinometer. A variant of this method is to use a stack of short cylindrical sections of tube, capable of relative shear.

(*v*) Pegs buried in the soil and subsequently re-excavated; positions determined by survey; these have been called *Young-pits* (after Young, 1960, 1962) (Fig. 21*B*, *C*).

(*vi*) Buried coloured beads or sand.

(*vii*) Buried plates, with electrical connections to a reference shaft; movements determined by resistance measurements.

(*viii*) Buried cones, connected by wires to a reference shaft (Fig. 21*D*).

Surface markers are affected by wash as well as creep, and are more suitable for recording movement on screes. The movements of surface pegs are relatively simple

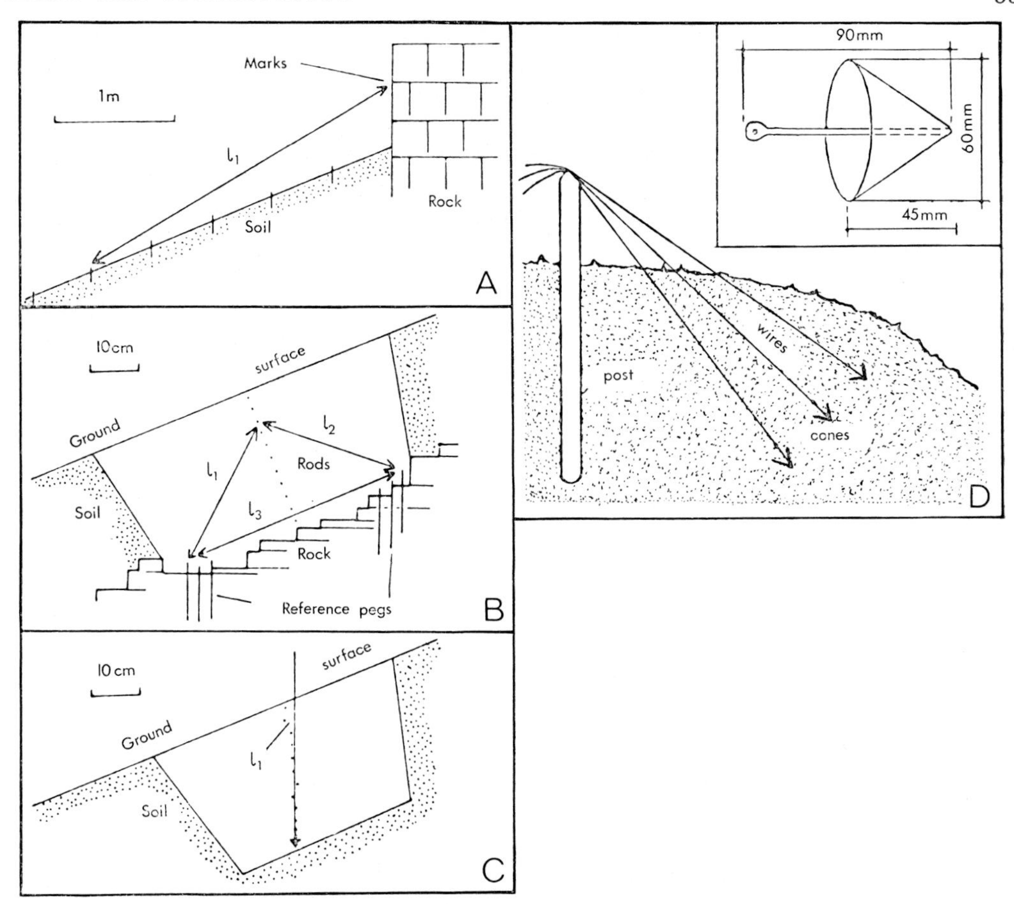

FIG. 21. Methods of measuring soil creep. *A-C* after Young (1960), *D* after Selby (1969).

to record, but the interpretation of the results raises difficulties. The pegs become tilted downslope, and to convert the movement of their tops into volume of soil movement requires various uncertain assumptions. Further, since the amount of movement decreases with depth approximately in a negative exponential manner, it is basically unsatisfactory to record such movement by means of rigid pegs. The same objections apply to tilt-bars, which have the added disadvantage that the above-ground weight of the bar exerts a force upon it. Tilt-bars can record short-term movements, in single cycles of wetting and drying, but are unsatisfactory for long-term measurements (Kirkby, 1967). Plastic tubing overcomes the difficulty of non-linear depth variation in movement; the difficulty is to record deformations with the high precision required. This method is mainly of use for recording the more rapid movements of solifluction and landslides. Buried pegs can provide a record of variation of movement with depth to an accuracy of about 1 mm, provided that accidental disturbance of the pegs on re-excavation can be avoided. A 'screen' of permanently-stained sand sprinkled around the ends of the pegs prior to burial is an aid to relocation. The specific gravity of the peg material should not exceed the bulk density of the regolith. Measurement may be with reference to either posts fixed into the rock beneath, or to a suspended plumb-line (Fig. 21*B*, *C*). This is the best method for long-term measurements.

Measurements of the rate of contemporary soil creep are summarized in Table 1.

Table 1. Observed rates of soil creep

Source	Method	Climate	Movement of surface or upper 5 cm, mm/yr	Volumetric movement, cm³/cm/yr
Young, 1960, 1962	Surface and buried pegs	Temperate	1–2	0·5
Everett, 1963	Buried plates	Temperate	1	—
Kirkby, 1964, 1967	Surface and buried pegs	Temperate	1–2	2·1
Slaymaker, 1967	Buried pegs	Temperate	—	2·8
Owens, 1969	Buried tubes	Temperate	11	3·2
M. A. J. Williams, 1969	Buried pegs and tilt-bars	Warm temperate	—	1·9–3·2
		Savanna	—	4·4–7·3
Schumm, 1964	Surface pegs	Semi-arid	6–12	—
Leopold *et al.*, 1966	Surface pegs	Semi-arid	5	4·9

They suggest that for humid temperate climates, the upper 2-5 cm of mineral soil experiences a downslope movement of the order of 1 mm per annum. The unit for measuring volumetric downslope movement of soil is the volume moved annually across a plane perpendicular to the ground surface and parallel with the contour of the slope, for unit horizontal distance along the plane; it is expressed in cm³/cm/yr. Present evidence indicates that in humid temperate climates the movement is of the order of 0·5-3·0 cm³/cm/yr. It is probably higher in semi-arid climates, perhaps due to crumbling of the soil and the formation of cracks as a result of desiccation.

Further understanding of the mechanism of soil creep, and better knowledge of its rate, requires improvements in instrumentation. To study the mechanism it is necessary to track, under laboratory conditions, displacements of the order of 0·1 mm of individual particles. To measure the rate of creep on natural slopes it is desirable to be able to make repeated measurements, to an accuracy of 1 mm, of the three-dimensional position of small objects permanently buried in the regolith.

ROCK CREEP AND TALUS SHIFT

A group of processes involves the slow downhill movement of rock fragments. Sharpe (1938) described as rock creep the movement of boulders away from an outcrop on the surface of, or embedded within, the soil; it is unnecessary to class this as a separate process, as the cause is creep of the soil. The term rock creep, or to avoid ambiguity *surficial rock creep* (Schumm, 1967) will be reserved for the movement of rock fragments across an inclined surface. The surface need not necessarily be of consolidated rock, the essential feature being that movement takes place by intermittent slip along a single plane between the fragment and the surface. When a symmetrical rock fragment resting on an inclined surface expands on heating, the position of the plane of no movement lies upslope of its centre; on cooling and contraction the reverse is the case, giving a net downslope movement of the whole coherent block. At an early date this was both shown theoretically (Moseley, 1855, 1869) and demonstrated experimentally (Davison, 1888). Movement may also be caused by ice forming beneath the rock and, in the case of small fragments only, by needle-ice. For this distinct, if uncommon, process, one of the few cases of experimental verification of the variation in rate of creep with the sine of the slope angle has been obtained; Schumm (1967) measured movements of fragments on shale slopes in a semi-arid climate over seven years, finding a significant correlation (Fig. 22).

Talus shift refers to the slow downslope movement of any mass of rock fragments in which the interstices are predominantly hollows rather than fine soil particles. The term talus shift (Gardner, 1969) is preferred to talus creep, as movements of individual stones are highly variable, and only a proportion of them are caused by the mechanism of creep. The separation by Sharpe (1938) of rock-glacier creep, in which ice was assumed to be present in addition to rock, is rejected, as this begs the question of the origin of rock glaciers. Talus shift may be initiated by heating and cooling, ice-wedging, slippage on moistened surfaces, and the impact of falling rocks. Movements in one part may be transmitted by gravity through the scree, causing rearrangement of the packing of rocks, possibly with mass slippage downslope. A theoretical discussion of heat-cool movement is given by Scheidegger (1961c); the total volumetric change is very small, however, and this is probably not an important cause except in conditions of extreme aridity. Caine (1963), studying screes on slates in northern England, found that movement was mainly caused by thin ice layers which formed parallel to the surface. A further cause of disturbance of equilibrium between rocks is the loss of material by weathering and solution.

CONTINUOUS CREEP

Continuous creep, as defined by Terzaghi (1950), is creep produced by the force of gravity, unaided by other agents. This is the normal engineering use of the word creep, as first recognized by Preston (1913). Continuous creep may affect both consolidated rock and the deeper parts of the regolith. The applied stress is produced by the weight of the overlying material, which causes molecular rearrangements and, in regolith, inter-particle movements. Under experimental conditions with constant applied stress the amount of creep strain increases approximately linearly with time. Granites undergo negligible creep, sandstones are moderately susceptible, and shales are subject to considerable creep (Farmer, 1968).

The main direct effects of continuous creep occur where weak beds, usually shales, are overlain by stronger beds such as sandstones. In valley centres, contortions and

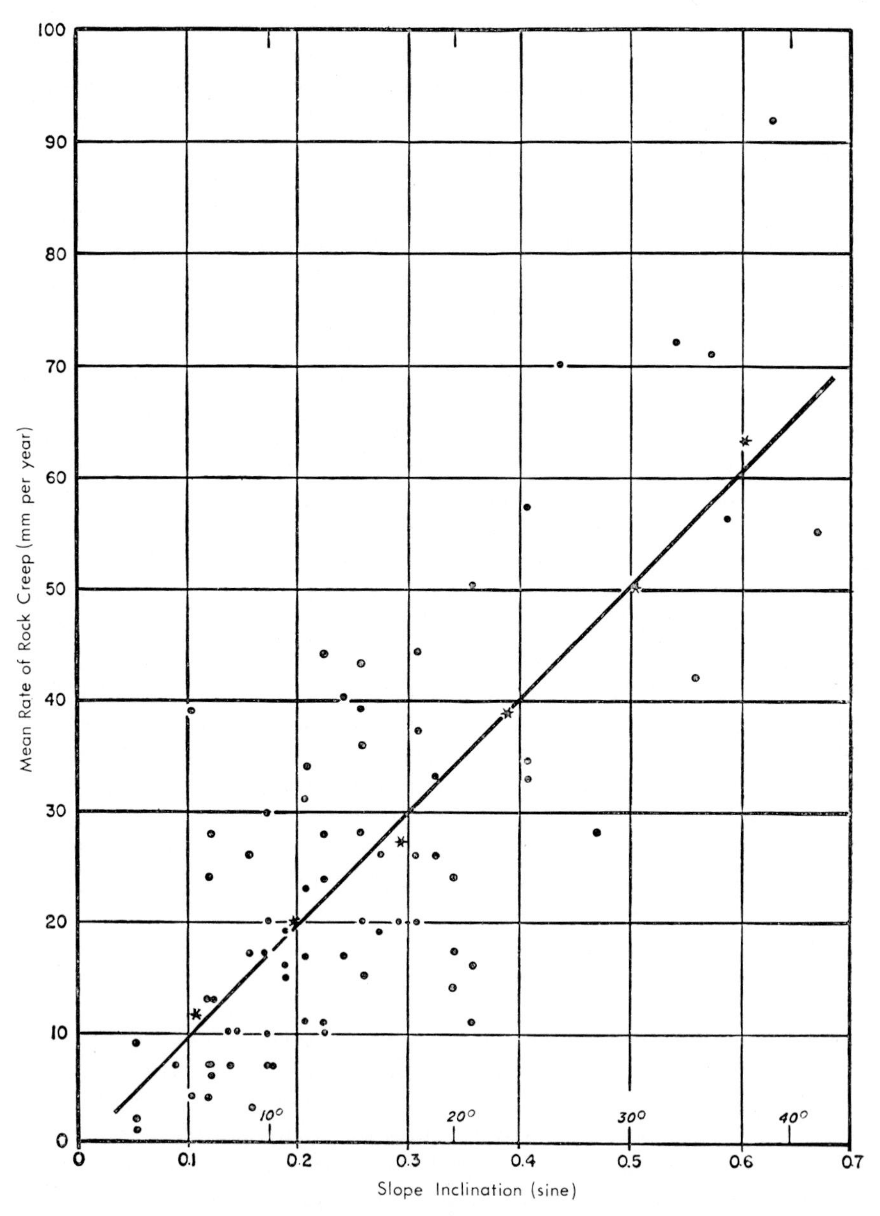

FIG. 22. Relation between sine of slope angle and rate of surficial rock creep on shale slopes in Colorado, observed over seven years. Standard error, 11·8 mm, $r = 0{\cdot}7$. After Schumm (1967).

bulging occur in the weaker rocks. Where the weak beds outcrop on valley sides the overlying rocks are subject to *cambering*, changes of dip in directions towards the valley centres (Stearn, 1935; Hollingworth *et al.*, 1944). In terms of volume-distance of material moved, continuous creep is only a minor surface process. Its main geomorphological importance is as a precursor of rapid mass-movements. Most landslides (except those following earthquakes) are preceded by slow creep deformation; the rate of creep remains slow for a relatively long period, then accelerates exponentially shortly before rapid movement under shear failure occurs (Terzaghi, 1950; Ackermann, 1959; Schumm and Chorley, 1964).

SOLIFLUCTION

Solifluction was defined by Andersson as 'the slow flowing from higher to lower ground of masses of waste saturated with water' (1906, p. 95). Its importance was recognized by Eckblaw (1918), and its part in forming patterned ground features noted by Beskow (1930) and Hollingworth (1934). There has been some variation in the use of the term (Dylik, 1967), but solifluction is here confined to phenomena of cold climates; slow flowage not associated with frozen ground is termed earthflow. Following Washburn (1967), movement caused by expansion and contraction of freeze-thaw is termed *frost creep*, and that due to flow *gelifluction*. Thus solifluction comprises both frost creep and gelifluction. The origin of patterned ground will not be considered; a comprehensive review is given by Washburn (1956).

The mechanism of solifluction is consequent upon an annual cycle of freezing in autumn and thawing in spring. On freezing, additional water is drawn up into the frozen layer, forming discrete ice bands or individual crystals, so increasing the amount of frost heave perpendicular to the ground surface. On thawing there is subsidence, which may initially be at some angle between perpendicular and vertical, as in soil creep. Gelifluction occurs when the thawed soil attains moisture contents close to the Atterberg liquid limit; flow is augmented by a loss of cohesion in the soil, resulting from the previous separation of particles by frost heave. In different areas, the major component of total regolith movement may be either frost creep (Washburn, 1967) or gelifluction (Rapp, 1966), and their relative importance may vary between different parts of a single slope (P. J. Williams, 1959, 1966; Washburn, 1967).

The rate of movement of the ground surface has been observed by distance measurements or theodolite readings to marked stones, pegs, or cone targets (Michaud and Cailleux, 1950; Pissart, 1964; Rapp, 1960, 1966; Rudberg, 1958, 1964; Washburn, 1967). The variation of movement with depth has been investigated using superimposed plastic cylinders (Rudberg, 1958, 1962), probes with attached electrical resistance strain gauges, either buried directly in the soil or inserted into buried plastic tubing (P. J. Williams, 1957, 1962, 1962b), and linear-motion transducers (Everett, 1966b).

Investigations of the variation in amount of downslope movement with depth in the regolith give consistent results (Rudberg, 1958, 1962, 1964; P. J. Williams, 1966). Displacements are very small below 50 cm depth, and almost nil below 1 m. Upwards from 50 cm the rate increases at a linear or greater than linear rate; typical results are shown in Fig. 23. On slopes with recorded solifluction, a marked orientation of stones in the direction of slope has been observed (Rudberg, 1958, 1962).

Some recorded rates of movement are summarized in Table 2. All investigators have found that even on apparently uniform slopes, rates vary considerably between individual observations, often by a factor of 10 or 100. Caine (1968b) found that movements approximated to a log-normal distribution, and suggested that for statistical

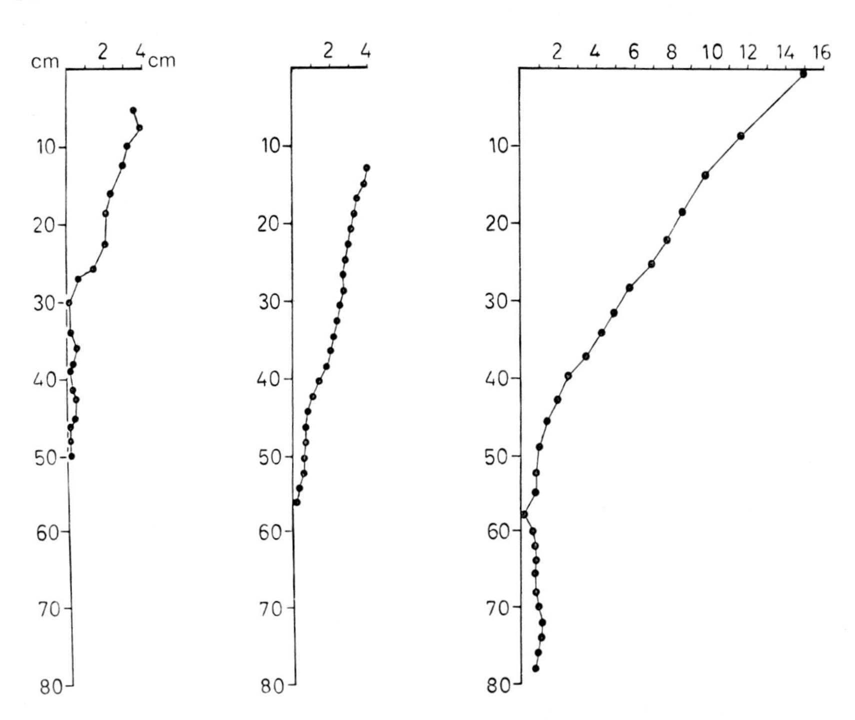

FIG. 23. Deformation of buried test pillars, by solifluction, Swedish Lappland. The left and central diagrams show total movements over five years in solifluction lobes on 10° slopes; the diagram to the right shows three-year movement in a lobe on a 30° slope. After Rudberg (1964).

comparison between different sets of results, logarithmic transformation should be applied to the data. Mean values for movement of the surface are of the order of 0·5-5·0 cm. Assuming movement in the upper 50 cm is, on average, at a quarter of this rate, this gives a volumetric downslope movement of 6-60 cm³/cm/yr. Movements on solifluction lobes and terraces and on polygonal and striated ground may be substantially more rapid. The high rates of movement on such features as solifluction lobes should not be taken as typical for slopes as a whole in the polar zone; to obtain mean rates of overall slope denudation they must be combined with estimates of the proportion of the total area occupied by these features, a proportion which in many areas is quite low.

Table 2. Observed rates of solifluction

Source	Method	Location	Movement of surface or upper 5 cm, mm/yr: Approximate mean	Range	Volumetric movement, $cm^3/cm/yr$
Rapp, 1960	Surface pegs	Lappland	40	0–80	50
Rapp, 1966	Surface pegs	Scandinavia	—	10–200	—
Rudberg, 1964	Surface marks and buried cylinders	Lappland, frost-debris zone	4	0–10	5
		Lappland, tundra zone	20	0–60	25
Pissart, 1964	Surface stones	Alps	10	2–60	—
Everett, 1966	Linear-motion transducers	Alaska	—	10–100	—
P. J. Williams, 1966	Buried tubes	Canada	—	0–100	50
Washburn, 1967	Surface pegs	Greenland, dry sites	9	—	—
		Greenland, wet sites	37	—	—
Caine, 1968b	Surface stones	Tasmania	25	0–337	—

There appears to be little correlation between rate of movement and slope angle. Pissart (1964) gave 16-year movements for over 100 marked stones; they show no relation to angle between 6° and 22°, but the median value on 32° slopes is four times greater. From several thousand measurements Washburn (1967) found a correlation with the sine of the slope angle only for sites that remained moist throughout the year. Moist sites show considerably greater movement than freely-drained sites. Büdel (1948b) distinguished between the frost-debris zone, with a stony regolith and little vegetation, and the tundra zone with finer soil and a vegetation cover; Rudberg (1964) found that, contrary to Büdel's supposition, movement in the frost-debris zone was five times slower than in the tundra zone. It is possible that movement on a footslope with fine-grained soil may be as great as on a steep slope with a scree or other coarse regolith (Pissart, 1966). Thus moisture supply is more important than slope angle in determining rate of movement. Largely because of moisture characteristics, movement is greatest in regolith with a high silt content, and least on sands and gravels (Washburn, 1967).

As frost creep is also a contributory agent to soil creep, transitions between soil creep and solifluction occur. Movement may be attributed to solifluction where the upper regolith remains frozen for a substantial part of each year, and where a flow component on thawing can be demonstrated. An empirical distinction is that only in solifluction are movements exceeding 1 cm/yr commonly found. P. J. Williams (1961) gives the limit of solifluction as a mean annual ground temperature of 0°C., corresponding to a mean air temperature of 3°C. At high altitudes in warmer latitudes there is a high frost-change frequency (Troll, 1943), and it is probable that regolith movement is mainly by frost creep.

VII WASH

SURFACE WASH

Surface wash is the downslope transport of regolith material across the ground surface, through the agency of moving water. There are two distinct processes involved, the impact of rain on the ground surface and the flow of water across it; these are called *raindrop impact* and *surface flow* respectively. Surface flow may be divided into *sheetwash*, when the ground is entirely or largely covered by a moving layer of water, and *rillwash*, when the water flows mainly as micro-channels. The channels in rillwash frequently change their location on the slope, causing the process to act areally; where channels are constant in location, the process is termed *gullying*. Besides the above distinctions between the active agents of surface wash, two separate effects of processes may be defined: *soil detachment*, the removal of particles from their initial positions in the regolith, and *wash transport*, the carrying of the detached particles downslope.

Recognition that rain, running off the ground surface, carried material to rivers is found in some of the earliest geomorphological writings. The first detailed description of sheetwash was by McGee (1897), and of rillwash by Fenneman (1908). Other papers of historical interest are by Jutson (1919) and Lawson (1932). The principal recent field studies of surface wash are those of Schumm (1956 *et seq.*), Leopold *et al.* (1966) and Ruxton (1967). More than twelve independent studies of the rate of wash under natural conditions have been made since 1960.

Knowledge of surface wash has been considerably increased from hydrological investigations of catchment runoff and studies of accelerated erosion, two fields closely related to geomorphology. The hydrological work is partly associated with accelerated erosion, and partly concerned with runoff and flood prediction. It is summarized, together with some aspects of its application to geomorphology, by Horton (1945). Pioneer studies on accelerated erosion were made in Germany by Wollny (see Baver, 1938). Experimental work on a large scale was carried out in the United States from 1929. Among discussions of this work the symposium of the American Geophysical Union (1941) may be noted. In 1953 all the U.S. data was assembled at the Runoff and Soil Loss Data Laboratory, Lafayette, Indiana, where it has since been correlated and analyzed (Smith and Wischmeier, 1962, and bibliography therein). The relation of normal to accelerated erosion has been discussed by Bailey (1941), Sharpe (1941), Schultze (1951) and Jahn (1963).

HYDROLOGY AND ACCELERATED EROSION

Some results of work on hydrology and accelerated erosion will first be outlined, followed by a consideration of surface wash under natural conditions. The symbols used in this section are as follows:

d_x	depth of flow at distance x from watershed, cm
e_r	erosion rate, expressed as depth of soil removed, cm/hr
f	infiltration capacity, cm/hr
i	rainfall intensity, cm/hr
l_o	total slope length, cm
n	roughness factor, determined empirically
v_x	velocity of flow at distance x from watershed, cm/sec
x	distance downslope from watershed, cm
K_1, K_2	constants
S	slope, expressed as fall/horizontal length, i.e. tan slope angle; most of the equations are also applicable to S expressed as sine slope angle
θ	slope angle, degrees
σ	supply rate, cm/hr

The basic equation for determining whether surface runoff occurs is

$$\sigma = i - f \qquad (7.1)$$

This states that water is supplied to a point on the ground surface, and in the absence of runoff would accumulate at a rate equal to the rainfall intensity less the infiltration capacity of the soil. When rain falls on dry soil the rate of infiltration is initially high, but falls to an almost constant value, which is the infiltration capacity. Actual capacities vary widely as between clays and sands, from almost nil to over 50 cm/hr. For comparison, rainfall intensities exceeding 2 cm/hr are uncommon in temperate climates, and the heaviest recorded one hour falls are less than 6 cm for most counties of Britain; in the tropics rates exceeding 2·5 cm/hr are common, and many stations record over 5 cm/hr at least once per year. If a soil horizon at some depth has a lower permeability than the surface horizon, the transmission capacity of the lower horizon replaces the infiltration capacity as the limit to infiltration after prolonged rain. Thus the occurrence or frequency of surface runoff on a given slope can be investigated by measurements of the infiltration capacity and comparison with records of rainfall intensity. Runoff without rainfall exceeding infiltration capacity may occur when the soil becomes saturated up to the surface, the limit to infiltration being either a less permeable subsurface layer or the permanent water table; runoff under such conditions is termed *saturation overland flow* (Kirkby, 1969), and is most common near the base of slopes. The phenomenon of water-repellent soils has occasionally been recorded, particularly on dry soils following bush fires (Debano and Krammes, 1966).

Equations 7.2 and 7.3 give the depth of flow at a distance x from the watershed*:

$$d_x = \left(\frac{\sigma}{K_1} \cdot \frac{x}{l_o}\right)^{0.6} \qquad (7.2)$$

where

$$K_1 = \frac{3600 S^{0.5}}{n . l_o} \qquad (7.3)$$

Where slope and other factors are held constant the depth increases at the 0·6 power of the distance. The constant K_1 is shown by equation 7.3 to be partly dependent on

* Equations 7.2 and 7.3 are from Horton (1945), converted to metric units.

slope. For a steeper slope K_1 increases and therefore the depth of flow becomes less. The flow of water may be turbulent or laminar; on natural slopes it is almost always turbulent. The equation for the velocity of turbulent flow in channels is known as the Manning formula. For the case of sheetwash, the depth of water effectively replaces the hydraulic radius in this formula:

$$v_x = \frac{1}{n} . d_x^{0.67} . S^{0.5} \tag{7.4}$$

The Manning formula shows that the velocity of flow increases with the two-thirds power of the depth, and with the square root of the percentage slope. Thus for a given rainfall intensity, infiltration capacity and surface roughness, (*i*) an increase in slope angle decreases the depth of flow, but increases its velocity, and (*ii*) both depth and velocity increase with distance from the slope crest, but at a less than directly proportional rate.

The force exerted on the ground surface by flowing water is proportional to both its depth and velocity. The relations that result are given by Horton (1945), whose final equation for the rate of erosion may be simplified to

$$e_r = K_2 . \frac{\sin \theta}{\tan^{0.3}\theta} . x^{0.6} \tag{7.5}$$

in which the constant K_2 is dependent upon the supply rate and surface roughness. This indicates that the rate of erosion varies with slope at somewhat less than the sine of the angle, and varies with the 0·6 power of distance from the crest.

On the basis of hydrological theory and experimental findings, Horton (1945) put forward the concept of the belt of no erosion. This assumes that soil possesses a finite 'resistivity' to erosion; where the eroding force is less than the resistivity, there will be no erosion. This will necessarily be the case on the watershed; erosion will commence beyond a certain distance down the slope, giving the belt of no erosion. For a given slope the factors of angle, infiltration capacity and surface roughness remain constant, therefore the width of the belt of no erosion varies with the intensity of individual falls of rain. The frequency with which erosion occurs will decrease toward the crest, owing to the decreasing frequency of higher rainfall intensities. This concept may be criticised, first, because it ignores erosion by raindrop impact, and secondly, because the notion of a fixed resistivity is based on experiments concerned with accelerated erosion, in which erosion rates typical of natural conditions are so much smaller that they may be taken as nil. Under natural conditions, there is no evidence for a fixed margin between no erosion and erosion.

The above equations may be compared with the empirical results obtained from analysis of all available U.S. accelerated erosion experiments (Smith and Wischmeier, 1962). The experiments are mainly on runoff plots with an area of 0·004 ha (0·01 acre) and a length of 22·2 m (73 ft). By isolating the effects of single variables, the following equations for the relative rate of erosion are obtained:

$$e_r = K_3 . S^{1.3} \tag{7.6}$$

or

$$e_r = 0{\cdot}43 + 0{\cdot}30S + 0{\cdot}043S^2 \tag{7.7}$$

and

$$e_r = K_4 . x^{0.5} \tag{7.8}$$

With respect to slope angle, the exponent in equation *7.6* varies between different investigations, but is more commonly above than below unity. An alternative form which provides a good statistical fit to a wide range of data is equation *7.7*. With respect to distance from the crest, the exponent in equation *7.8* varies considerably, from 0·3 to 0·9, but 0·5 ± 0·1 is suggested as the best working hypothesis. Where gullying occurs it rises to 1·0-2·0 (Kirkby, 1969). With increase in particle size of the soil the relative importance of slope angle decreases, and that of volume of flow, and therefore distance from the crest, increases.

Two observed phenomena, rillwash and surges, complicate the attempt to relate surface wash to the laws of water flow. Where flow takes place as rillwash there is a temporary increase in available force of flow along rill channels; the mechanism whereby such temporary channels do not become eroded, and so made permanent, requires investigation. Surges refer to the series of uniformly-spaced waves, approximately parallel to the contour, which are sometimes seen in sheetwash; the concentration of flow into surges causes temporary local increases in available force, moving down the slope.

The effects of raindrop impact are most apparent when bare soil containing stones is exposed to a heavy storm. The stones are subsequently found resting on miniature earth pedestals, several centimetres high, as a result of the lowering of the surface of the surrounding unprotected soil. Raindrop erosion has been the subject of a series of studies by Ellison (1944 *et seq.*; cf. also Lehmann, 1931; Rose, 1960; Smith and Wischmeier, 1962).

The kinetic energy in falling raindrops varies with drop size, terminal velocity and rainfall intensity. Drop diameter is found to vary approximately with rainfall intensity, but is about half for orographic *vis-à-vis* non-orographic rain. Drop size in orographic rain rarely exceeds 2 mm, corresponding to a terminal velocity of 6·5 m/sec. Raindrops attain close to their terminal velocity in a fall of 8 m or less. A high proportion of the kinetic energy of fall is dissipated on impact as heat, very little being transmitted to surface flow.

In cases of accelerated erosion, the following effects of raindrop impact have been distinguished: detachment of soil particles by impact; downslope transport of particles thrown above the surface; sealing of the soil surface through dispersion of fine particles and clogging of pores, hence increasing runoff; deterioration of soil structure; and removal of fine particles with relative accumulation of sand. On bare soil, raindrop impact is considerably more effective than surface wash in causing soil detachment. This has been demonstrated by covering the ground with a layer of mulch; the effect of the mulch in reducing erosion is little altered if it is raised above the surface, permitting runoff but preventing raindrop impact. As well as detaching particles which are then transported by surface flow, raindrop splashing may act directly as an agent of transport; this is caused partly by the angle of impact on a slope, but mainly because particles thrown downslope travel farther before landing. An approximate rule is that the proportion of all splashing particles that travel downslope is 50% plus the percentage slope. De Ploey and Savat (1968) found the proportion rose to 75% on 20° slopes and 80% on 30°.

SURFACE WASH UNDER NATURAL CONDITIONS

Accelerated erosion takes place under conditions which geomorphologically are highly unnatural—cleared vegetation, bare ground, and a weakened soil structure. Results based on such conditions are not necessarily applicable to natural slopes as far as either the mechanism or the manner of action of surface wash is concerned. Among these results, the following are potentially of particular geomorphological significance:

(*i*) Raindrop impact is a more powerful agent of soil detachment than surface flow, but is relatively less important as an agent of transport. The detachment caused by raindrops does not vary with position on the slope nor, substantially, with slope angle.

(*ii*) The lowering by surface wash varies directly with the slope angle. The precise relation varies with soil and surface conditions, but a linear proportional increase with the sine of the slope angle is the best general approximation.

(*iii*) The ground loss caused by surface wash varies with approximately the 0·6 power of distance from the slope crest.

In respect of surface wash, a distinction may be made between *control by detachment* and *control by transport*. If the agents of transport are more than able to carry away downslope all the material supplied by detachment, the rate of ground loss is determined by the rate of detachment. If control is by wash transport, detached material is supplied faster than it can be transported, so that the transporting media are then always fully loaded (Fenneman (1908) described this condition as overloaded). This distinction affects the manner of action of surface wash. Let it be assumed as a simplification that detachment is caused by raindrop impact, and transport by surface flow. Under control by wash transport, the rate of ground loss would increase toward the base of the slope, but under control by detachment, ground loss would be independent of distance from the crest. The effect of slope angle would also be less for the latter than for the former case.

Conditions intermediate between those of accelerated erosion and most natural slopes exist on badland relief, where, due to a combination of weak bedrock and vigorous gully erosion, ground loss is too rapid for vegetation to gain a hold. Schumm (1956, 1956b, 1962) found that ground loss on badlands varied approximately with the sine of the slope angle, but with respect to distance from the crest, the null hypothesis that loss is independent of such distance was not disproved. Rectilinear slopes therefore undergo parallel retreat. Slope evolution, largely through surface wash, takes place by closely-spaced gullying, parallel retreat of steep slopes, and the formation of pediments; similar conclusions were reached by Smith (1958). Natural badland relief is rare, but this rarity must in part be due to the rapidity with which it is eliminated. It is possible that, in terms of volume of bedrock removed, badland erosion is of some geomorphological significance where unconsolidated rocks, particularly clays, are first exposed to rapid stream erosion.

Evidence of surface wash under natural conditions may take the form of direct observations of turbid flowing water, accumulations of material above obstacles such as tree roots, and measurements over a period of time. Wash probably reaches its optimum effectiveness in semi-arid climates, due to the sparse vegetation cover, the absence of a surface soil organic horizon, and the occurrence of convectional rainfall of high intensity. Schumm (1963, 1964) has described a seasonal cycle of surface processes in Colorado. During winter the surface is loosened by frost action; raindrop impact in spring and summer compacts the soil and increases runoff; surface wash occurs and rill channels are formed; the channels are then destroyed during the succeeding winter. Net ground loss occurred on convex and straight slopes; on concavities there were irregular alternations between exposure and burial of stakes, suggesting that transport was occurring, but no net ground loss. On unconsolidated Pleistocene beds in New Mexico, Leopold *et al.* (1966) recorded substantial ground loss by wash, but found no significant variation with either slope angle or distance from the crest. There was some indication that ground loss may be influenced less by slope angle as conventionally measured than by micro-relief of the order of a few centimetres. In a Mediterranean climate Gabert (1964) found that the rate of wash varied greatly with the density of vegetation. Arid climates lead to exposure of the ground surface, but wash is limited by the infrequency of rain and, in many areas, high infiltration capacities. Grove (1960) and Savigear (1960) describe surface flow in arid regions. The flow takes place in braided, anastomosing rills; it is turbulent, and carries much suspended material.

Surface wash occurs under tropical evergreen rainforest (Rougerie, 1956; Ruxton, 1967). Raindrop impact occurs, not only through temporary gaps in the forest canopy caused by treefall but also because the canopy height is sufficient for drops falling from leaves to attain nearly their terminal velocity. After the first few minutes of a heavy storm, the rain experienced beneath the canopy is as great as in the open. The leaf litter is frequently thin, and is subject to transport by wash. Miniature earth pillars and root-impounded earth steps occur. Flow takes place, not in rills, but as diffuse sheetwash. Substantial amounts of water reach the ground by running down tree trunks, causing local intensification of erosion. In savanna climates conditions are also favourable for wash. Savanna grasses grow in tufted forms, leaving bare soil exposed between. Vegetation protection is removed by dry-season fires; in Malawi a vernacular term for the start of the wet season means 'the rains that wash away the burnt bush'. Earth brought to the surface by termites is particularly affected by wash (Alexandre, 1967; M. A. J. Williams, 1968). In a part of the Mato Grosso, Brazil, with sandy soils and very high infiltration capacities, the soil surface becomes closely pitted by raindrop craters, and sand grains are thrown several decimetres (Fig. 24); it is possible that in such circumstances the transporting effect of wash is mainly by raindrop impact, surface flow being infrequent owing to the rapid infiltration.

Polar climates provide the condition of a sparse vegetation cover, but rainfall intensities are low. Surface wash takes place mainly after the spring snow melt (Jahn, 1960). Sites in which snowdrifts persist late in the year, inhibiting vegetation, are particularly susceptible.

FIG. 24. Sandy soil under a savanna climate pitted by raindrop impact. Sand grains were thrown horizontally the full width of the wash-trap, adhering to the back. Mato Grosso, Brazil.

It is mainly in humid temperate climates that the efficacy of surface wash is in doubt. Under the normal climax vegetation of deciduous or coniferous forest, the surface of mineral soil is protected by a closed tree canopy, layers of living and partly decomposed leaf litter, and a soil horizon composed mainly of humus. Most temperate climate grasses are of the turf-forming type, giving a dense root mat over the humus horizon. Runoff is inhibited not only by the protection of the surface, but because the humus layer acts as a sponge, capable of absorbing a 2-5 cm depth of water. If excavations are made during a heavy storm, slow seepage at the base of the humus horizon may sometimes be observed. Surface wash occurs on river bluffs and marine cliffs, kept sufficiently steep by undercutting to inhibit vegetation, on sites where bare ground is temporarily exposed by rapid mass movement, and on land under agricultural use. But there is no evidence that it is an important denudational process on slopes under natural conditions.

There are two methods of measuring the rate of surface wash. The first is to embed stakes deeply into the ground, and record the lowering of the ground surface around them; washers are sometimes placed over the stakes, and the falls of these measured. (Schumm, 1956 *et seq.*; Leopold *et al.*, 1966.) The other method uses wash traps, collecting the material in sedimentation chambers. Sediment collection was extensively used in U.S. work on accelerated erosion. Simple troughs for installation on natural slopes (Figs. 24, 25) have been described by Young (1960) and Gerlach (1967).

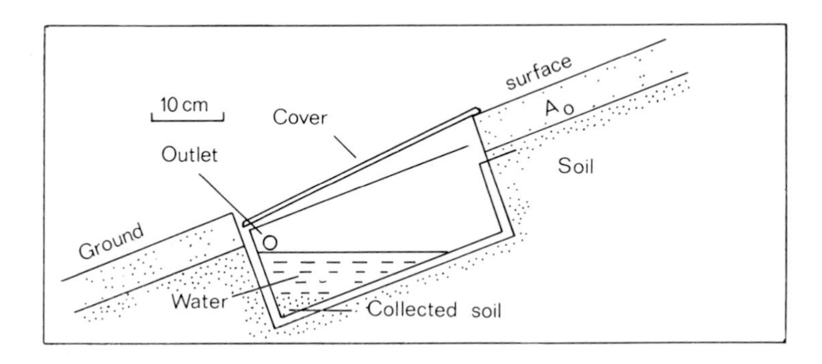

FIG. 25. A wash-trap. After Young (1960).

Further descriptions are given in *Revue de Géomorphologie dynamique* (1967). The nature of the information given by the two methods differs, in that stakes record net ground loss, whilst traps record the rate of downslope transport. If the area supplying the traps is enclosed, a conversion of collected sediment into ground loss is possible, but the enclosure introduces an artificial condition. Where no enclosure is used it is difficult to estimate the slope catchment from which the trapped sediment has been derived. Pairs of traps, placed a fixed (downslope) distance apart, have been used to indicate net soil loss or gain over the intervening portion of the slope (Townshend, 1970). On any part of natural slopes except the crest, a high rate of transport does not necessarily indicate ground loss. Since both methods are relatively simple, they are best used in conjunction; a series of stakes and traps installed from the crest to the base can in theory provide the data necessary to obtain the sediment input and output budget at each point, which can then be compared with the observed ground loss. Particles thrown by raindrop impact can be collected in shallow horizontal troughs, compartmented and partially shielded. Additional information on height of trajectory is provided by vertical plates coated with an adhesive substance, e.g. groundnut oil.

Measurements of the rate of contemporary surface wash under natural conditions are summarized in Table 3. Extreme values for individual studies are omitted, yet a wide range of values is still apparent. The very much slower rates under humid temperate climates as compared with all others are confirmed. Several investigators record rates of the order of 10-100 times higher on sites where the vegetation is locally less dense. Working on grassed slopes in the Carpathians, Gerlach (1967) obtained the

Table 3. Observed rates of surface wash

Source	Method	Climate	Ground lowering mm/yr	Volumetric movement, cm³/cm/yr
Young, 1960	Traps	Temperate	—	0·08
Starkel, 1962	Traps	Temperate	< 0·005	—
Gerlach, 1963	Traps	Temperate	< 0·005	—
Gerlach, 1967	Traps	Temperate	0·03	—
Kirkby, 1967	Traps	Temperate	—	0·09
Smith and Stamey, 1965	Experimental plots	Temperate	0·01–0·06	—
Soons and Rayner, 1968	Traps	Temperate montane	0·01	—
M. A. J. Williams, 1969	Traps	Warm temperate	0·05–0·10	—
Gabert, 1964	Traps	Mediterranean	0·09	—
Schumm, 1964	Stakes	Semi-arid	2·00	—
Leopold *et al.*, 1966	Stakes	Semi-arid	6·40–8·20	—
M. A. J. Williams, 1968, 1969	Traps	Savanna	0·039*	—
Rougerie, 1956	Stakes	Rainforest	5·0–15·0	—

* A correction (personal communication) of the value given in Williams (1968).

following variation in volume carried by wash: on rectilinear slopes the volume was greatest on the central part, decreasing towards both upper and lower ends; on convexities, volume continuously increased downslope, and on convex-concave slopes there was net accumulation on the concavities. At any given point there may be an alternation of loss and gain of material over short periods.

The mechanism of surface wash, and its empirical relation with slope parameters under experimental conditions, is relatively well-established. A greater range of comprehensive measurements under natural conditions is now required. For geomorphological purposes it is essential that measurements should be made under the nearest possible approximation to the natural climax vegetation. Among questions requiring investigation under natural conditions are: (*i*) the relative importance of raindrop impact and surface flow for soil detachment and for transport; (*ii*) whether ground loss is subject to control by detachment or by wash transport; (*iii*) the applicability of hydrological equations to natural slopes; (*iv*) quantitative relations between the rate of wash and vegetation cover, using established ecological methods to record the latter; (*v*) whether, as has formerly been supposed, the rate of surface wash (with respect to both transport and ground loss) increases downslope, or whether, as some present evidence suggests, the rate is relatively independent of distance from the slope crest.

SUB-SURFACE WASH AND GULLYING

Sub-surface wash is here employed as an *ad hoc* heading to cover a group of processes associated with water flowing within the regolith. These processes include lateral eluviation and tunnelling. Slope retreat through gullying is also conveniently treated in this section.

LATERAL ELUVIATION

Throughflow (also termed interflow) is the flow of water downslope through the regolith. It occurs where soil permeability decreases with depth. Water falling on a slope may be removed from it by evaporation and transpiration, surface flow, throughflow, and downward percolation through the bedrock. The ratio of throughflow to surface flow is highest where the regolith is thick and has a high infiltration capacity; that of throughflow to downward percolation depends on the relative permeabilities of regolith and bedrock. Water carried downslope by throughflow may emerge at the surface lower down the slope and continue as surface wash. The existence of throughflow invalidates the Horton (1945) erosion model, which assumes that all downslope water movement is by surface runoff. On an experimental plot in Provence, Clauzon and Vaudour (1969) found that throughflow was the normal means of drainage, surface runoff being exceptional. A comparison with stream hydrographs shows that in most environments, throughflow exceeds surface runoff (Whipkey, 1965; Kirkby and Chorley, 1967; Kirkby, 1969).

The term throughflow refers to water movement. The downslope transport of fine particles through the regolith, by the agency of throughflow, will be described by the pedological term *lateral eluviation.* The existence of lateral movement through the soil of material in solution is a recognized pedological process. Whether lateral eluviation of clay particles on an appreciable scale also takes place is unproven. Doubts have recently been cast on the more limited phenomenon of clay translocation from the upper to the lower horizons of a soil profile (Oertel, 1968). Lateral eluviation has been cited as a denudational process by Hauser and Zötl (1955), Hadley and Rolfe (1955), Ruxton (1958), and Mabbutt (1966), although most of the evidence given is indirect. Direct measurement of this process, although more difficult than that of surface wash, is necessary before it can be accepted as a process of quantitative importance.

TUNNELLING

There is more evidence for the existence of sub-surface wash in the form of concentrated lines of seepage or flow. The possibility of such processes was first recognized by Rubey (1928), who suggested that certain gullies on the U.S. Great Plains were formed by the washing out of fine particles along seepage lines, the resulting voids being filled by subsidence of the coarser soil fragments. This led to a sinking of the turf and so to the formation of a surface gully. The formation of subsurface channels is termed *tunnelling.** Examples of it have been described by Gibbs (1945), Buckham and Cockfield (1950), Schumm (1956), Zeitlinger (1959), and Bishop (1962). In Britain, the presence of tunnels within the regolith, covered over by turf, soon makes itself known to

* The term piping has also been used, but this is better reserved for the phenomenon of tongues of a drift deposit filling solution hollows in limestone.

walkers in moorland country. Parker (1963) considers that semi-arid regions are particularly subject to tunnelling; the regolith contains montmorillonitic clays which crack on drying, giving initial conditions for the concentration of subsurface flow; once initiated, flow lines converge towards tunnels, which are subsequently enlarged by mechanical corrasion.

An early stage in the development of lines of concentrated drainage on slopes was demonstrated by Bunting (1960, 1961, 1964). Many upland slopes in Britain show frequent *seepage lines*, in which the soil is normally waterlogged and carries *Juncus* or other hydrophytic vegetation. These may or may not have surface expression as shallow linear depressions, and they frequently pass downslope into gullies with permanent channels. By means of a closely-spaced grid of auger observations Bunting discovered integrated seepage lines without ground surface or vegetational expression, identified as lines along which the regolith/bedrock junction lay at greater depths. These are termed *percolines*; they pass downslope into visible seepage lines.

Spring sapping is a phenomenon in which the denudational processes of lateral eluviation, slumping, and possibly tunnelling pass downslope into the erosional process of channel formation (e.g. Hauser and Zötl, 1955; Small, 1964). It operates more as a process of channel extension and thus of slope dissection than as one of slope formation.

Seepage steps are micro-scarps, 1-60 cm high, formed within the regolith, sometimes capped by turf. They may be crescent-shaped, concave downslope, or run parallel to the contours. The steps retreat progressively upslope, causing in effect the removal of a layer of regolith. Hadley and Rolfe (1955) demonstrated that in the case described on the semi-arid Great Plains, there was a decrease in the permeability at the regolith/bedrock junction, and hence seepage and undermining. Vita-Finzi (1964) attributed microscarps in Jordan to sheetwash.

GULLYING

Percoline formation, tunnelling and spring sapping lie on the boundary between surface processes and linear erosion processes. The process of gullying is primarily one of linear erosion. It may have quasi-areal effects where a slope of relatively large dimensions is scored by numerous sub-parallel gullies. The floors and interfluve crests of the gullies both have similar angles to that of the main slope. In detail, the operative processes are linear channel erosion combined with surface processes on the steep gully sides. The net effect is that the main slope retreats considerably faster than if it were undissected, since its mean slope angle is increased by the presence of the gully sides. Steep scarps in the savanna and semi-arid tropical zones not infrequently retreat by this means.

Bryan (1940b) described a cyclic operation of processes, comprising gully formation, the infilling of gullies with permeable debris, and the cutting of fresh gullies on the less permeable former interfluve sites. The whole slope would at some time be liable to be the locus of a gully floor, and an apparently linear process would be acting areally. The occurrence of such inversion of relief is unlikely on *a priori* grounds, and has not subsequently been reported. It is a possible mechanism to account for the fact that pediments are not normally subject to gully dissection. Beaty (1959) described a more plausible cycle (Fig. 26); variation in weather conditions causes alternations between channel filling by denudation from gully-side slopes, and channel deepening. The whole slope retreats, but the gullies remain relatively unaltered in location.

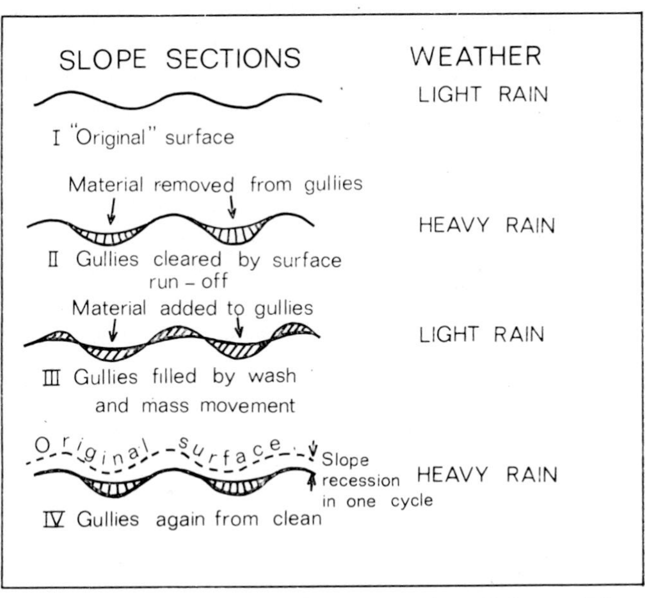

FIG. 26. A mechanism of slope retreat by gullying. After Beaty (1959).

Gullies are often initiated by seepage lines close to a main river channel, often combined with debris avalanching or other forms of mass movement; superimposition from a peat cover is another mechanism (Thomas, 1956). Gully extension may follow some or all of the stages percoline—seepage line—tunnel—gully. Gullying is particularly likely to affect slopes relict from different environmental conditions, for example boulder clay deposits and the steep sides of glaciated valleys.

Gullies caused by accelerated erosion, due to man's activities, will not be discussed; the classic description is by Ireland *et al.* (1939). In some cases doubt exists as to whether gullies are man-induced or natural (Bryan, 1925; Bailey, 1935, 1941; Antevs, 1952; Schumm and Hadley, 1957; Höllermann, 1963; Szupryczyński, 1967; Denevan, 1967). The trend of recent findings is to attribute more features than formerly to accelerated erosion following vegetation clearance by man. The same tendency is found in pedological studies in Britain and Europe, in which it is thought that Neolithic or Bronze Age forest clearance may have substantially altered the dynamics of soil moisture (e.g. Dimblebey, 1962).

SLOPE DISSECTION

The density of slope dissection, the mechanisms by which it is brought about, and the factors which affect it, are questions of geomorphological importance which have received little systematic study. The pioneer work on this topic was done by Morawetz (1937, 1944, 1950, 1962). Some areas possess broad, smooth slopes with lateral extents of the order of a kilometre, free from tributary valleys, gullies or seepage lines; in other areas the valley sides of the main rivers are dissected by numerous channels. It is relief of the former type, expressed morphometrically as having a low drainage density, that is the most noteworthy phenomenon. Any chance concentration of water flow is known to cause an increase in erosive and transporting capacity by the cube or higher power of the discharge; the fact that all slopes do not become dissected into a closely-spaced network of minor channels therefore calls for explanation.

The most important factor controlling dissection is probably infiltration capacity, either of the regolith or the bedrock. Badlands, the most closely-dissected type of relief, are formed from unconsolidated or weak materials with low permeability; conversely, limestones and permeable sandstones frequently have undissected slopes. Relatively close dissection may occur in semi-arid climates with a sparse vegetation cover, and even in deserts (Fig. 27). But well-dissected landscapes also occur under rainforest and

FIG. 27. Fine-textured gully dissection in an arid climate. Northern Iran.

in some temperate regions (pp. 228, 236). The density of slope dissection is a prominent feature in the appearance of the landscape as a whole, and affects its hydrological characteristics. There is opportunity for further systematic study in this field.

VIII RAPID MASS MOVEMENTS; COMPARISON OF SURFACE PROCESSES

RAPID MASS MOVEMENTS

CLASSIFICATIONS of mass movements are based on three properties: type of material, type of movement and rate of movement. The materials involved are rock, regolith, water, ice and snow, and the movements comprise fall, slip, flow and combinations of these. The principal classifications are those of Heim (1882, 1932), Sharpe (1938; with bibliography of earlier classifications), Ward (1945), Varnes (1958) and Hutchinson (1968). A modified form of Sharpe's classification is used here.

RAPID FLOW

Sharpe (1938) placed earthflow, mudflow and debris avalanche under the category of rapid flow, and debris slide in the group of slide movements. His descriptions of earthflow appear to include two substantially different speeds of movement, distinguished by Varnes (1958) as slow and rapid earthflow; movements of the latter type are here grouped with mudflow. In movements of broken masses of regolith, called debris, elements of both flow and slip are always present; Rapp (1960) noted the difficulty of distinguishing between Sharpe's classes of debris slide and debris avalanche, and criticised the application of 'avalanche' to movements not involving snow. Nevertheless, no better word to describe the mixed slip, flow and rolling movement involved has been found, and so the established term debris avalanche will be retained.

The following definitions are therefore adopted. Mass movements in which movement is largely or entirely by flow are designated *mudflow* if the moving material is almost entirely in a fluid state, and *earthflow* if it is partly solid; mudflows move at more than 10 m/hr, and earthflows at lower rates. No connotation is implied by 'earth', nor is the turf cover in earthflow necessarily unbroken. Mass movements of regolith which disintegrates substantially during movement will be termed *debris avalanches*; movement is necessarily very rapid. Examples intermediate between mudflow and debris avalanche occur. *Debris slide* will be reserved for cases in which regolith moves largely along a single slip plane, with relatively little disintegration and flow movement in the slipped mass.

In *earthflow*, movement is normally imperceptible to the eye, and the turf cover may remain intact. In the source area crescentic scars appear in the turf followed by slumping, due to flowage out of the underlying soil; the flow track is normally short, and leads to an area with bulging lobes of soil. Clays are most affected. (Blackwelder, 1912; Ackermann, 1959; Crandall and Varnes, 1961; Crozier, 1969.) Earthflows of this localized form are quantitatively a minor denudational process. It has been suggested that where a moist regolith underlies a strong root mat, as in tropical rainforest, earthflow may act areally over slopes, the regolith flowing slowly downslope whilst the turf remains intact but subsides (Freise, 1935; Sapper, 1935). White (1949) described in Hawaii the movement of a 'muddy sludge' below and through the root mat, occasionally breaking out onto the surface as mudflows. This phenomenon is not visibly apparent in most rainforest areas, and has not yet been quantitatively measured.

Mudflows move at a visibly perceptible rate, varying from viscous material moving at the order of 10 m/hr to flow at rates comparable to that of water on a similar slope, and capable of destroying buildings. A basin-shaped source area, a long and relatively narrow flow track, and expanded depositional toe zone can usually be distinguished. Mudflows are particularly associated with slopes lacking a coherent turf cover, and therefore with semi-arid and montane climates, but are not confined to such zones. They frequently follow tracks of existing gullies, and may act as a means of headward extension of such channels. (Blackwelder, 1928; Cailleux and Tricart, 1950; Grove, 1953; Rapp, 1960; Prior *et al.*, 1968.) In sandy materials lying below the water table at the time of movement, spontaneous liquefaction may be a cause of the initiation of rapid flow (Terzaghi, 1950).

Debris avalanches also follow relatively long and narrow tracks. They may occur at gully heads, and are a common mechanism by which seepage lines on steep slopes are converted into incipient gullies dissecting the slopes. Selby (1966b) gives examples of valleys believed to be formed entirely by such mass movements. Debris avalanches can also occur on unchannelled slopes, stripping the regolith and so permitting weathering to recommence on exposed bedrock (Gifford, 1953; Rapp, 1960). In Romania, dissected country formed of intercalated clays and soft sandstones is subject to numerous shallow mass movements (Morariu and Gârbacea, 1967). Boulder clay and periglacial deposits on steep slopes are also commonly affected (Fig. 28). On steep slopes under tropical rainforest high mean rates of surface lowering by debris avalanches have been recorded (Wentworth, 1943; Simonett, 1967). A thick regolith is not a necessary condition for debris avalanches; many recorded instances, both in temperate and tropical climates, affect 20-100 cm of soil overlying bedrock. Detailed descriptions and photographs of typical debris avalanches are given by Hack and Goodlett (1960).

Snow avalanches may incorporate some rock debris (Rapp, 1959). Of greater importance as a denudational process are *slush avalanches* (slushers), formed when melting snow saturates the underlying regolith and a mixture of snow, water and regolith flows rapidly downslope (Washburn and Goldthwait, 1958; Rapp, 1960; Jahn, 1967; Caine, 1969).

Earthflows, mudflows and debris avalanches invariably occur when the regolith becomes exceptionally wet, often when a heavy storm follows a period of unusually high rainfall. The onset of movement is not due to lubrication, since only a thin film of water, such as is nearly always present, is required to produce the maximum decrease in cohesion between particles. The cause is a rise in the water table to a position above the material affected, which produces an increase in pore water pressure, and a consequent decrease in shearing strength (p. 79). The capacity of a soil to resist flow

FIG. 28. Debris avalanches on Plynlimon, Wales. The upper movement affected a slope deposit of congelifractate (p. 242), the lower, a boulder clay deposit in a moist site. An earlier avalanche has occurred on a similar site to the right.

becomes very low at the moisture content denoted by the Atterberg liquid limit. The susceptibility of clays to rapid flow varies with the *activity* of the clay, where activity = plasticity index/percentage clay fraction, and plasticity index = Atterberg liquid limit − Atterberg plastic limit. The activity of clay minerals increases in the order kaolinite, illite, montmorillonite (Skempton, 1953).

Rapid flow mass movements are therefore associated with exceptional meteorological conditions. Thick superficial deposits laid down on slopes under environmental conditions different from those of the present, for example boulder clay, may be unstable with respect to contemporary extreme-weather conditions, and so are often affected. A feature of the manner of action of most rapid flow mass movements is their tendency to dissect slopes. Occurrences of large numbers of closely-spaced debris avalanches have been reported, giving a quasi-areal effect (Fig. 29), but in many cases most of the movements occur in wet, seepage sites, suggesting a tendency towards dissection rather than areal slope retreat. They are sometimes localized in valley-head slopes,

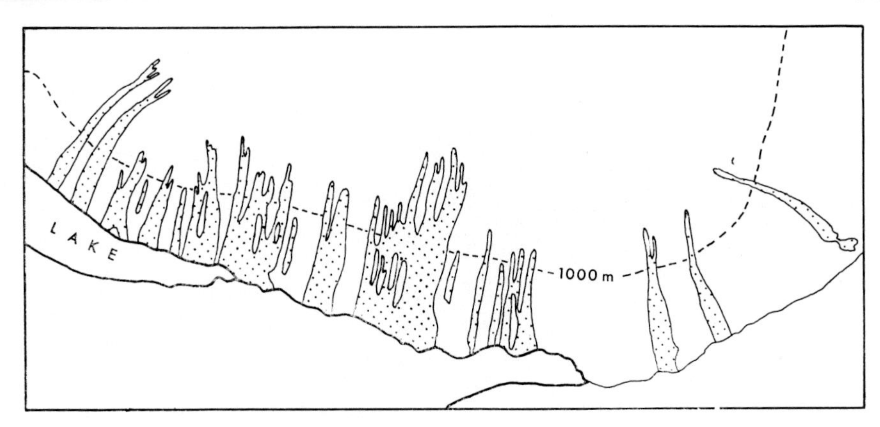

FIG. 29. Tracks of debris avalanches at Ulvådal, western Norway, 26th June 1960. After Rapp (1963).

acting as a means of headward extension of drainage channels (Hack and Goodlett, 1960).

A manner of action that is in the long term areal may also be produced if individually-localized mass movements occur over a slope for a long period of time. A *regolith-stripping cycle* (landslip-healing cycle) has been suggested for tropical rainforest environments (Freise, 1938; Wentworth, 1943; White, 1949), and more recently reported on very steep slopes in a cool temperate climate (Schweinfurth, 1966). The cycle involves the progressive deepening of the regolith by weathering, followed by its stripping down to bedrock by landslides after reaching a thickness critical for stability under the local conditions of soil strength, vegetation cover and storm intensity. Such a cycle would give a landslide distribution over a period of time more regular than random. Scars of recent mass movements are common in steeply dissected rainforest relief; to prove a truly areal effect, as opposed to a dissecting tendency, it would need to be shown first, that landslides occurred where the regolith was thickest, and secondly, that sites with exceptionally thick regolith were more or less randomly distributed on the slope, and not localized in linear patterns.

LANDSLIDES

The feature distinguishing landslides from other mass movements is that movement occurs mainly along a discrete failure surface. Initial movement is by slip, the material above the plane being internally undeformed. The slipped mass partially disintegrates, and subsequent movement may include an element of flow. In *rockslide* and *debris slide* failure takes place along a relatively flat plane approximately parallel to the ground surface, movements of consolidated or relatively coherent rock being rockslides, and those of regolith or other incoherent material debris slides. Rockslide and debris slide are relatively superficial movements. Where the failure surface penetrates deeply into the slope, the failure plane may take various forms. For a rectilinear slope of finite height and homogeneous material, failure takes place along a curved surface, concave to the slope and passing through or below its base; this is the ideal case of *rotational slip*, or slump. Where structural conditions are non-uniform, movement may take place along failure surfaces not readily placed into the preceding classes; these are referred to as undifferentiated landslides. The process of *rockfall* is discussed in relation to cliff evolution in Chapter XI.

Skempton (1953) characterized landslides by their depth/length (D/L) ratio, the ratio of the maximum thickness of the moving mass to its length up the slope, expressed as a percentage. Mass movements affecting the regolith only, or *surface landslides*, have D/L ratios of the order of 2-5%; in *deep landslides* the ratio is commonly 10-30%.

The theoretical basis and empirical methods for the study of the causes of landslides were developed within the science of soil mechanics, particularly by Karl Terzaghi and D. W. Taylor. Much of this theory was developed with respect to the artificially constructed slopes of cuttings and earth fills. It has been modified to apply to the long-term stability of natural slopes by Terzaghi (1950) and Skempton (1953, 1964; Skempton and Delory, 1957). Numerous case studies of landslides are available in engineering journals, and the mechanism of the process is fully treated in textbooks on soil mechanics (e.g. Taylor, 1948), so a brief account only of this mechanism is included here.

Slip failure occurs within a material, rock or regolith, when the shearing stress exceeds the shear strength along any plane through the material. The shear strength, equal to the shear stress at failure, is dependent on the normal stress, the pressure in a direction perpendicular to the plane. Let

s = shear strength
σ = normal stress on the shear plane
c = cohesion
ϕ = friction angle
h = head of water
w = unit weight of water
$u = hw$ = pore water pressure

The shear strength of a freely drained material is given by

$$s = c + \sigma . \tan \theta \qquad (8.1)$$

For sands and silts the value of cohesion c is nil. Fig. 30*A* and *B* illustrate this equation. Where the water table rises above a potential failure plane the normal pressure is reduced by an amount u, equal to hw; in effect the intergranular pressure is reduced by an element of 'flotation' imparted to the overburden through its immersion in water. The normal stress is then reduced to the effective normal stress σ', equal to $(\sigma - u)$; the values of cohesion and of friction angle are also affected, the altered values being termed the effective cohesion c' and effective friction angle ϕ'. The modified equation for shear strength, commonly called Coulomb's Law, is then

$$s = c' + \sigma' . \tan \phi' \qquad (8.2)$$

The shear strength is at a minimum when the water table is at its highest, therefore landslides almost invariably occur following periods of prolonged or exceptional rainfall. If active flow of water occurs in the material its strength is further reduced by seepage forces.

On diagrams like Fig. 30 a rectilinear slope of angle θ may be represented by a line at this angle passing through the origin. The slope will be stable, i.e. not subject to landslide failure, where the line representing the slope lies below that representing the shear strength. Cohesionless materials are stable if the slope angle is less than the friction angle, irrespective of the height of the slope (Fig. 30*C*); under natural conditions this applies to well-drained fragmental slopes, such as sand dunes and screes. Materials possessing cohesion can stand at an angle above that of the friction angle up to a certain height, dependent on the relation between c, ϕ and θ (Fig. 30*D*). On clays and other cohesive materials, high-angle slopes can exist for limited periods of time up to a certain critical height. An example of *high-cohesion cliffs* are those formed in loess, from which vertical bluffs up to 30 m high have been reported (Smalley and Taylor, 1970).

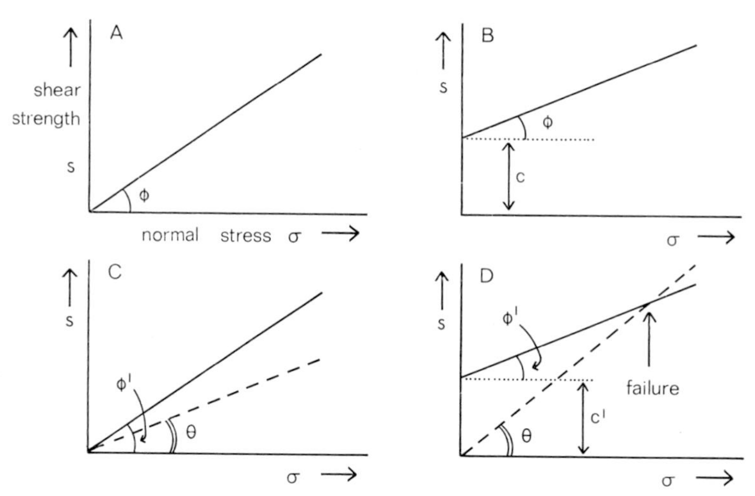

FIG. 30. Relations between normal stress and shear strength. *A* and *C* = cohesionless materials; *B* and *D* = materials possessing cohesion.

Most natural clays have been subject in the past to greater overburden pressures than at present. This applies both to clays from which overlying strata have been removed by denudation and to boulder clays formerly bearing an ice load. These are known as *overconsolidated clays*. Owing to their lower water content and closer packing of particles, overconsolidated clays initially possess a greater strength than normally consolidated material; when the overburden is removed, overconsolidated clays undergo a progressive decrease in shear strength over a period of the order of 10-50 years, principally by the opening of fissures. Recently-eroded cliffs or valley sides in soft strata may therefore retain steep angles for a short period, after which they will be subject to landslides. From an analysis of records of slope failure in London Clay, Skempton (1948) found that for slopes 6 m in height, a 25° slope is liable to fail after 10-20 years and an 18° slope after 50 years. The equations and parameters in normal use in soil mechanics thus refer to stability conditions over very short periods of time in terms of landform evolution.

A slope possesses *long-term stability* when it is not subject to landsliding within 100 years or more. From stability analyses based on samples taken from natural clays, the angle at which slopes in such clays should remain stable may be calculated; but the observed steepest natural slopes unaffected by landsliding are found to be less than this angle. Thus for the case cited above of London Clay, stable natural slopes rarely exceed 10° (Hutchinson, 1967). Skempton (1964) has investigated this discrepancy, and proposed that analysis of the long-term stability of clays should be based on their residual strength. A typical result from a large-strain shear-box test on clay is shown in Fig. 31. With a constant normal stress, only slight strain occurs initially, and the

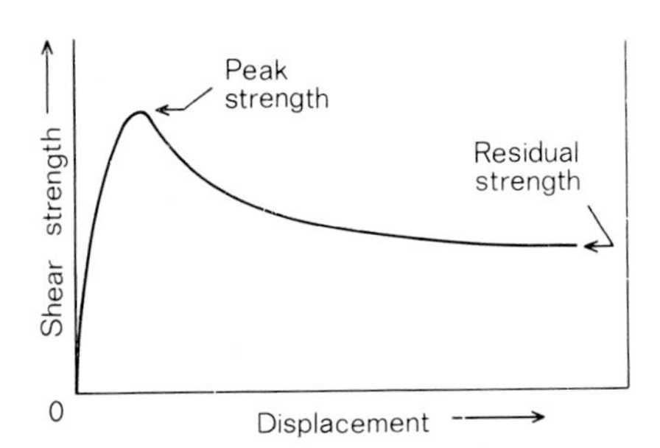

FIG. 31. Residual strength of clays, as shown by the variation of shear strength with displacement in a shear-box test. Based on Skempton (1964).

shear strength rises to a maximum value, the *peak strength*; following the commencement of shear, the shear strength drops off rapidly at first, ultimately settling down to a constant value, the *residual strength*. From a series of such tests at different normal pressures, the values of residual cohesion c_r and the friction angle with respect to residual strength ϕ_r may be obtained. Residual cohesion is in practice found to be almost zero, so Coulomb's Law with respect to residual strength s_r becomes

$$s_r = \sigma' . \tan \phi'_r \qquad (8.3)$$

The physical cause of the difference in values may be the formation of domains of oriented clay along the shear plane. For example, different strength envelopes obtained for London Clay with respect to peak and residual strength. Using the latter, the calculated stable slope when the effects of a high water table are allowed for is about 10°, which agrees with the observed slopes.

Further advances can be made in the application of the methods of soil mechanics to natural slopes, but there are reservations regarding such applications. First, the standard techniques of soil mechanics were developed with respect to, and are intended to apply primarily to, artificially cut and constructed slopes. Secondly, even with respect to the short-term stability, which is the concern of the engineer, the calculated values of stability are known to be only rough approximations to the observed behaviour of slopes. In practice an arbitrary but substantial safety factor is incorporated, probably to allow for the undetected presence of localized fractures or other planes of weakness in the material. Thirdly, as indicated above, the value of strength used must be that which the material retains after existing close to the ground surface, under low overburden pressures and affected by weathering, for long periods. Finally, analysis must be in terms of extreme hydrological conditions, such as may have a recurrence interval of 100 years or longer. Data on such conditions may not be directly available, and consequently an element of uncertainty in extrapolating known data is introduced.

The causes of landslides under natural conditions have been listed by Sharpe (1938, p. 84). They are grouped into *passive conditions*, favouring landslide occurrence, and *active conditions*, which initiate movement. The main passive conditions are:

Lithological — weak or unconsolidated rocks, especially clays.
Structural — permeable beds overlying impermeable.
— dips toward a slope.
Topographic — cliffs or steep slopes formed by basal erosion.
Hydrological — conditions causing surface or subsurface concentration of water.
— seepage within the rock or regolith.
Climatic — liability to high-intensity rainstorms.

The main active causes of movement are steepening of the slope by basal erosion, earthquakes and, the most frequent immediate cause, the occurrence of exceptionally heavy rainfall, especially following a prolonged wet period when the water table has reached a relatively high level.

Rotational slip in its theoretically ideal form is approximated to in natural landslides mainly in cases of steep slopes cut in thick clay strata, including boulder clay (see Skempton, 1953). Various forms of shallow slips also occur in unconsolidated material, including shallow rotational slips sometimes occurring in series up a slope (Hutchinson, 1967), and slides along a more or less plane surface parallel to, or forming a low angle with, the ground surface (Skempton, 1953; Hutchinson, 1961). Even in the latter cases there is usually some element of rotation close to the toe. Movement is usually very rapid.

A common structural situation for the occurrence of deep landslides is where permeable beds (sandstone or limestone) overlie weak, impermeable strata, the latter outcrop on the lower part of a slope, and the beds dip gently towards the slope. The theoretical curved failure plane of rotational slip is much distorted to pass through a greater thickness of the weak beds, but the slipped masses normally show substantial backward rotation, finishing with steep dips towards the slope from which they have come. A succession of such landslides may occur, working progressively backwards into the slope until the structural conditions favouring movement have been eliminated. Landslide complexes of this type occur in the Millstone Grit strata of the English Pennines (Bass, 1956; Johnson, 1965), where in most cases slipping is no longer active, the slopes having attained under periglacial conditions a form that is stable with respect to the present climate. An active landslide in this type of structure is the Mackenröde Spitze, near Göttingen, where Muschelkalk limestone overlies red marls (Mortensen and Hövermann, 1956; Mortensen, 1960b). The main currently active slide began in 1880. The moving material consists of a series of limestone blocks, at the foot of which more disintegrated material moves as an earthflow. Active movement is evidenced by tilted and damaged trees, and ravines covered by turf. In 1962 the late Professor Mortensen took an excursion of the I.G.U. Slopes Commission over the landslide; a root was stretched tautly across a ravine behind a slipped mass; on cutting the root, so releasing part of the force holding back the mass, the two halves could no longer be rejoined. Between 1880 and 1952 movement of the main slide averaged 25 cm/yr. The toe of the earthflow attained a speed of 4 m/yr in 1944-7, but slowed to 0·5 m/yr in 1952-8. Deep slips of this nature frequently move at visibly imperceptible rates, the speed of movement showing some correlation with rainfall (Mortensen and Hövermann, 1956; Schumm and Chorley, 1964; Merriam, 1960).

Slopes at steep angles in high cohesion materials, such as loess, may not fail by rotational slip but along plane fractures passing through the base of the slope. The theory of *high-angle plane failures* is given by Lohnes and Handy (1968). The angle of the failure plane, A, is given by

$$A = \frac{i}{2} + \frac{\phi}{2} \tag{8.4}$$

where i = angle of initial slope and ϕ = angle of internal friction. From calculations based on shear strength tests on loess, Lohnes and Handy found three angles would theoretically be expected: slopes of 77°, up to a critical height of 4·7 m, resulting from vertical cleavage failure; 51°, up to a critical height of 7 m, resulting from shear failure; and 38°, the angle of repose of failed loess. Observed angles confirmed these theoretical calculations, having a tri-modal frequency distribution with an absence of slopes at about 60° and 45°, as predicted.

Soil slips are surface landslides affecting the regolith only (Bailey and Rice, 1969; Rice *et al.*, 1969). They include a *breakaway microscarp* left in the turf above, a *slip plane*, a *toe* of relatively coherent material that has moved above the slip plane, and sometimes a *tongue* of broken or semi-fluid material below. Soil slips occur on steep slopes and indicate regolith instability (cf. p. 165).

The net geomorphological effect of landslides is to reduce slopes to angles at which they possess long-term stability, with respect both to their internal strength after long periods near the surface, and to the most extreme groundwater conditions occurring over long periods. If marine or river erosion produces slopes steeper than such critical angles, these slopes will be rapidly eliminated. An individual landslide does not always produce a gentler slope than that which preceded it, but if a steeper slope is produced it is less high. In rotational slip, an immediate decline in slope angle results, through removal of material from the upper part of the slope and its deposition on the lower part; in cases where slipped masses break away to leave a steep cliff, the retreat of the cliff is usually accompanied by a progressive rise of the slope at its foot. Thus landslides alone can cause a type of slope evolution of the nature of either slope decline or slope replacement. The fact that areas in which landsliding is the dominant process of slope evolution are of limited extent is a necessary consequence of the rapidity with which such evolution occurs. When a slope has attained the condition of long-term stability, landslides cease.

OTHER PROCESSES

Penck (1924) identified a process of *corrasion*, defined as the freeing of rock fragments from their *in situ* positions by the force exerted by regolith moving over them. The significance he gave to this process is linked with the assumptions of his deductive system (p. 28); Penck held that corrasion loosened fragments before they had reached the degree of reduction necessary to move by spontaneous mass movement on a given slope angle. The evidence cited as showing the existence of corrasion is outcrop curvature. Penck also believed that this process could produce corrasion valleys, lacking stream channels but containing a 'stream' of regolith; unlike most dry valleys, they are found on non-calcareous rocks. Several German-language studies of such valleys (*Dellen*) have appeared (Lehmann, 1918; Schmitthenner, 1925; Stratil-Sauer, 1931; Klatkowa, 1967).

It does not seem necessary to distinguish corrasion beneath a soil cover as an independent process. Outcrop curvature is sufficiently considered as a consequence of soil creep or solifluction. Dry valleys in non-calcareous rocks are now usually attributed to periglacial action. The existence of corrasion has been justified in the special circumstance of the abrasive action exerted on a bare rock slope by rapidly rolling or sliding boulders, originating from rockfall or rockslide (Blackwelder, 1942; Sparrow, 1965).

Mineral accumulation by plants has the net effect of downslope transport. Mineral matter is taken up in an approximately vertical direction into the plant body, and returned to the soil downslope following leaf fall and plant death and decay. Most plants take up substantial weights of calcium, magnesium and potassium, and some grasses and tropical rainforest trees accumulate silica; this material originates from rock weathering, subsequently cycling between plant and soil (Lovering, 1959). The amount of silica removed from the soil by tropical rainforest has been estimated at 1·0 t/ha/yr (McKeague and Cline, 1963), equivalent to a layer 38 mm thick per 1000 years. The manner of action of this process is progressive downslope transport, and it may in some respects be regarded as an agent of soil creep.

Violent winds cause *tree uprooting*, in the course of which substantial amounts of soil are loosened and uplifted (Lutz, 1960). Downslope transfer is brought about either by free fall on the decay of the tree, or by surface wash on the exposed bare soil. In the short term the process is highly localized, but if randomly distributed it will have an areal effect over a long period comparable with that of soil creep. In some circumstances gully dissection is initiated. Tree uprooting in temperate woodlands follows infrequent and localized gales; in tropical rainforest it is a normal and widespread phenomenon.

Root growth of trees on rock cliffs can act as a denudational agent, penetrating and widening fissures, and detaching slabs of rock. Roots may also exert a binding action temporarily stabilizing the cliff (Jackson and Sheldon, 1949).

Deflation, the movement of rock particles by wind, requires the exposure of loose, dry fragments at the surface, and is therefore significant mainly in arid, semi-arid and polar climates. It is now generally held that the erosive action of sand blast in deserts accounts only for minor details in the shaping of landforms, but as an agent of transport, wind is of substantial importance, removing fine particles as soon as they are brought by weathering into a loose state. The efficacy of wind transport under periglacial conditions is testified by loess deposits. Among studies of polar environments, Everett (1966) considered deflation to be a major contributor to surface movement in Alaska; Pissart (1966b), specifically investigating wind action in Prince Patrick Isle, Canada, found little evidence for it, and Rapp (1960) considered its quantitative importance to be very small. The manner of action of deflation is intermediate between downslope transport and direct removal from the slope, the latter predominating for fine particles.

THE RELATIVE IMPORTANCE OF SURFACE PROCESSES

Relative importance refers in this context to the proportion of original rock material, by volume, transported across and away from slopes by the agency of different surface processes. There are two main questions: the relation between catastrophic and continuous denudation, and the relative importance of the individual processes in each. The former is closely related to that of the relative importance of rapid mass movements, although the two questions are not identical since catastrophic denudation may also involve surface wash. Only present-day processes are considered in this section; the extent to which existing slopes are relict features is discussed in Chapter XIX.

Catastrophic denudation takes place after intense storms and other exceptional meteorological conditions, and is effected largely, but not necessarily entirely, by rapid mass movements. The area subject to catastrophic denudation on any single occasion is localized; conversely a given locality is affected only very infrequently. *Continuous denudation* is caused by processes which affect the whole or large parts of a slope simultaneously, principally soil creep, surface wash, solution loss and solifluction; the intensity is not uniform, but the frequency interval for the occurrence of substantial denudation is not more than one year. Thus catastrophic denudation is spatially localized in the short term, and highly irregular in time, while continuous denudation acts more uniformly in space and is less variable in time.

On Friday 13th December 1946 an intense depression moved inland from the Moçambique Channel, reached Zomba, the capital of Malawi, at midday, and remained stationary for 36 hours, drawing in large amounts of moist unstable air. 70 cm of rain fell in 40 hours, including 12 cm in one half-hour. The capital lies on the lower slopes of a steep-sided plateau. Debris avalanches and mudflows coursed down the slope, damaging buildings and blocking roads, and parts of the slope were stripped bare of regolith down to bedrock, the debris accumulating at lower levels (Talbot Edwards, 1948). The scars and deposits left by this storm have not been substantially modified during the succeeding 20 years. There are many accounts of similar catastrophic denudation, in temperate latitudes (e.g. Gifford, 1953; Common, 1954; Hack and Goodlett, 1960; Tricart *et al.*, 1961; Meynier, 1965), in polar environments (Rapp, 1963) and in the tropics (Cooray, 1958). The process most often involved is debris avalanching, together with other types of rapid mass movement, and surface wash at rates greatly above normal. The volume of material moved may be of the order of 100 000 m^3. The contrast between events of this kind and the instrumental refinements needed to detect any movement at all by continuous processes raises doubts as to whether the latter are of substantial importance in the long term.

The theoretical basis for investigating this question is a comparison of the magnitude and frequency of geomorphological events. In Fig. 32 the abscissa represents the intensity of events, or net applied stress. Possible units are rainfall intensity, as cm/hr, or some integral of rainfall amount over a given period. The ordinate scale represents for curve A the volume-distance of material moved in unit time for an event of given

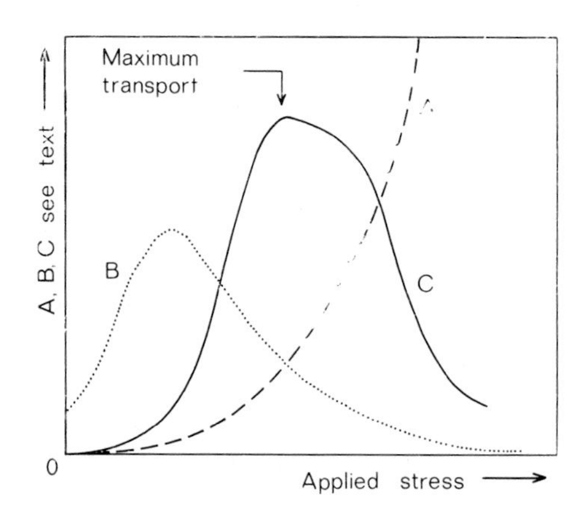

FIG. 32. Magnitude and frequency in surface transport. A = work accomplished by individual events of a given magnitude; B = frequency of events of a given magnitude; C = total work accomplished by events of given magnitudes. Based on Wolman and Miller (1960).

intensity, and for curve B the frequency of occurrence of such events. It is possible that A is exponential in form, and B approximately log-normal. Curve C is the product of A times B, and shows the total volume of surface transport accomplished by events of given intensity. For the analogous case of fluvial transport, Wolman and Miller (1960) found that the maximum work was accomplished by floods with a recurrence interval of two years or less; the rapidity of transport during very high flows was more than compensated for by the extreme infrequency of such flows. With respect to surface processes, there is at present no data available in the form required for this model.

A single minor landslide moving a volume of 10 m^3 a distance of 10 m is equivalent to a continuous process with an average rate of 10 $cm^3/cm/yr$ acting for one million years. This order-of-magnitude comparison suggests that if the whole of the surface of a slope is affected by rapid mass movements with even a very low frequency they will outweigh the effects of continuous processes. Rapid mass movements probably account for a high proportion of total denudation on slopes above the angle of long-term stability (p. 81), and in all areas where visible signs of mass movement are common, for example in steeply dissected rainforest relief. On most slopes, however, neither of these conditions apply, and for such slopes, an indirect argument may be advanced. Assume the hypotheses (*i*) that the manner of action of most rapid mass movements is to produce irregular topography, frequently with a linear, dissecting tendency; (*ii*) that rapid mass movements are the main cause of denudation. It then follows that slopes will be irregular in profile and highly dissected in plan. This is contradicted by observation. If the first hypothesis is accepted, on the basis of studies of actual mass movements, the second must be false. Thus on undissected slopes lacking marked irregularities of form, which constitute a high proportion of all valley slopes, the relative importance of rapid mass movements must be low.

Outstanding among observational comparisons of surface processes is the study of the Kärkevagge, a formerly glaciated trough valley in northern Lappland, by Rapp (1960). All forms of denudation were recorded for eight years (Table 4). Solution loss

Table 4. Denudation in the Kärkevagge, northern Lappland, 1952-1960. The values are converted into ton-metres moved vertically and horizontally. Surface wash is excluded, but is believed to be of minor importance. After Rapp (1960).

Process	Average gradient	Vertical transport, ton-metres	%	Horizontal transport, ton-metres	%
Rockfall	45°	19 565	7	19 565	4
Slush and snow avalanche	30°	21 850	8	37 820	8
Debris avalanche, debris slide, mudflow	30°	96 375	34	166 630	34
Talus creep	30°	2 700	1	4 700	1
Solifluction	15°	5 300	2	19 800	4
Solution loss	30°	136 500	48	236 500	49
Totals		282 290	100	485 015	100

accounts for nearly half the total denudation, despite this being an area of steep slopes and a cool climate. Movements of a rapid flow nature come next in importance, and solifluction is relatively insignificant. For a polar environment in Greenland, Everett (1967) found solifluction to be the predominant process, followed by rapid mass movements, with wind and possibly surface wash also significant. On Flysch sandstone in the Carpathian mountains, Gerlach (1967) estimated the relative importance of processes as, in descending order, rapid mass movements, surface wash, solution loss, soil creep, needle-ice, and tree uprooting.

Some comparative measurements of soil creep and surface wash on the same slopes are available. For cool temperate conditions in Britain, two independent studies have shown that creep is respectively ten and twenty times faster than wash (Young, 1960; Kirkby, 1967). In semi-arid conditions in New Mexico creep accounted for less than 1% of total denudation, and wash for 98%, despite an absolute rate of creep ten times faster than in Britain (Leopold *et al.*, 1966). In Australia, wash has been recorded as seven times faster than creep in a warm temperate climate, and five times faster under savanna conditions (M. A. J. Williams, 1969). There have been negative findings for both creep and wash; in warm temperate southeastern U.S.A., Parizek and Woodruff (1956) found that evidence for the existence of creep, such as stone lines and tree curvature, were either absent or could be explained by other means, and in the cool temperate Carpathians, Starkel (1962) recorded extremely low rates of surface wash on grass slopes.

Conclusions as to the relative importance of the continuous processes can only be provisional, owing to the small amount of data. In cool temperate climates soil creep is more important than surface wash, the absolute rate of wash being very low, but this is the only climate, with the possible exception of rainforest, in which creep is predominant. In semi-arid, savanna and possibly warm temperate climates surface wash predominates over creep; this contrast with cool temperate climates is due to higher absolute rates of wash, not lower rates of creep. Solifluction in polar climates is at least ten times faster than soil creep in other climates. It follows from the above that surface transport is slower in cool temperate climates than in most other zones. For the rainforest environment there is insufficient data to estimate the relative importance of creep and wash. These results refer only to downslope transport; in all climates, solution loss probably accounts for 10-50% of the total removal of rock material from a slope.

ABSOLUTE RATES OF GROUND LOWERING

The order of time required for the evolution of a slope fundamentally affects consideration of its origin. It is significant to know whether the present features of slope form could have originated largely in post-glacial time; whether one or more glaciations and inter-glacial periods have been involved; or whether the origin of any feature dates back into Tertiary time. The history of ideas on this subject has shown a progressive decrease in the time scale attributed to landforms; Davis estimated 20-200 million years as the time required for an erosion cycle to proceed to old age, whereas in 1954 Thornbury gave as one of the fundamental concepts of geomorphology that 'Little of the earth's topography is older than Tertiary and most of it no older than Pleistocene'.

A suitable unit for an order-of-magnitude answer is the average rate of lowering of the ground surface. This may be expressed as mm/1000 yr, equivalent to metres per million years; in terms of volume of material it is equivalent to $m^3/km^2/yr$. Evidence of four main types is available: first, measurements of river load; secondly, the sediment accumulated in reservoirs; thirdly, measurements of surface processes on slopes; and fourthly, what may be termed geological evidence, the comparison of known geological or radiocarbon dates with landform changes identifiable as subsequent to them. The two first types include the work of river erosion as well as surface denudation. Results from unconsolidated rocks, which may be eroded 10-1000 times faster than consolidated, have been excluded.

There are no clear correlations with either climate or rock type. Previous studies of the climatic variable reached inconsistent conclusions (Corbel, 1959; Schumm, 1963; Ahnert, 1970c). A marked grouping appears, however, if the results are divided into two relief classes: 'normal' relief, including plains, moderately dissected areas, and gentle to moderate slopes; and steep relief, including mountainous areas and individual steep slopes. The results are shown grouped in this way in Fig. 33. For data of such a heterogeneous nature the median and inter-quartile range are appropriate measures of

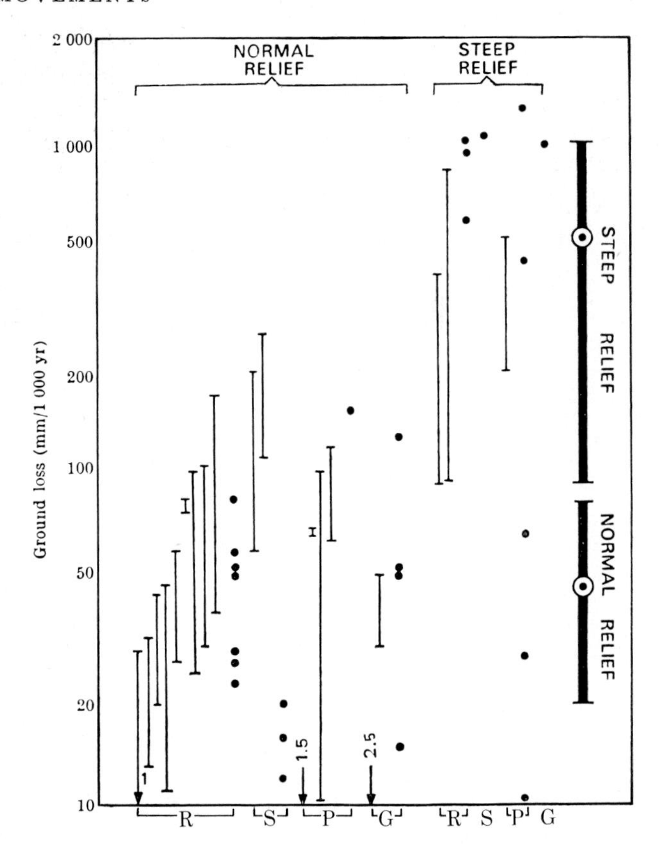

FIG. 33. Reported rates of lowering of the ground surface. To the right are shown median values and inter-quartile ranges for steep relief (mountain areas, individual steep slopes) and normal relief. Types of evidence: *R* = river load; *S* = sedimentation in reservoirs; *P* = surface process measurements; *G* = geological evidence. After Young (1969).

central tendency. The values are, in mm/1000 yr, normal relief: median 46, range 20-81; steep relief, median 500, range 92-970. Thus ground loss from mountainous areas and steep slopes is of the order of ten times faster than from other landforms.

Table 5. Time required for slope retreat, based on median values in Fig. 33

	Slope retreat, metres	
Time, years	Steep slopes	Gentle slopes
10 000	5	0·5
100 000	50	5
1 000 000	500	50

An extrapolation in time of the median values is given in Table 5. There are clearly considerable reservations over the validity of such a procedure. If, however, the orders of magnitude are correct, the results suggest that slope form has not been greatly modified in post-glacial time; that a steep slope could have been appreciably modified, and a small gully completely formed, during the last inter-glacial and glaciation; but that the evolution of extensive, relatively gentle, slopes may have extended over a substantial part of Pleistocene time, and possibly into the late Tertiary.

IX THEORY

THE practical usefulness of an hypothesis, or other abstract concept, depends upon four considerations: its logical validity, including whether it is precisely formulated and internally consistent; the degree of correspondence between its basic assumptions and the observed facts; whether it can be tested under field conditions; and how widely it is applicable. Most discussions have given too little attention to the three latter aspects, which require comparison with field and experimental data. Nevertheless, an examination of the deductive basis of concepts formulated in abstract terms is prerequisite to consideration of their usefulness as explanations of observed facts. It is in particular necessary, in order to compare hypothesis with reality, to make explicit all of the assumptions contained in the former, and to deduce the consequences.

Most hypotheses concerning slopes involve relations between process and form. Such hypotheses may arise either inductively, out of generalizations from field studies, or deductively, from a consideration of the forces and properties concerned. The hypothesis that the rate of downslope transport by surface wash increases linearly with distance from the crest could have been reached by recording wash, and finding that such a relation appeared to be a good approximation to the data. Alternatively, it could have been formulated from a consideration of the factors which determine the volume of wash, its speed of flow and transporting power. If proposed on the latter basis (as was historically the case in the example given), one or more of its deduced consequences must be tested experimentally before the hypothesis can be considered as established.

Principles in science exist on a series of levels of differing scale; the sub-atomic, atomic and molecular levels; the level of physical properties, such as force, pressure and work; and what may be termed the environmental level, at which forces cannot be examined in isolation, under controlled conditions, but must be considered as they occur in nature, ineluctably combined with other variables. Geomorphology is concerned primarily with the environmental level, supported in many cases by explanations in terms of the action of physical forces.

Explanations intended to account for general features of slopes are most firmly based if they are suggested initially by field or experimental observations, of either process of form. However, in the period 1920-1950 a substantial proportion of contributions to the geomorphology of slopes was of a dominantly theoretical nature, and such discussions still appear. Most are based on qualitative observations of slope form or process, and attempt to explain features of form in terms of the assumed manner of action of processes. The following discussion is not confined to concepts proposed on theoretical grounds, but includes those initially arising out of field studies. However, many of the main concepts were first put forward in primarily theoretical discussions. In addition to the work of Davis, Penck and King, which is essentially of this nature, the principal theoretical studies are those of Maw (1866), Gilbert (1877), de la Noë and de Margerie (1888), Hicks (1893), Marr (1901), Chamberlain and Salisbury (1904), Fenneman (1908), Gilbert (1909), Jeffreys (1918), Lehmann (1922), Lawson (1932), Morawetz (1932), Louis (1935), Baulig (1940), Bryan (1940), Meyerhoff (1940), Blache (1942), Wood (1942), Challinor (1948), Birot (1949), Cotton (1952), Ahnert (1954), Holmes (1955), de Béthune (1967) and Jahn (1968b).

The first part of this chapter discusses hypotheses of the origin of slope form and certain aspects of slope evolution; the treatment is disproportionately brief in comparison with the volume of published work on this subject. The succeeding section brings together concepts associated with equilibrium and uniformity. Models expressed in numerical terms are discussed in Chapter X.

SLOPE EVOLUTION: QUALITATIVE AND SEMI-QUANTITATIVE HYPOTHESES

Much discussion of slope form and evolution has centred upon the problems of the origin of the convexity, the origin of the concavity, and the evolution of the steepest (or any rectilinear) part of the slope. These problems will be considered in the abstract, assuming homogeneous rock but without reference to specific conditions of climate or vegetation.

Explanations of the convex-concave form of slopes

(*i*) The convexity and concavity are formed by different processes. Suggested hypotheses are: the convexity is produced by weathering and the concavity by surface wash (de la Noë and de Margerie, 1888; Hicks, 1893, Marr, 1901); the convexity is produced by soil creep and the concavity by surface wash (Davis, 1892; Gilbert, 1909; Birot, 1949); and the convexity is produced mainly by creep but also by diffuse sheetwash, and the concavity by concentrated wash (Baulig, 1940).

(*ii*) The convexity and concavity correspond respectively to periods of accelerating and decelerating basal river erosion (Penck, 1924).

Hypothesis (*i*) is based on the observation that river longitudinal profiles are concave, and the reasoning that running water in the form of wash will also produce concavities, and so the convexity must be formed by some other process. This is argument by analogy, and in view of the hydrological differences between water flow in river channels and as surface wash it cannot be accepted. As a statement about process-form associations, it may be directly investigated by measurements of processes on convexities and concavities; no such associations have been demonstrated. If the dominant surface process changes between convexity and concavity, some corresponding difference in the properties of the regolith might be expected; such a difference has not usually been reported, although the 'junction' described by Furley (1968) could have such an origin. This hypothesis is neither proven nor disproven. (*ii*) is now generally discredited (p. 29).

Explanations of the origin of the convexity

(*iii*) The convexity is formed by weathering. It results from attacks by weathering proceeding inwards from both sides of an initial angular break of slope (de la Noë and de Margerie, 1888; Hicks, 1893; Marr, 1901).

(*iv*) The convexity is formed by soil creep. The amount of soil to be removed increases downslope, requiring a progressively steeper slope removal (Gilbert, 1909; Baulig, 1940; King, 1953, 1957; de Béthune, 1967).

(*v*) The convexity is formed by soil creep and weathering. It is produced by (*va*) the removal of soil from the upper part of the slope, without compensating replacement from upslope, and (*vb*) the thinning of soil consequent upon this, with a resulting increased exposure to weathering (Götzinger, 1909; Lehmann, 1918; Penck, 1924; Blache, 1942; Birot, 1949).

(*vi*) The convexity is formed by surface wash. The volume of wash increases downslope from the crest, with a consequent increase in its power of detachment or transport (Maw, 1866; Chamberlain and Salisbury, 1904; Fenneman, 1908; Lawson, 1932).

(*vii*) The convexity is formed by surface wash. Extending a certain distance downslope from the crest there is a belt of no erosion, in which the force of wash is insufficient to overcome the resistance of the soil to erosion; fluctuation in the width of this belt, caused by rains of varying intensity, produces the convexity (Horton, 1945; King, 1957).

(*iii*) implies that the convexity is a denudation slope, subject to control by weathering. Qualitatively it is valid, and the ability of this mechanism to produce short convexities can be seen in the formation of corestones from initially angular rocks. For it also to produce long convexities of low curvature requires improbable quantitative assumptions. A necessary condition for the mechanism is that the products of weathering do not accumulate; it might therefore be applicable to limestones, where material is lost by direct removal in solution rather than by surface transport. (*iv*) is based on Gilbert's argument that the quantity of material that must pass each point on the profile increases downslope, therefore 'on a mature or adjusted profile, the slope is everywhere just sufficient to produce the proper velocity . . . and therefore increases progressively with distance from the summit'. This leaves unanswered what would happen if the slope were not so adjusted, a question which may be investigated by a process-response model. The implied assumptions are probably the same as those of (*va*). (*vb*) further assumes that the rate of weathering varies inversely with regolith depth; since weathering rate is a dependent variable, slope form is still subject to control by removal. As a statement about how processes operate, (*vii*) has been discussed above (p. 64), where its applicability to natural slopes is doubted. If it exists, however, its effect on form is similar to that of (*vi*), which states that for a constant angle there is a downslope increase in transporting power. If, but only if, the increase in transporting power with distance from the crest is greater than linear, an initial rectilinear slope may be transformed into a convexity. (*iii*), (*iv*) and (*v*) can only form a convexity by rounding an initial angular break of slope (Fig. 34*A*); (*vi*) and (*vii*) permit additionally the transformation of an initially rectilinear part of the slope into a convexity (Fig. 34*B*). (*iv*) to (*vii*) all imply that the convexity is a denudation slope, subject to control by removal.

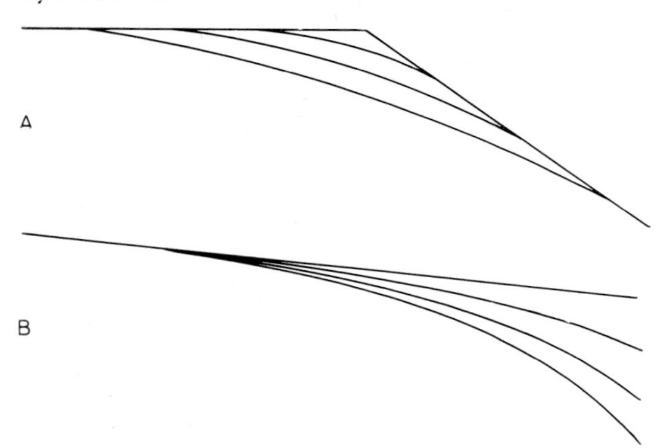

FIG. 34. Formation of a convexity: *A* = from an initial angular break of slope; *B* = from an initial rectilinear segment.

Explanations of the origin of the concavity

(*viii*) The concavity is essentially due to the restricting influence of the unchanging height of the base of the slope (implied in many explanations, and stated explicitly by Fenneman, 1908; Penck, 1924; Wood, 1942, Challinor, 1948, and Birot, 1949).

(*ix*) The form of the concavity represents the minimum slope across which material brought in from upslope can be transported (Chamberlain and Salisbury, 1904; Meyerhoff, 1904; Wood, 1942; Horton, 1945; Macar, 1955; Holmes, 1955).

(*x*) The concavity is formed by surface wash. The volume of wash increases downslope, with a consequent increase in its transporting power. It is thus due to the same mechanism that produces the concave form of a graded river profile (Tylor, 1875; Gilbert, 1877; de la Noë and de Margerie, 1888; Hicks, 1893; Marr, 1901; Jeffreys, 1918; Lake, 1928; Baulig, 1940; Meyerhoff, 1940; Horton, 1945; Birot, 1949; King, 1953, 1957; Holmes, 1955).

(*xi*) The regolith becomes finer-textured downslope, therefore a progressively gentler slope is required for it to be transported. This may be applied to transport by either creep or wash (Davis, 1932; Birot, 1949; Derruau, 1956; King, 1957; cf. also Penck, 1924.)

These explanations are not mutually exclusive. (*ix*), (*x*) and (*xi*) imply that the slope of any given section of the concavity is at the minimum angle necessary for the material arriving at its upper end to be transported to its lower end. If a section is gentler than such an angle, material will accumulate; if steeper, the regolith will be thinned and the slope lowered. In either event the tendency is to restore the condition of a transportation slope. The form of the whole of a concavity produced in this way is dependent on the position of its lowest point. (*viii*) and (*ix*) do not alone cause a concave form; if they are combined with an assumption of unchanging rate of transport with distance downslope, the resultant transportation slope is rectilinear. Only the assumption of a downslope increase in the rate of transport, as by (*x*) and/or (*xi*), causes the transportation slope to be concave. Attainment of a transportation slope implies cessation of further ground loss only for so long as the amount of material arriving at its upper end remains unchanged. If the rate of ground loss from the upper parts of the slope decreases with time, the concave transportation slope will in the long term undergo a gradual decrease in angle. That is, over the relatively short period of time sufficient for material to be carried from the upper to the lower end, the concavity is practically a transportation slope, undergoing no ground loss, but over the much longer period necessary for the upper parts of the slope to be lowered, there is ground loss from the concavity. This is the probable mechanism of pediment evolution (p. 207). All of the above explanations imply first, that the concavity is a transportation slope, subject to control by removal, and secondly, that a concavity would not be formed but for restriction in the rate of lowering of the base of the slope.

Explanations of the evolution of the steepest, or any rectilinear, part of the slope

(*xii*) Conditions are uniform on all parts of a rectilinear slope, therefore it retreats parallel (Penck, 1924; Bryan, 1940; Wood, 1942).

(*xiii*) More soil has to cross the lower part of a rectilinear slope than the upper part, and the regolith is consequently thicker on the lower part, giving more protection from weathering. The rate of weathering therefore decreases downslope, so the slope declines in angle (Götzinger, 1907; Davis, 1930; Morawetz, 1932; Louis, 1935; Rich, 1938; Baulig, 1940; Cotton, 1941; Birot, 1949; Savigear, 1952, 1956).

(*xii*) is valid where loss of regolith from the slope is by direct removal, as in solution loss, or where surface transport at a very rapid speed, as in rockfall, produces an effect of quasi-direct removal. If any part of regolith loss is by normal, slow, transport, the hypothesis is only applicable if the rate of weathering is assumed to be independent of regolith thickness, and the rate of transport is assumed to increase with regolith thickness. In these circumstances, a rectilinear slope with a downslope increase in regolith thickness could retreat parallel. (*xiii*) applies if loss is in whole or part by surface transport, the rate of which does not vary with distance from the crest, and if the rate of weathering varies inversely with regolith thickness. Both the above cases refer to a denudation slope, subject to control by weathering in (*xii*) and control by removal in (*xiii*).

Some of the above explanations are examined further in the following section. It is clear, however, that for any given slope form, or type of evolution, a large number of alternative, internally consistent, mechanisms exist. To illustrate this further, consider two alternative hypotheses for the evolution of a rectilinear part of a slope that cause it to remain rectilinear but to steepen with time:

(*a*) The slope is subject to control by weathering: the rate of weathering increases downslope linearly with distance from the crest.

(*b*) The slope is subject to control by removal; removal is by downslope transport, the rate of which varies as the square of distance from the crest.

Evolution in this manner has never been proposed, nor is it likely that either of these sets of assumptions corresponds with real conditions. The hypotheses are given to show how mechanisms for almost any type of evolution can be constructed in the abstract. All abstract explanations for slope evolution, including hypotheses (*i*)-(*xiii*) above, remain as statements of pure theory unless it can be shown that their assumptions correspond to observed conditions on slopes.

CONCEPTS OF EQUILIBRIUM, GRADE AND UNIFORMITY

A variety of concepts concerned in some way with equilibrium, grade or uniformity of form or evolution have been proposed, some of which have already been noted. These are grouped below, and certain of their features are summarized in Table 6. This section is based on Young (1970).

Existing terminology and definitions are adhered to, with one exception. In the interests of giving a precise meaning to an otherwise diffuse concept, the definition of a graded slope is restricted to only one of its several original attributes. The previous definition of time-independent form is also slightly qualified. The first six concepts described refer to ground loss; *7* and *8* refer to static properties of the regolith, and *9* to change in the regolith with time; *10-14* are related to the cause and origin of slope form. It is suggested that concepts *1-10* are valid and useful, but that the utility of *11-14* has not been satisfactorily demonstrated.

1. Transportation slope (Holmes, 1955). Part of a slope on which no ground loss occurs because at each point the material brought in from upslope is equal to that carried away downslope.

This concept implies an equality between the capacity of transporting agents to move material and the amount of material supplied from upslope. It is equivalent to the most frequently employed definition of a graded reach of a river. For no ground loss to occur there is a further necessary condition, namely that there shall be no direct removal of material. (Chamberlain and Salisbury, 1904, p. 58; Bryan, 1925; Gilluly, 1937; Meyerhoff, 1940; Wood, 1942; Macar, 1955; Holmes, 1955; Souchez, 1961, 1966.)

If surface wash is assumed to be the only process operative, this can be tested by a series of wash traps and pegs spaced at intervals down a slope profile. The hypothesis is proven if all traps collect equal amounts of soil and the pegs show no net lowering of the surface. The concept is necessarily inapplicable to the uppermost part of slope,

Table 6. Concepts of equilibrium, grade and uniformity. After Young (1970)

Concept	Name	Aspect of slopes to which definition refers	Applicable to whole or part of slope	Variation in rate of ground loss	Change in regolith thickness with time
1.	Transportation slope	Ground loss	Part	Nil	Nil or increase
2.	Uniform ground loss	Ground loss	Whole or part	Uniform, as measured perpendicularly to surface	Not specified
3.	Parallel retreat	Ground loss	Whole or part	Uniform, as measured horizontally	Not specified
4.	Time-independent form	Ground loss	Whole or part	Uniform, as measured vertically	Not specified
5.	Stable slope	Ground loss by rapid mass movements	Whole or part	Not specified	Not specified
6.	Static slope	Ground loss	Whole or part	Nil	Nil
7.	Graded slope	Regolith cover	Whole or part	Not specified	Not specified
8.	Uniform regolith thickness	Regolith thickness	Whole or part	Not specified	Not specified
9.	Equilibrium balance of denudation	Change in regolith thickness with time	Whole or part	Not specified	Nil
10.	Non-relict slope	Origin of slope form	Whole or part	Not specified	Not specified
11.	Profile of equilibrium	Origin of slope form	Whole	Not specified	Not specified
12.	External equilibrium	Origin of slope form	Whole	Not specified	Not specified
13.	Endogenetic equilibrium	Origin of slope form	Whole	Not specified	Not specified
14.	Open system in a steady state	Relation between energy, process and slope form	Whole	Not specified	Not specified

and has been applied mainly to concavities and pediments. It cannot remain true over long periods of time, since lowering of the sector upslope must ultimately reduce the debris supply. Over shorter periods of time it is a valid concept, and may be applicable to pediments.

2. *Uniform ground loss.* A manner of slope evolution in which ground loss measured perpendicularly to the surface occurs at the same rate over the whole slope, or the part involved (*A* in Fig. 35).

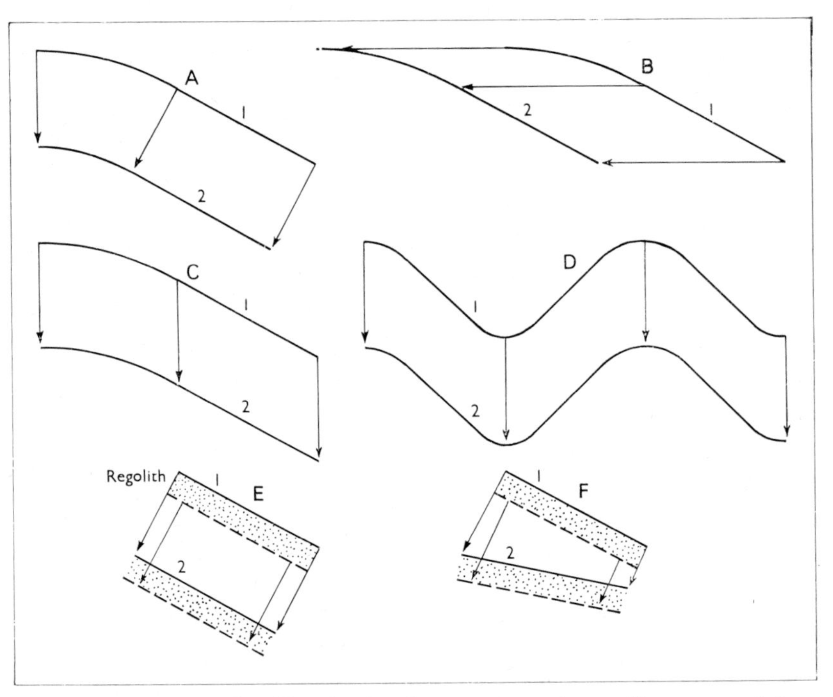

FIG. 35. Concepts of uniformity in slope retreat: *A* = uniform ground loss; *B* = parallel retreat; *C* and *D* = time-independent form; *E* and *F* = equilibrium balance of denudation. After Young (1970).

Possible reasons why such a condition may exist are: (*a*) on a free face, control of slope retreat is by weathering, to which all parts of the cliff are equally exposed (Penck, 1953, p. 135); (*b*) on a convexity or rectilinear slope on which regolith thickness is uniform, weathering and ground loss are also uniform (Gilbert, 1909; Lawson, 1932; Louis, 1932); (*c*) a uniformly retreating slope, progressively extending in length, develops above a river eroding vertically at a uniform rate (de le Noë and de Margerie, 1888, pp. 77-9; Penck, 1953, pp. 143-8; Ahnert, 1954; cf. *12* below); (*d*) on a concavity, the slope has become so adjusted that the amount of material removed by wash, the volume of which is increasing downslope, is everywhere the same (Jeffreys, 1918; Lake, 1928); (*e*) on rocks, such as limestones, on which all denudation is by direct removal, the rate of ground loss is not influenced by angle, distance from the crest, or other factors, and is therefore uniform.

Whether uniform ground loss is occurring at the present can be tested by measurements of surface processes; whether it has occurred in the past, by inductive comparison of slope profile form, making assumptions that observed slopes represent different stages of an essentially similar evolution. The concept is logically valid. It is frequently assumed to apply to cliffs, although this has been disputed (Mortensen, 1960). For a rectilinear part of a slope, uniform ground loss is identical to the two succeeding concepts, parallel retreat and time-independent form. For non-rectilinear slopes, the three concepts are virtually identical in the short term, i.e. for an amount of retreat small in relation to the dimensions of the slope. For greater retreat of non-rectilinear slopes, the three concepts are different.

3. Parallel retreat. A manner of slope evolution in which ground loss measured in a horizontal direction occurs at the same rate over the whole of the slope, or of the part involved (*B* in Fig. 35).

The main reason given is that the balance between surface processes and slope form is such that the several hillslope elements (in the sense of King, see p. 36) maintain the same relative proportions (Fair, 1947, 1948; King, 1951, 1953, 1957). Cotton (1941, p. 160) pointed out the surprising consequence, that parallel retreat defined in this way implies that the rate of ground loss as measured perpendicular to the surface is proportional to the sine of the slope angle.

This can be tested by the same two methods as *2* above. Inductive evidence suggests that parallel retreat may be applicable to the upper part of the slope (i.e. excluding the concavity or pediment) in areas of structurally controlled landforms, such as escarpments, plateaux and mesas with a caprock.

4. Time-independent form. The relative form of the slope, that is, the position of each point on the profile relative to the other points, remains unchanged with time; the altitude of the entire slope may be lowered (*C* and *D* in Fig. 35).

Since the position of the slope with respect to base-level is one property of form, absolute time-independence of form would imply a static slope. The term is therefore defined in such a way as to permit uniform lowering of the ground surface; consequently, loss measured perpendicular to the ground surface is proportional to the cosine of the slope angle.

This can only be tested by the same two methods as *2* above. In addition, the existence of time-independent form is suggested if inductive studies indicate that parameters of slope form show a good correlation with lithology, stream gradient and other external variables, but no correlation with assumed age. In the sense of relative form, as defined here, time-independent form is a valid and useful concept. It implies that neither evidence of past epicycles of erosion nor relict features from different climatic conditions exist in the landscape. Time-independent form is probably present in badlands and other closely-dissected (feral) relief, and may be more widely applicable. This manner of slope evolution is sometimes called a condition of *dynamic equilibrium*, but in view of the possible confusion with the concept of an open system in a steady state (see *14* below), to which it is not equivalent, the designation time-independent is preferred. The relation of time-independent form to available relief and valley spacing has been noted above (p. 21).

5. Stable slope. A slope, or part of a slope, on which rapid mass movements do not occur.

The applied stresses, tending to cause slope failure, are less than the strength of the materials (rock and regolith) forming the slope. The stability of a slope may be estimated by laboratory tests of the strength of materials. Inductive evidence of the maximum angle at which slopes on a given rock type are stable is provided by field observation of the angles of slope respectively affected and unaffected by rapid mass movements (p. 165).

6. Static slope. A slope, or part of a slope, on which neither addition nor removal of material occurs.

On a static slope the forces tending to carry away material fail to overcome the resistance of the material to removal. Reasons that have been given are that the slope is protected by the vegetation cover (de la Noë and de Margerie, 1888, p. 46); that surface wash does not detach soil because the applied force is less than the resistance of the soil to detachment (Horton, 1945); that plastic flow does not occur because the plasticity threshold is not exceeded (Souchez, 1961, 1963, 1966); and that the slope is stable (see *5* above). In general, the slope is below the limiting angle for all processes of removal.

This may be tested by field observations of surface processes. Positive results disprove the hypothesis; negative results fail to disprove it, but it cannot be proven. In an absolute sense, this condition can never be attained in the natural environment.

7. Graded slope (Davis, 1898). A slope possessing a continuous regolith cover, without rock outcrops.

Davis's use of the term carried the further implications that angular breaks of slope are absent, and that there is an adjustment between rate of weathering and rate of removal such that at no point does removal exceed weathering and so expose bedrock (p. 26). Owing to the difficulty of defining when a break is angular, and of testing the relation between weathering and removal, it is here suggested that the meaning of the term be limited to that of absence of rock outcrops.

In this restricted definition, whether a slope is graded can be directly ascertained by field observation. As such, it is a precise and useful term; the elimination of rock outcrops, and thus the attainment of grade, marks a stage at which there is a qualitative difference in the surface processes active. If the other conditions given by Davis are included in the definition, the concept becomes imprecise and untestable.

8. Uniform regolith thickness. A slope, or part of a slope, over which the regolith thickness is uniform.

The existence of uniform regolith thickness does not necessarily imply uniformity in the rates of either weathering or ground loss, although it has been argued that such a relationship exists. Whether this condition occurs may be ascertained directly by field observation.

9. Equilibrium balance of denudation (Jahn, 1954, 1968). A condition on a slope, or part of a slope, in which the regolith cover remains unchanged in thickness with time (E and F in Fig. 35).

Equilibrium is one of three possible states of the balance of denudation, as defined by Jahn. Let A = addition of material to the regolith by weathering of the rock beneath, S = subtraction of material from the regolith by direct loss, and M = net loss of material from the regolith as a result of excess of the amount carried away downslope over the amount brought in from upslope. If $A > S+M$, the active balance of denudation, the regolith thickens with time; if $A < S+M$, the passive balance, the regolith becomes thinner. The equilibrium balance of denudation occurs when $A = S+M$. The balance is only likely to remain active or passive for short periods, since a decrease or increase in regolith thickness is assumed to cause, respectively, an increase or decrease in the rate of weathering, tending to restore the equilibrium balance. No particular pattern or rate of ground loss is necessitated by the existence of an equilibrium balance. The concept is equivalent to *denudational equilibrium* as defined by Ahnert (1967), and also to one definition of a mature soil (Nikiforoff, 1949).

This concept is virtually untestable by direct methods, since extremely refined measurements of weathering and ground loss would be required. The finding of a correlation between regolith thickness and slope angle would provide indirect evidence suggesting that the hypothesis might be true, but not proving it. The concept implies that regolith thickness (but not the position of the ground surface) is time-independent. It is conceptually valid, and can be reproduced in process-response models of slope evolution (p. 103), but the difficulty of field testing limits its usefulness. The suggestion that the balance acts as a control on the rate of ground loss rests on the assumption, widely made but unproven, that the rate of weathering of rock into soil decreases with increasing soil depth. Ahnert's formulation, although later, is more elegant than Jahn's.

10. Non-relict slope. A slope, or part of a slope, the form of which is related entirely to processes that are similar to those of the present.

This implies that there has been no change of climate over a period long enough for all features of form produced under different, past, conditions to have been obliterated.

The hypothesis that a slope is non-relict can be disproved by demonstrating the presence of relict features, but the absence of such features does not prove it to be true. It can be supported by accumulating evidence that present processes, as measured in the field, can satisfactorily account for existing form. A slope with time-independent form is necessarily non-relict, but the reverse is not the case.

11. Profile of equilibrium (Baulig, 1940). A slope on which the position of each point of the profile depends at any moment on that of all the others.

A disturbance of equilibrium (i.e. change in external variables) at one point is transmitted to all the other points by the moving regolith. In the formulation of this concept, Baulig was in part anticipated by Gilbert (1877, p. 118): 'Every slope is a member of a series, receiving the water and waste of the slope above it, and discharging its own water and waste upon the slope below. . . . And as any member of the system may influence all the others, so each member is influenced by every other."

This hypothesis is untestable in the field. It can be tested in a process-response model or a scale model, by disturbing conditions at one point and observing whether the resulting adjustment of form affects the entire slope. The validity of the concept depends on the interpretation of 'at any moment'. Taken literally, it is manifestly untrue; the excavation of a quarry has no observable effect on the rest of the slope in 100 years. Hence it must be taken to refer to a period of time long enough to achieve substantial ground loss from the entire slope. The analogy with the links of a chain hung in catenary curve does not hold, the forces and processes being of a quite different nature. If slopes were in fact profiles of equilibrium in this sense, polycyclic valleys could not exist, and no evidence of past base-levels would be preserved; this is clearly not the case.

12. External equilibrium (Ahnert, 1967). A slope, the form of which expresses an adjustment between the action of surface processes on the slope and the rate of river erosion at its base.

This concept was previously termed 'uniform development' (Penck, 1953, p. 143) or 'dynamic equilibrium' (Ahnert, 1954). It may be held either that the rate of river erosion is an independent variable, as by Penck, or that this rate is partly dependent on the amount of debris supplied from the slope, and there is consequently a feedback effect (Strahler, 1950; Ahnert, 1954). The arguments of Penck and Ahnert also differ in respect of the relative importance attributed to past changes in river erosion and its present rate, and in the mechanism of slope adjustment. The 'adjustment' is a complex matter of relations between slope form and processes of linear erosion and surface denudation. This is a similar concept to the profile of equilibrium, with emphasis placed on the effects of changes in one variable, that of river erosion.

The hypothesis is untestable, directly, in the field. Indirect inductive evidence for it would be the finding of a correlation between stream gradient (or volume) and slope form. It can be reproduced in process-response models. The suggestion that the debris supplied from slopes appreciably influences the rate of river downcutting is unlikely, in view of the magnitude of the quantities and forces involved. Some dependence of slope form on rate of river erosion is clearly present, but the precise meaning of the 'equilibrium' involved in such dependence has not been demonstrated. Both Penck and Ahnert claim that with a uniform rate of river downcutting, the resulting slope will be rectilinear; this does not necessarily follow.

13. Endogenetic equilibrium. A slope, the form of which expresses an adjustment between the endogenetic forces of uplift and the exogenetic forces of denudation.

This hypothesis, proposed by Penck (1924) and also discussed by Lehmann (1922), is untestable and is now generally discredited.

14. A slope as an open system in a steady state. A slope possessing the properties of an open system in a steady state (Von Bertalanffy, 1950). This concept is theoretically sound, but cannot be practically tested (p. 20).

X MODELS

THE systems of Davis, Penck, and King, described in Chapter IV, are models of slope evolution. Other non-numerical models, concerned primarily with form, are discussed in the section on slope classification in Chapter XIV. This chapter is concerned with models in which the variables are expressed in numerical terms. Quantitative models may refer to process, form, or to the interaction of the two. Examples of process models are equations for the manner of action of surface wash, whether based on empirical or rational grounds (pp. 63–5), and the numerically-based hypothesis of soil movement by plastic flow proposed by Souchez (p. 111). Models of form include the fitting of mathematical curves to surveyed slope profiles (p. 153). This chapter deals with process-response models.

Process-response models relate the action of processes to the resulting change of form over a period of time. Two sets of assumptions are made, concerning the initial form, and the manner of action of processes. The form at subsequent periods of time, and therefore the evolution, are then computed. This may be illustrated by reference to a simple case (Fig. 36). The initial form is taken as a rectilinear slope of 35°, above which is a horizontal plateau. This might result from the initial dissection of an uplifted plateau by a stream cutting rapidly downwards, in conjunction with the elimination of slopes steeper than 35° by rapid mass movements: slopes approximating to such a form are in fact commonly observed. The lithology is taken to be uniform.

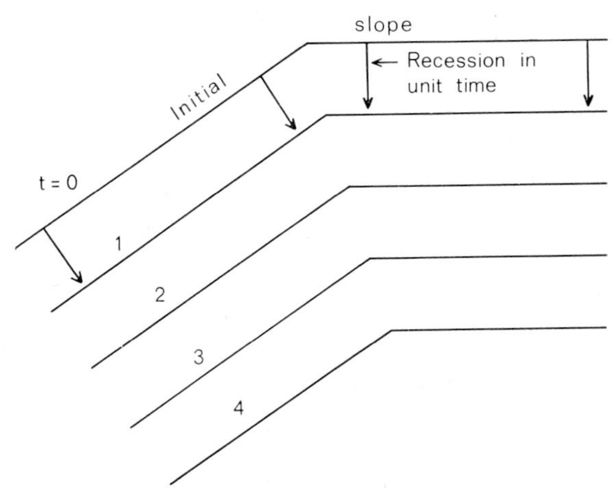

FIG. 36. A process-response model of slope evolution.

No assumptions regarding climate or vegetation are necessary. Let it now be assumed that the only process acting upon the slope is solution loss. An assumption about the type of process is insufficient to construct the model; it is necessary further to specify its manner of action. In this instance, it will be assumed that solution causes ground loss, by direct removal, in a direction normal to the surface of the slope, and that the rate of loss is uniform on all parts of any slope. It is also assumed that no erosion occurs at the slope base. Given these assumptions, it is clear by graphical construction alone that slope evolution will follow the successive stages 2, 3, 4 . . . shown in the figure, i.e. will be by parallel retreat. To complete the quantification, assumptions are made of the height of the initial slope, in metres or in dimensionless units, and of the rate of ground loss by solution, e.g. in mm/1000 years. The number of years required to reach any later form can then be obtained.

It is necessary to note which of the assumptions are essential features, and which give additional information. With respect to the initial conditions, only the actual form is part of the model, not its origin. The circumstances suggested for the origin of this form, by valley downcutting and mass movements, provide grounds for the selection of initial conditions that will subsequently be useful in matching the model to reality, but are not part of the model itself. Similarly, solution loss is used above as a rational basis for selecting a certain manner of action of process, but the existence of this process as a mechanism is extraneous; the model is applicable to any process which acts in such a manner as to produce uniform ground loss irrespective of slope angle or other variables. Thus the matching of a model with reality in respect of process requires not only the identification of the processes operative on a slope but also information on their manner of action. Models are initially constructed using simplified or idealized conditions; this permits the subsequent incorporation of complications, whereas the reverse would not necessarily be the case. The above model could readily be adapted to initial conditions of varying lithology, by adding to the assumptions about process the different rates of ground loss on each rock type.

In matching a process-response model to real conditions, circumstances are most favourable if observations of process, form and form change can all be made. The model then acts as a verification that an assumed relation is valid. In practice, one or more variables are usually unknown; in slope models, the change of form with time is virtually unobservable except in special circumstances. Where there is only one unknown a process-response model can be used to obtain it (Whitten, 1964). If, however, either two or more variables are unobserved, or the magnitudes of certain parameters are known only with a large margin of error, then the number of solutions to the model becomes infinite. This is a serious limitation to the use of process-response models as a means of prediction.

Many process-response models of slopes involve feedback between variables. A common example is found in cases involving the removal of regolith by downslope transport. In this instance, the relation between form and process differs according to the time interval under consideration (cf. Schumm and Lichty, 1965). Over a short period of time, slope form may be considered as an independent variable, upon which the rate of downslope soil movement is dependent. Over a considerably longer period, however, slope form is itself dependent upon the changes brought about by downslope soil movement. This inter-dependence is handled computationally by a method of iteration. Events are considered separately for short time intervals, within each of which process is held to act as if dependent on conditions at the beginning of that interval. From such process behaviour, the form change over the interval is then calculated. The new form obtained is the basis for process operation during the succeeding time interval. All models so far developed refer to slopes considered in profile only, the third dimension being assumed to be uniform and of unit dimension. The profile is represented by a series of points, the positions of which are defined by rectangular co-ordinates. Examples of the detailed method of calculation are given in Young (1963c). Having first calculated such models before the use of computers became common, the present author appreciates the suitability of computers for work of this type.

Two main classes of process-response models have been developed, the first concerned with the evolution of a free face with scree accumulation at the base, and the second with the more general case of regolith-covered slopes. Penck's system is unusual in that it attempts to combine both of these cases, a fact which in part accounts for its fundamental error (p. 31). The models of cliff recession were developed earlier, in part, perhaps, through historical accident, but mainly because the processes involved act relatively rapidly and so are more readily observed. These classes will be considered separately.

MODELS OF CLIFF RECESSION

The first type of process-response model is concerned with the retreat of a free face with scree accumulation at the base. The essential features of this were deduced by Fisher in 1866, and independently by Lawson in 1915. Lehmann (1933, 1934) added to this analysis the effects of the variables cliff angle, scree angle, and rock/scree volume ratio. The above discussions refer to the case of parallel recession of the free face, which will be termed the *Fisher-Lehmann model*. Gerber (1934) extended the model to instances where the cliff increases in height, through fault movement or river erosion at its base, contemporaneously with its denudation. Bakker and Le Heux (1946) further refined the Fisher-Lehmann model. A similar theoretical approach was subsequently extended to cases of non-parallel cliff recession (Bakker and Le Heux, 1947/50, 1952; Van Dijk and Le Heux, 1952; Bakker and Strahler, 1956).

The Fisher-Lehmann model assumes an initial rectilinear free face of height h and angle β, with level ground above and below (Fig. 37). It is assumed that the whole of the free face is equally exposed to weathering, and that in unit time a uniformly thick layer of rock falls away from it. This rock accumulates at the base as a scree, with a rectilinear surface slope of angle α. Cliff recession combined with scree accumulation causes the formation of a rock core, buried beneath the scree; an important assumption of the model is that this core is protected from further change of form. The height of the remaining free face is progressively reduced by scree accumulation. The

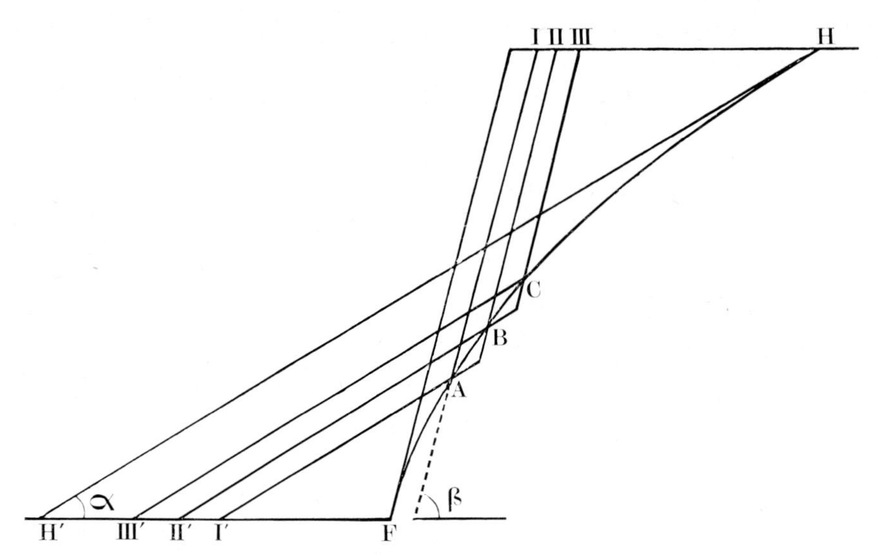

FIG. 37. Parallel retreat of a free face (I, II, III .... *H*) with scree accumulation (I′, II′, III′ .... H′) and the formation of a buried rock core (*F*, *A*, *B*, *C*, *H*). β = angle of cliff, α = angle of scree. After Lehmann (1933), modified.

ratio between the volume of rock falling V_r and the volume of scree accumulating V_s is expressed by

$$V_s = \frac{V_r}{1 - c} \qquad (10{\cdot}1)$$

In the literature, c is treated as a single constant. It is in fact influenced by two variables: an increase in volume consequent upon the existence of voids in the scree, and a decrease in volume if part of the scree is removed (whether by erosion at its base or by loss through weathering and surface wash). If the increase, expressed as a proportional gain in volume, is denoted by c_1, and the proportion of scree remaining by c_2, then

$$V_s = \frac{V_r}{1 - c} = V_r . c_1 . c_2 \qquad (10.2)$$

and

$$c = 1 - c_1 . c_2 \qquad (10.3)$$

The value of c_1 is always positive, approaching zero for sands with initial loose packing. If there is no scree removal $c_2 = 1$, therefore c is positive and $V_s > V_r$. With small amounts of scree removal c may be positive or negative, $c = 0$ occurring when the volume gain through looser packing is just matched by the proportion removed. With substantial scree removal c becomes negative, and the scree volume is less than the rock volume. For the limiting case where all scree is removed, $c_2 = 0$ and $c = -\infty$.

Let x and y be respectively the horizontal and vertical co-ordinates of the buried rock core. For parallel cliff recession the shape of the rock core is given by

$$\frac{dx}{dy} = \frac{h.\cot\beta - (\cot\alpha - c.\cot\alpha - \cot\beta)y}{h - cy} \qquad (10.4)$$

An alternative assumption is that recession takes place faster on the upper than on the lower part of the cliff. There are an infinite number of possibilities concerning the manner of this variation; one for which the consequences have been deduced is termed *central rectilinear recession* in which the free face is assumed to remain rectilinear but to decline in angle, rotating about its initial base (Fig. 38).

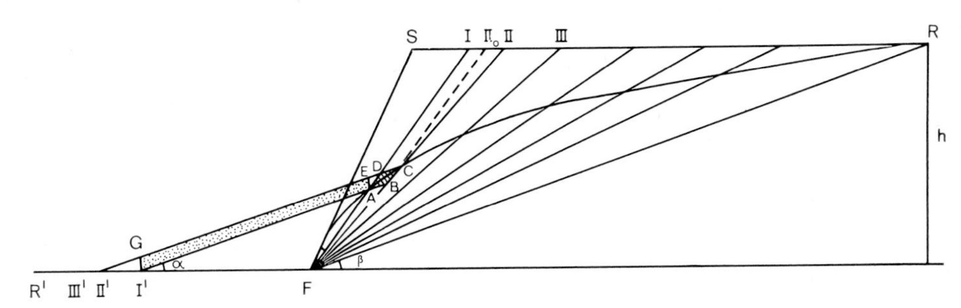

FIG. 38. Central rectilinear recession of a cliff with scree accumulation and the formation of a buried rock core (*F, A, C, R,*). After Bakker and Le Heux (1947).

All models of this type give a similar general result: the buried rock core is convex, with its lowest point tangential to the initial cliff angle β, and its uppermost point, formed at the moment when the whole free face is eliminated by scree accumulation, tangential to the scree angle α. Variation in β, even down to 45°, has relatively little influence on form; if the angle of the initial cliff is made less than 90° the lowest part of the core becomes gentler in angle. Variation in α has a slightly larger effect; the steeper the angle at which the scree stands, the steeper the slope of the core, particularly in its upper, last-formed, parts. Positive and low negative values of c have only small effects. The results from the model are most substantially affected if a high proportion of the scree is removed; c_2 is then small, c acquires high negative values, and the curve of the rock core becomes considerably flatter and less curved (Fig. 39). The effects of α, β and c as stated above apply to both parallel and central rectilinear recession; for any given values of these variables, the rock core is steepest for parallel recession, becoming less steep for central rectilinear or any other type of recessional decline.

For the limiting case in which all the scree is removed, the residual rock surface is a rectilinear slope at the angle at which scree would accumulate if present, α, irrespective of whether cliff recession is parallel, central rectilinear, or in some other manner (Fig. 39). This result, stated as a 'geomorphological law', was obtained by Bakker and Le Heux (1952). The rectilinear slope is termed a *Richter denudation-slope*,

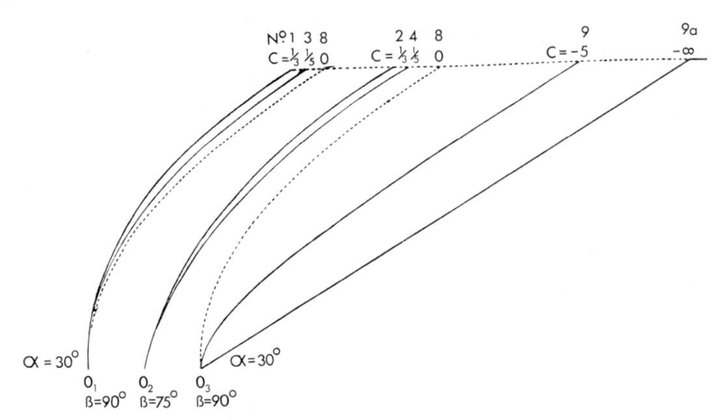

FIG. 39. The influence of the constant c and the cliff angle β on the form of the buried rock core. After Lehmann (1933).

after Richter,* who drew attention to the existence in the Alps of frequent rock slopes at angles close to those of screes (Fig. 40). The formation of such a rectilinear slope is a correct deduction from the model; it rests, however, on the assumption that the residual slope remains unmodified subsequent to its formation, which in this limiting case is unrealistic. Nevertheless this is the most significant result obtained from cliff recession models.

FIG. 40. Frost-shattered rock being replaced by rectilinear slopes at the angle of repose of screes. Arid montane environment, near the Sallang Pass, Hindu Kush, Afghanistan.

* 'Der Grat zersplittert und zerfällt, sein Fuss aber wird allmählich vom Firn verschlungen. Denn die schiefe Denudationsfläche, oberhalb derer der Fels abgetragen wird, gibt Raum zur Schneeauflagerung' (Richter, 1900, p. 58).

Mortensen (1960) noted that the Fisher-Lehmann model gives a convex rock core, whereas Penck's model of cliff recession results in a concave basal slope. The reason lies in the assumptions: the Fisher-Lehmann model assumes that the rock core, once formed, remains unmodified while Penck's model assumes that the basal slope is continuously altered, concurrently with cliff recession. Thus two models, both deductively sound, produce opposite results; yet adequate field evidence to decide between the two sets of hypotheses is not available. No purpose is served by further refinements to this type of model until more evidence on which to base its assumptions, from field and experimental studies, is obtained.

MODELS OF SLOPE EVOLUTION

The concept of process-response models of slope evolution is not new. The Davisian cycle, based on a discussion of waste movement, is one such, very versatile in its capacity to incorporate structural and cyclic complications, but quantitatively so ill-defined as to be of limited value. The main precursor of quantitative models is Penck's system, which employs graphical construction as a form of analogue computation, combined with rigorous logical reasoning. Penck's explicit aim was to find the unobservable variable, which he believed to be endogenetic processes. Lehmann (1922) further enunciated the principle of matching observed landforms to deduced model forms. Lawson (1915) and Putnam (1917) obtained models for the special case of progressive burial of a suballuvial bench below a retreating mountain wall, as believed to occur in arid basins. It is mathematically similar to the Fisher-Lehmann model of cliff recession, an alluvial fan sloping at only a few degrees replacing the scree slope. Morawetz (1932) introduced the concept subsequently employed in many models, that the amount of slope retreat is determined by the difference between the volume of regolith material brought into a portion of slope from upslope, and the amount carried away from it downslope. This was extended by Ahnert (1954), who showed how a change in angle initiated at the slope base could be transmitted upslope by regolith movement.

In the following discussion, the initial conditions are assumed to be a steep rectilinear slope above which is a level plateau, as in Fig. 36, formed of homogeneous rock. At the foot of the slope there is unimpeded basal removal (p. 7), but no erosion. Given the behaviour of a model under these conditions, its modifications with non-homogeneous rock, vertical or lateral basal erosion, and impeded basal removal can be readily obtained; such complications are treated in many of the publications cited. In cases involving erosion, a constraint is usually added that if a part of the profile with an angle greater than 35° is produced, this is instantly converted to a 35° slope by retreat of its upper point, the reasoning being that such transformation takes place on real slopes by rapid mass movements, which on the time scale of the model can be assumed to act quasi-instantaneously. The models are grouped into four main classes: models of direct recession, models based on creep and flow mechanisms, models involving form interaction, and special cases based on process measurements.

The first group of models was developed by Scheidegger (1960, 1961, 1961b,

1964) and Hirano (1968, summarizing previous publications in Japanese). Direct assumptions about the rate of slope recession are made. Scheidegger makes the alternative assumptions that degradation (equated with weathering) acts vertically (the 'linear theory'), and that it acts normal to the slope surface (the 'non-linear theory'). He analyzes three cases for each theory:

Case 1—that degradation (i.e. ground loss) is uniform.
Case 2—that degradation under all conditions is proportional to altitude.
Case 3—that degradation under all conditions is proportional to slope angle.

Hirano presents the same concepts as equations giving the change in elevation of points on the slope with time:

$$\frac{du}{dt} = -cu \text{ (the 'denudational coefficient')} \qquad (10.5)$$

$$\frac{du}{dt} = -b\frac{du}{x} \text{ (the 'recessional coefficient')} \qquad (10.6)$$

$$\frac{du}{dt} = a\frac{d^2u}{dx^2} \text{ (the 'subduing coefficient')} \qquad (10.7)$$

where u = elevation, t = time, x = horizontal distance and a, b and c are constants of proportion. Equations *10.5* and *10.6* are identical to Scheidegger's Cases 2 and 3 respectively for the linear theory. The subduing coefficient states that ground loss is proportional to convex curvature. Case 1 for the non-linear theory was that used in Fig. 36. Some patterns of slope evolution based on these assumptions are illustrated in Fig. 41. Equation *10.5* results in central rectilinear recession, *10.6* in parallel retreat, and *10.7* in a progressive decrease in curvature and increase in length of convexities and concavities. By varying the proportions of the constants a, b and c, Hirano demonstrates that a wide variety of convex-concave profiles, similar in general form to Fig. 41*D*, may be developed.

Both authors take the general correspondence between the deduced evolution and observed slope form as evidence of the correctness of the assumptions. With respect to the correspondence of a single profile this is not justified. In any system, a given state of form can be arrived at in an infinite variety of ways; this has been called the principle of equifinality (Von Bertalanffy, 1950). All that has been demonstrated is that the assumptions provide one such way. The argument for confirmation of assumptions would be somewhat stronger if a complete set of profile forms during evolution could be obtained by observation, such as can only be done directly in small-scale experiments; indirect means, by scaling and superimposition of observed profiles, involve additional and uncertain assumptions. Hence models of this type cannot be used to derive the manner of action of processes. Approached from the point of view of process mechanism, all of the above models, except for those involving the

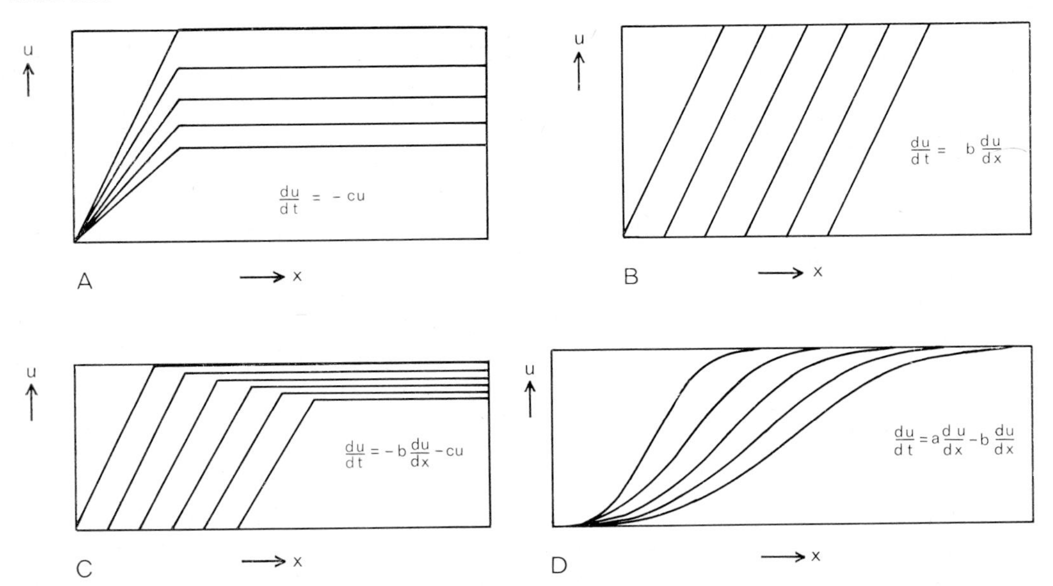

FIG. 41. Hirano's models of slope development. *A:* effect of the denudational coefficient. *B:* The recessional coefficient. *C:* the denudational and recessional coefficient combined. *D:* The recessional and subduing coefficients combined. After Hirano (1968).

subduing coefficient, demonstrate the slope evolution that would follow in the absence of any restraint upon ground loss caused by the need to remove weathered material. This situation could occur either with denudation entirely by direct removal, as in solution loss, or under the conditions of control by weathering.

The second class of models starts from analysis of the manner of action of processes of downslope regolith movement. Culling (1963, 1965) bases his analysis on an argument that soil creep is a stochastic process, produced by random movements of individual particles (see p. 52), a concept involving the use of the mathematics of molecular diffusion. He assumes that ground loss is subject to control by removal, that removal is entirely by downslope transport, and that the rate of transport is proportional to the slope gradient. This leads to the same relation as given in equation *10.7*, except that the constant a represents the coefficient of diffusion. It follows that the rate of ground loss at a point is proportional to its convex curvature; no ground loss occurs on a rectilinear slope, whilst a concavity results in accumulation, more soil being brought in from upslope than is removed downslope. Souchez (1961, 1963, 1964, 1966) explicitly relegates soil creep to a minor role in denudation, and assumes that downslope transport is by plastic flow. Such flow is governed by the relations:

$$F = \frac{\tau - \tau_c}{\eta} \qquad (10.8)$$

and

$$\tau = W.\sin\theta \qquad (10.9)$$

where F = rate of flow, τ = tangential stress, τ_c = critical stress, η is the coefficient of viscosity of the regolith, W = specific gravity of the regolith, and θ = slope angle. Below a certain threshold angle, governed by η, no flow occurs. For the special case of viscous flow, where $\tau_c = 0$, an equation is obtained identical in form to *10.7*, except that the constant is equal to $W/3\eta$.

Thus two different approaches to situations of denudation subject to control by removal, with downslope transport as the agent of removal, both reach the conclusion that if the rate of downslope regolith movement is proportional to slope angle, then the rate of ground loss is proportional to convex curvature (as in equation *10.7*). The resulting slope evolution given by Culling for an initial vertical slope is similar to that shown here as Fig. 42, Model 1. Its features are that the slope remains entirely convex at all stages; that the curvature of the convexity progressively decreases; and that equal amounts of ground loss require considerably longer periods of time in the later stages, or with low curvature, than in the earlier stages with relatively high curvature.

It is open to doubt whether either of the above process mechanisms, soil creep by stochastic diffusion or by plastic flow, is a correct description of the behaviour of

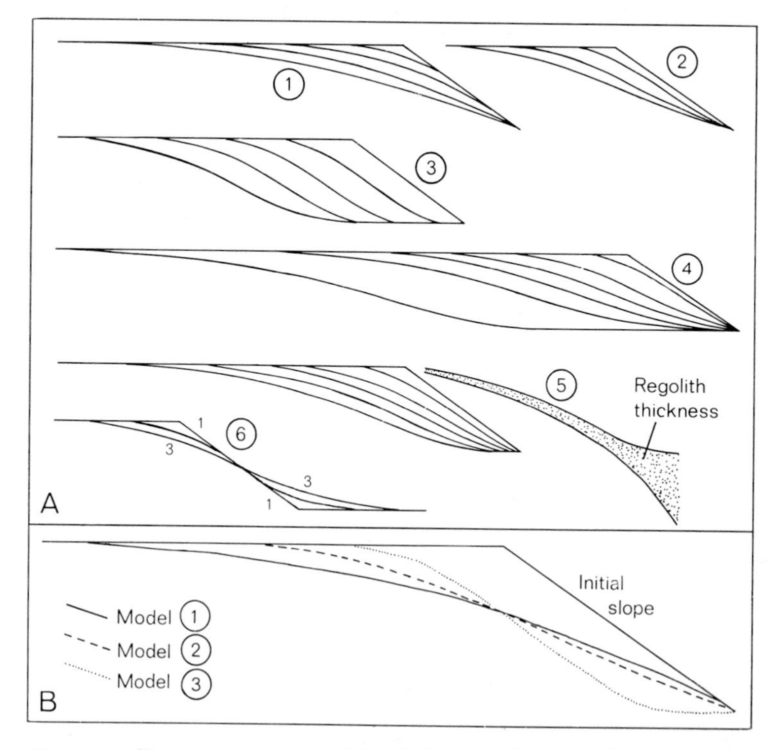

FIG. 42. Process-response models of slope evolution. A = successive profiles developed by models 1-6; B = comparison of profiles developed by models 1, 2 and 3. For assumptions on which models are based, see text. Based on Young (1963c).

regolith on a slope. This does not invalidate the later, form-evolution, part of this model, which is applicable to any mechanism that leads to the assumption of downslope regolith transport at a rate proportional to the slope angle. Thus processes acting in such a manner cause a decline in the maximum angle of the slope and the development of a long convexity, but such processes are unable (except by deposition) to produce a concavity.

The third class of model was developed by Young (1963c) and Ahnert (1966, 1970b) who, working independently, arrived at almost identical techniques but employed different assumptions. The models are based on consideration of the interaction of process and form, calculated by a method of iteration. The assumptions on which the 6 models shown in Fig. 42 are based are as follows:

Model 1. (*i*) Unimpeded basal removal, but no basal erosion
(*ii*) Slope retreat subject to control by removal
(*iii*) Removal of regolith entirely by surface transport
(*iv*) Rate of surface transport proportional to sin θ

Model 2. (*i*), (*ii*), (*iii*) as in Model 1
(*iv*) Rate of surface transport proportional to sin θ and distance from crest

Model 3. *Either* (Model 3*a*):
(*i*), (*ii*) as in Model 1
(*iii*) Removal of regolith entirely by direct removal
(*iv*) Rate of direct removal proportional to sin θ
Or (Model 3*b*):
(*i*) as in Model 1
(*ii*) Slope retreat subject to control by weathering
(*iii*) Rate of weathering proportional to sin θ. (Rate of surface transport immaterial)

Model 4. (*i*), (*ii*) as in Model 1
(*iii*) Removal of regolith partly by surface transport and partly by direct removal
(*iv*) Rate of surface transport and rate of direct removal both proportional to sin θ

Model 5. (*i*), (*ii*), (*iii*) as in Model 1
(*iv*) Rate of surface transport proportional to sin θ and to regolith thickness
(*v*) Rate of weathering (conversion of rock to regolith) proportional to sin θ and inversely proportional to regolith thickness
(*vi*) Regolith thickness proportional to total regolith weathered upslope of a given point, and inversely proportional to rate of downslope transport at the point

Model 6. (*i*) Impeded basal removal
(*ii*) Where ground loss occurs, slope subject to control by removal; where ground gain occurs, slope subject to control by accumulation
(*iii*), (*iv*) as in Model 1

The assumptions of Model 1 were suggested by analogy with the supposed manner of action of soil creep, and Models 2 and 3 similarly with surface wash and solution loss respectively. It may again be stressed that the profile evolution computed is dependent only on the assumptions and not on the actual process. Thus Model 1 would only show evolution under soil creep if it could be shown that the rate of soil creep is proportional to the sine of the slope angle; if it were found by field measurement that the rate of creep showed a downslope increase, then Model 3 would apply.

The evolution of Model 1 initially shows a rounding of the angular junction at the crest, but no ground loss on the lower part of the slope. Subsequently the convexity extends to the base, after which the evolution shows a decreasing maximum angle, and the development of a convexity of progressively increasing length and decreasing curvature. Even at late stages of evolution no concavity is formed. In Model 2, retreat of the whole of the initial rectilinear slope commences immediately, with the later stages being partly similar to Model 1, except that a concavity is formed. The evolution in Model 3 is largely by parallel retreat; the simpler case in which the rate of direct removal is uniform has been given in Fig. 36 (p. 103). Model 4 combines the assumptions of Models 1 and 3, by analogy with a slope influenced by both soil creep and solution loss. The addition of an element of direct removal radically alters the evolution from that of Model 1, permitting the formation of a long, pediment-like concavity. Model 5 incorporates an interaction between regolith thickness, and the rates of weathering and transport. It is found that regolith thickness increases downslope, gradually on the convexity and rapidly on the concavity; where rate of surface transport is assumed to increase with regolith thickness, this has the effect of causing a downslope increase in rate of transport. Thus the profile evolution is similar to that of Model 2. Model 6, based on identical assumptions to Model 1 except that basal removal is impeded, demonstrates the formation of a concavity by accumulation.

Ahnert's models are closely similar in technique; the assumptions generally resemble those of Model 5 above, but there are differences which are of some methodological importance. Ahnert computes the regolith thickness for each point on the slope, obtained by reference to its previous thickness and to the amounts of soil brought to a point from upslope, and removed downslope away from it. The rate of surface transport varies with sin θ. The rate of weathering is assumed to vary inversely with regolith thickness. In case *A*, suggested as being applicable to chemical weathering, weathering rate decreases linearly with regolith thickness. In case *B*, suggested as applicable to physical weathering, the decrease is exponential. There follows a critical assumption: that the amount of slope retreat in a time interval is equal to the amount of weathering. This differs from Model 10 above, in which the amount of slope retreat is held to be ultimately dependent on the ability of processes of transport to remove the weathered regolith. In Ahnert's system, slope retreat is ultimately determined by weathering, although weathering is itself limited by regolith thickness and so indirectly by processes of removal. The resulting slope evolution (Fig. 43) shows a decline in maximum angle, with the development of a convexity. Only in case *B* is a concavity formed.

Although constructed with dimensionless parameters, actual slope dimensions may be applied to slope models, together with observed or assumed rates of action of processes. The time required for a given amount of profile evolution may then be calculated. Table 7 shows the results for Models 1 and 5 of Fig. 42, for two slope heights and two assumed rates of action of processes. It shows that a process of downslope transport can fairly rapidly round off an angular junction into a short convexity, but for such a process to cause substantial modification to slopes of the order of 100 m high, either very rapid rates of transport or long periods of time are necessary. In

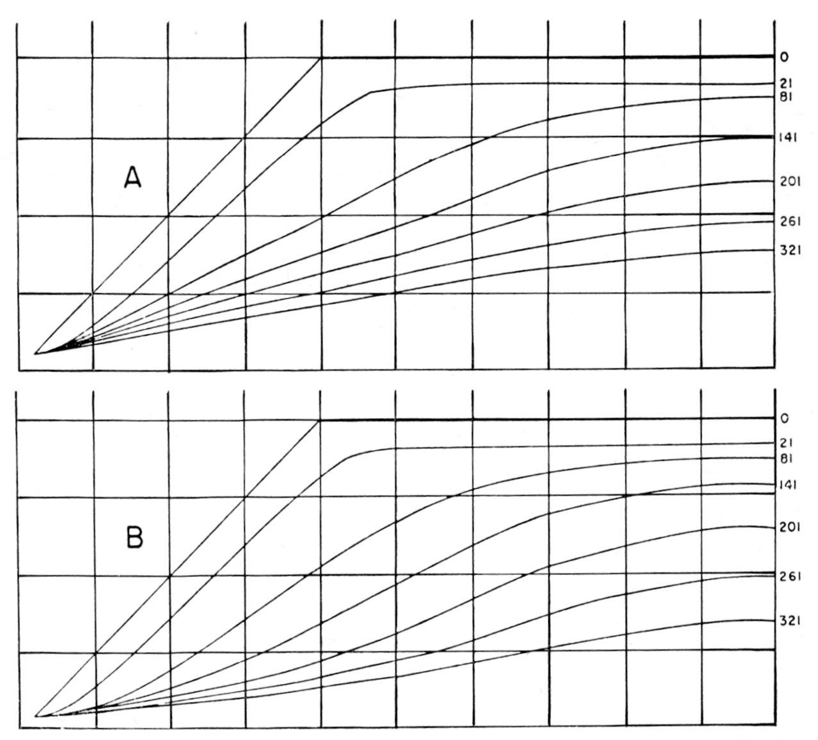

FIG. 43. Ahnert's models of slope evolution. Slope retreat is subject to control by weathering (p. 24). A = rate of weathering decreases linearly with increase in regolith thickness; B = rate of weathering decreases exponentially with increase in regolith thickness. After Ahnert (1966).

Table 7. Rates of development on Models 1 and 5. The number of years required to reach the final profile forms shown in Fig. 42 are given

Model	Assumed rate of downslope transport, cm³/cm/yr	Assumed rate of direct removal, cm/yr	Height of initial slope, 10 m	100 m
1	0·5	nil	1 500 000	150 000 000
1	3·0	nil	260 000	26 000 000
5	nil	0·005	100 000	1 000 000
5	nil	0·02	25 000	250 000

contrast, processes of direct removal can substantially modify a large slope in one million years, even if a very low rate of removal is assumed.

The last group of models apply to special circumstances only, but are constructed on the basis of observational evidence for the manner of action of process. De Ploey and Savat (1968) noted that, in a region of sandy soils under savanna in the Congo, rainsplash was evident, as shown by pitting of the ground surface. By an experimental study they obtained the relation between transport by splash and slope angle. The application of this relation in a process-response model showed that this process alone was capable of producing, from initial V-shaped valleys, the broad convexities of low curvature that characterize the region observed. Caine (1969) surveyed the profile form of scree slopes in New Zealand, and made field measurements of the rate of scree accumulation resulting from rock fall and slush avalanching. On the basis of empirically-obtained equations for the manner of action of these processes, he was able to deduce models which matched the observed profile forms.

The models provide a means of testing the validity of some of the hypotheses given in Chapter IX (pp. 92-5). Hypothesis *iii* corresponds to Fig. 36, which demonstrates that the assumption that weathering alone can form a convexity is false. Hypotheses *va* and *vb* correspond respectively to Fig. 42, Models 1 and 5; *vb* is also equivalent to Fig. 43. Although this shows that the proposed mechanisms are capable of producing a convexity, it does not prove that convexities are of this origin; thus Model 2 shows that hypothesis *vi* is similarly able to produce a convexity. Model 2 also shows that hypothesis *x* is one mechanism by which a concavity may be formed. In the early stages of Model 2 the concavity is a denudation slope, and in strict terms it remains so throughout; in the later stages of evolution, however, the rate of ground loss on the lower part of the concavity becomes very low, owing to the large amount of material brought in from upslope, and the condition approaches that of a transportation slope. All models confirm that hypothesis *viii*, a restraint in the rate of lowering of the base of the slope, is a necessary condition for the formation of a concavity; Model 1 shows that it is not a sufficient condition. Hypothesis *xii* corresponds to Model 3, and *xiii* to the early stages of Fig. 43, demonstrating that a rectilinear slope may either retreat parallel or decline, depending on the assumptions.

Process-response models are a useful means of testing the consequences of hypothesis. The matching of form, between model and observed slopes, does not, however, prove that the assumptions of the model are correct. Owing to the principle of equifinality, models can never be used to obtain process from a basis of observed form. At the present, very few of the assumptions concerning process used above have been confirmed by observation. Models become of considerably greater value if, as in the special cases of rainsplash and scree formation described, they are constructed on the basis of process observations. With observational evidence on both process and form, models could be used to obtain the unobservable variable, slope evolution. Until there is a better knowledge of surface process on regolith-covered slopes, the value of models will remain severely limited.

SCALE MODELS

Experimental work using scale models is employed in studies of marine and fluvial erosion, but has been little applied to slopes. The procedure in hydrological work is to reproduce actual conditions at a small scale, adjusting materials used and the intensity of applied stresses until the existing field behaviour appears to be simulated; and then to introduce changed conditions, for example harbour works or meander straightening, and record the model's subsequent behaviour. Other environmental phenomena which have been reproduced in scale models are mountain building and glacier flow.

The main advantage of scale models is an accelerated time scale; the smaller the dimensions of the model, the larger the ratio between real and model time. A subsidiary advantage is that conditions are controllable, enabling variables not under study to be excluded. These benefits are achieved at the cost of some sacrifice of reality. Reduction of the length dimension alters other constants by differing ratios; for instance, volumes are reduced by the third power of the length reduction. A table of model ratios for the main physical constants is given by Hubbert (1937). Some properties, however, cannot be altered, chief among which are the viscosity of water and gravitational acceleration (hence the need for film shots of explosions etc. to be shown in slow motion). It is impossible to match all constants, but in individual models many forces are so small that they can be ignored. In general, the strength of materials used should be very low.

The main application to slopes is the outstanding series of experiments by Wurm (1936). Mixtures of ground limestone and slag, sand, colouring material and water were placed on a 70 cm square table, and subjected to an even fine spray. The main features of slope evolution (Fig. 44), found with both cones and ridges, were that steep

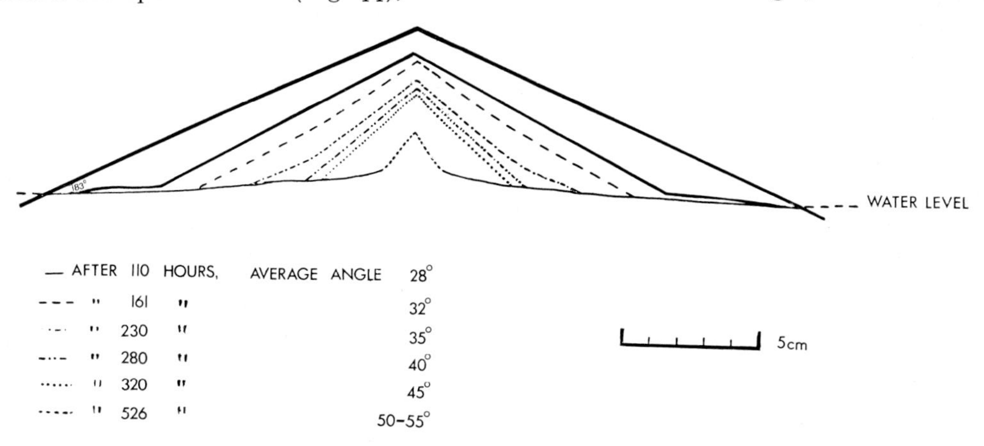

Fig. 44. Slope evolution in a sand model of a cone under fine spray. The number of hours and mean angles are given. After Wurm (1936).

slopes became progressively steeper, remaining rectilinear or acquiring a slight convexity, and were replaced by a gentle footslope. Wurm interprets the steepening as being caused by the downslope increase in volume, and acceleration, of wash, leading to a downslope increase in erosion. In the terminology used here, it is clear that the steep slopes and footslopes are respectively denudation slopes and transportation slopes. Twelve other experiments, involving cuesta topography, structural benches and concurrent uplift, are also described. It is noteworthy that the behaviour of these models does not correspond with any of the recognized theories of slope evolution.

Model reproduction of surface wash has subsequently been repeated, always with similar results: the parallel or steepening retreat of steep slopes, often with rill dissection, and pediment formation. Other processes are harder to reproduce. Flow has been simulated in kaolin glaciers (Lewis and Miller, 1955). An element of solution loss is readily obtained, by mixing salt or other soluble material with sand. The heave mechanism of creep can be obtained by wetting and drying small soil troughs (Young 1960; Kirkby, 1967), but not at a sufficiently accelerated speed for model slope evolution. It is further necessary to confine the heave to a surface layer of a few centimetres only. A possible means is to use loose grains of a material with a high coefficient of thermal expansion, subjected to intermittent brief periods of heating by electric plates suspended just above the surface. By applying greater technical refinements, and by a close comparison of experimental results with field measurements, there is the potential for useful work using scale models.

A link between models and work on natural slopes exists in studies of erosion on badlands, abandoned quarries, or other rapidly evolving landforms (Bradley, 1940; Schumm, 1956, 1956b, 1962; K. G. Smith, 1958). On most badland slopes surface wash is the main process. The results are similar to those from models; parallel retreat with pediment formation, the pediments being subject to regrading. In England, experimental earthworks, each consisting of a ditch and bank, were constructed in 1960, mainly from archaeological interest; there is a planned schedule of observations extending over more than a century. So far the face of the ditch has been replaced by scree (Jewell, 1963; Jewell and Dimblebey, 1966).

XI CLIFFS AND SCREES

CLIFFS

The terms *free face* and *cliff* are used synonymously, to refer to part of a slope which is steep and formed of bare rock. An arbitrary minimum angle of 45° is suggested. More gently-sloping rock surfaces, as found in deserts and on domed inselbergs, are termed *rock slopes*. Rocks fallen from a cliff accumulate at its base to form an approximately rectilinear slope at about 20°-35°; this is termed a *scree slope* (or talus slope) if comprised of rock fragments, or a *debris slope* if wholly or partly covered by regolith. There is no received term for a free face with its associated scree or debris slope; both Koons (1955) and Schumm and Chorley (1966) have employed *scarp* in this sense, and this usage is retained here, despite the alternative meaning of any structurally-controlled steep slope, whether cliffed or not. A scarp may be *simple* if formed of one main rock type; *compound* where formed of two rock types, usually a more resistant overlying a less resistant stratum; and *complex* where there are more than two strata (Schumm and Chorley, 1966).

A *caprock* is a relatively resistant stratum occurring at or close to the crest of the scarp. A cliff actively or recently undercut by the sea is a *marine cliff*. A *corniche* is an almost vertical free face below a horizontal projection, at or near the crest, formed by a resistant rock stratum.

Cliffs originate as a result of undercutting or undermining of a slope. Undercutting may be caused by a river, glacier or the sea. Undermining occurs when two superimposed strata outcrop on a slope, the lower of which is retreating more rapidly than the upper, thus removing basal support from the latter. In either case the base retreats faster than would the slope itself if its angle were moderate, causing a steep slope to be formed. A cliff results if this slope is steeper than the limiting angle for retention of a regolith.

The strength of hard rocks, such as most igneous and metamorphic rocks and many older sedimentaries, is sufficient for vertical cliffs over 1500 m high to be formed if the rocks are mechanically intact. Such cliffs rarely occur, indicating that it is the presence of mechanical defects, mainly joints, that limits the height and angle. This is confirmed by observations during engineering tunnel construction which show that jointing is invariably present.

The directions and frequency of joints may be ascertained by a *joint survey*. This usually reveals the presence of one or more *sets* of sub-parallel joint-plane directions, the sets constituting the *regional joint pattern*. In sedimentaries, the pattern comprises *bedding joints* together with one or more sets of *cross joints*, approximately perpendicular to the bedding. Close to the slope surface a local set of *sheeting joints* (dilation joints), approximately parallel to the slope, may also be present; these are thought to be caused by *unloading*, the release in a preferential direction of the high pressure under which the rocks previously existed (Bain, 1931; Bridgeman, 1936, 1938; Kieslinger, 1960; Ollier, 1969, pp. 5-9).

The following account of the theory of stability in jointed hard rocks is based on Terzaghi (1962). Any cross-section through rock will consist partly of joints and partly of *gaps*, the areas of intact rock separating joints. Along the joints, cohesion is zero. The *effective cohesion* c_i, along a section is given by

$$c_i = c \, . \, \frac{A_g}{A} \qquad (11.1)$$

where c = cohesion of intact rock, A = total area of section, and A_g = area of gaps in section. It is not in practice possible to determine the effective cohesion with any accuracy; thus the stability of a hard rock cliff is more difficult to determine than that of a slope in unconsolidated materials.

As the height of a cliff becomes greater, the shearing stresses on the rock close to its face increase, and more of the gaps are eliminated by splitting; weathering aids this process. The rock adjacent to the free face 'will be changed into a practically cohesionless aggregate of more or less irregular blocks which fit into each other', (Terzaghi, 1962, p. 254). Cliff retreat occurs by rock fall, in which individual joint-defined blocks or smallish masses of rock drop away from the face, and by deeper-seated movements which may include rockslides. Deeper movements develop by *progressive failure*: stress first exceeds strength at one point, cohesion at that point becomes zero, shearing stresses in the adjacent rocks therefore increase, and the failure spreads.

The *critical angle* for a cliff is the steepest angle at which it possesses long-term stability. If the rock retains effective cohesion, the critical angle is 90° up to a certain limiting height of cliff, above which it becomes less; the stability equation is similar to that for unconsolidated materials (equation *8.2*, p. 79), substituting effective cohesion as above defined. For jointed rocks close to the free face, the conservative assumption in an engineering stability analysis is that the effective cohesion is zero. The critical angle is then determined by the angle of friction along the walls of joints (normally about 30°) and the joint pattern. For a random joint pattern the critical slope is about 70°. With rectangular jointing, the critical slope ϕ_c depends on the angle of friction ϕ_f, the angle of dip α, the direction of dip, and the relative spacing and offset of bedding and cross joints. Based on this theory, possible cases are (Fig. 45):

			As in Fig. 45
Horizontal bedding	$\alpha = 0°$	$\phi_c = 90°$	*A*
Dip towards cliff	$\alpha < \phi_f$	$\phi_c = 90°$	*B*
	$\alpha > \phi_f$	$\phi_c = \alpha$	*C*
Dip into cliff	$(90-\alpha) < \phi_f$	$\phi_c = 90°$	*D*
	$(90-\alpha) > \phi_f$:		
Cross joints not offset:		$\phi_c = (90-\alpha)$	*E*
Cross joints offset, $C/D \geqq 1$:		$\phi_c = 90°$	*F*
Cross joints offset, $C/D < 1$:		$\phi_c = (90-\alpha) - \tan^{-1}(C/D)$	*G*

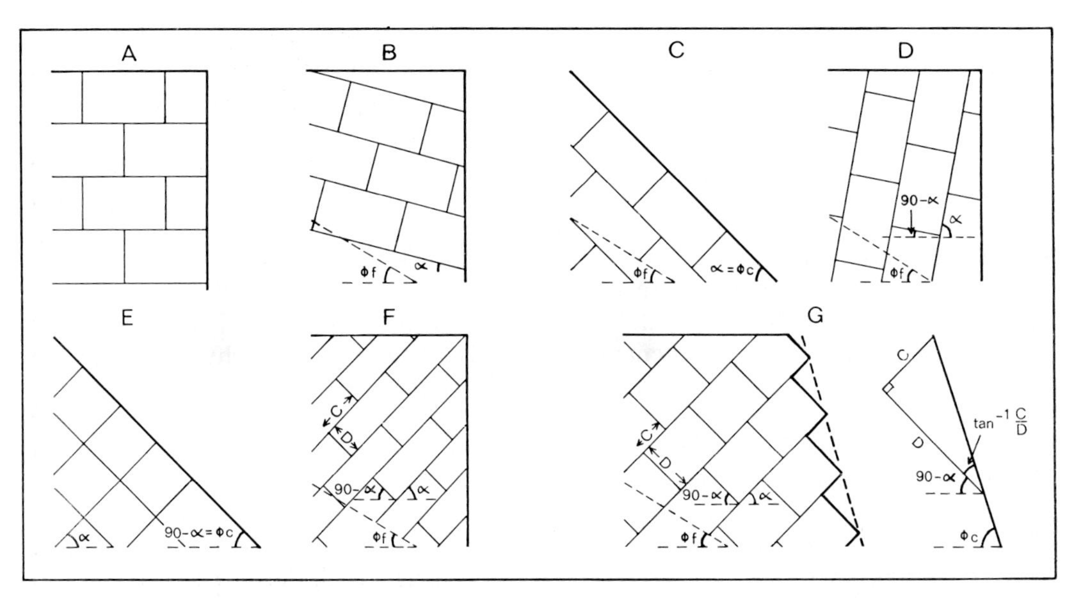

FIG. 45. The theory of critical angles in hard bedded and jointed rocks.

In the last two cases, C and D are as shown in Fig. 45*F*. For a given value of α, ϕ_c increases with increasing values of C/D. For closely-spaced bedding or widely-spaced cross joints, $\phi_c = 90°$.

Water percolating through hard rock moves largely through the joints; consequently, in contrast to unconsolidated materials, seepage forces are insignificant. Stability is lowered, however, by *cleft-water pressure*, the hydrostatic pressure of water accumulating in joints, and by *ice-wedging*, the pressure exerted by the further freezing of water sealed in joint cavities by the initial freezing of its own surface. A statistical study of the date and altitude of rockfalls in Norway showed maxima in April and October-November. This was interpreted by Terzaghi (1962) as being caused by cleft-water pressure, from spring snow-melt and autumn rainfall maximum; Bjerrum and Jörstad (1963) argued that it was caused by ice-wedging, showing that most falls occurred in periods when the mean diurnal temperature change crossed freezing point. Rapp (1960) found that a maximum rockfall frequency was produced by release on thawing following ice-wedging.

Bain (1939) describes the disintegration of glaciated cliffs in New England. Sheeting joints were present in nearly all rock types; these subsequently widened, and rock slips of distances up to 2 m occurred above them.

Little systematic quantitative data is available on the observed form and height of cliffs, for varying rock types and climatic conditions. A subjective impression is that vertical cliffs are relatively rare, and that the frequency distribution of angles would show a rapid decrease above 70°. It is necessary to distinguish between critical angles for short-term and long-term stability. The former is the angle at which a recently formed cliff will stand for a few months or years. The latter refers to stability after a period sufficiently long for joint-splitting to have taken place (through unloading and weathering), with consequent reduction or elimination of effective cohesion. This difference is well seen in marine cliffs, where the overhang associated with a wave-cut notch possesses short-term but not long-term stability.

Following Penck, it is frequently assumed that since all parts of a free face are equally exposed to weathering, a simple cliff will retreat parallel unless buried by scree. Mortensen (1960) suggests that because of the effects of weathering in progressively reducing the strength (cf. effective cohesion) of the rock, a closer approximation to the actual retreat of cliffs in homogeneous rock is central rectilinear recession (Fig. 38, p. 107). Actively undercut or undermined cliffs retreat parallel; this is shown by marine cliffs, and by the maintenance of free faces of similar height and angle on cliffed slopes with a caprock (e.g. Fair, 1947). Some oscillation is likely between steepening through undercutting and major rockfalls, and the restoration of a gentler angle by minor falls. In the retreat of cliffs not actively undercut there is probably an element of decline, although this has not been demonstrated.

In cases of compound scarps, formed for example by a sandstone or limestone overlying a shale, a convenient abbreviation is to refer to the upper rock as more resistant. The strict meaning of this is that, other conditions being equal, the upper rock tends to retreat less rapidly than the lower. The cause of this slower retreat may be greater rock strength (either strength of the intact rock or difference in effective cohesion due to joint spacing), greater permeability and hence less rapid attack by surface transport processes, or a slower rate of weathering. In compound scarps the base of the free face may lie at the junction between rock types, but it is also frequently below. In examples of the latter type, the formation has been observed of fractures dipping at about 45° towards the cliff in the lower rock; these are followed by vertical cracking of the caprock, leading to rockfall involving both rocks (Koons, 1955).

Estimates of the rate of cliff recession are summarized in Table 8. Soft shales and unconsolidated rocks retreat 10-1000 times faster than hard rocks. On the basis of

Table 8. Rates of cliff recession

Source	Type of evidence	Rock	Climate	Recession, mm/yr
Schumm and Chorley, 1966	Recorded	Shale	Semi-arid	2–13
Freise, 1932	Various	Granite, gneiss	Rainforest	2–20
L. C. King, 1956	Geological	Various	Humid sub-tropical	1·6
Rühl, 1914	Historical ruins	Sandstone	Humid temperate	·5
Schumm and Chorley, 1966	Geological	Sandstone	Semi-arid/arid	·6*
Mortensen, 1956	Geological	Sandstone	Semi-arid/arid	·2*
Rapp, 1957, 1960	Recorded	Schists	Polar	·01–·1
Mortensen, 1956	Archaeological monuments	Sandstone	Arid	·00004

* Both estimates of the retreat of the Grand Canyon, Colorado.

the limited data available, the following order-of-magnitude values for hard rocks are indicated:

Climate	Time required for retreat of 1 m
Rainforest	10^2–10^3 years
Humid temperate and sub-tropical	10^3 years
Semi-arid	10^3–10^4 years
Polar	10^4–10^5 years
Arid	10^7 years

Mortensen (1956) cites evidence for the extreme slowness of cliff recession in arid environments; he argues that the frequency of cliffs in deserts is associated with slowness of both cliff and scree weathering, due to the absence of chemical weathering. The low value obtained for a polar climate is noteworthy, as it is sometimes held that these are regions of rapid landform change.

A receding cliff no longer undercut will be progressively eliminated by the accumulation and upward growth of a scree slope unless the scree is removed (Figs. 46, 47).

FIG. 46. An early stage of the replacement of a cliff by a scree slope. Sandstone, Egypt. The material in the foreground is water-worn, and the cliff may be relict from a more humid period.

FIG. 47. Later stages in the replacement of a cliff by a scree. The site is adjacent to Fig. 46.

If weathering and removal of fine material from the scree occurs, a time-independent condition may be reached, in which the cliff and scree slope retain the same heights during retreat. Conditions apparently of this nature have been described by Fair (1947) and Schumm and Chorley (1966), in both cases for compound cliffs. To account for the volume of talus Schumm and Chorley proposed the *talus weathering ratio* $W = p/d$, where p = rate of talus production (i.e. supply, by rockfall etc.), and d = rate of talus destruction and removal. In the Colorado Plateaus the volume of talus is small, despite evidence of active rockfall; Schumm and Chorley explain this by supposing that $W \leqq 1$. This explanation is incomplete. For a time-independent condition to be attained, W must exactly equal 1. If $W < 1$ the only talus that can accumulate is that temporarily present, from individual rockfalls. To account for the relative heights of cliff and talus slope under time-independent conditions it is necessary that d is a function of the area of talus exposed to weathering; a rational assumption is that $d = kT$, where T = length of talus slope (measured downslope) and k is a constant. For example, let $p = 100$ m^3/1000 years, and $d = 2T$; a time-independent state $W = p/d = 1$ is reached when $T = 50$, i.e. when the talus slope is 50 m long. Among problems associated with cliffs, the determination of whether time-independent conditions exist, and if so the control of height relations between cliff and scree, are of much interest. Measurements of form, that of the existing cliff and talus slope, and of process, the rates of cliff retreat and talus weathering, are required.

The stability of cliffs in hard rocks, necessary information in, for example, the construction of excavations for dams, is not readily determined by the methods of soil mechanics. 'From an engineering point of view, even the results of a conscientious joint survey still leave a wide margin for interpretation' (Terzaghi, 1962). A systematic collection of data on cliff height, angle and stability, for different rocks and climates, would therefore be useful as a complement to calculated stability values. Observation of stability may be made by the presence or absence of signs of recent rockfall. The establishment of a precise classification of rock types, with particular reference to joint spacing, would be necessary.

MARINE CLIFFS

Marine cliffs differ from most other slopes in that the erosion at their base is entirely lateral, with no vertical component. In the short term, undercutting is intermittent, taking place mainly when storms coincide with high spring tides; between such times the cliffs are subject to rockfall, giving temporary protection to the base until the material is reworked and removed. In the long term, marine cliffs may be regarded as subject to unimpeded basal removal, with lateral basal erosion.

If undercutting is active, the steepness of a marine cliff may be that of short-term rather than long-term stability, the form being dependent on the relative rates of undercutting by the sea and slope retreat caused by subaerial processes. Consequently there is no correlation between rock hardness and cliff angle; among the steepest cliffs are those on chalk. Unconsolidated rocks such as boulder clays may be subject to rapid marine erosion, yet have a mean angle of 35°-45°, owing to the equally high rates of slope retreat by rapid mass movements. The influence of subaerial processes is clearly demonstrated where two or more rock types are present; for example, where drift deposits overlie chalk the former retreat more rapidly, giving an upper unit of gentler angle above the almost vertical chalk face. Details of geological structure, including bedding planes, joints and faults, may be picked out by both marine and subaerial processes (e.g. Wilson, 1952).

There appear to be three orders of magnitude for rates of retreat. Cliffs in most igneous and metamorphic rocks and some Palaeozoic sedimentaries retreat very slowly. Thus in Devon, no detectable difference between photographs taken at 50-year intervals has been found (Arber, 1949). In Mesozoic sedimentaries, cliffs typically undergo loss of a few metres at intervals of one or more years, with relatively little change between. The retreat of unconsolidated cliffs can average several metres per year, and the loss of 30 m in a single storm has been recorded (W. W. Williams, 1956).

Active erosion of a cliff base may cease as a result of progradation by coastal depositional features or, more gradually, a fall in sea-level due to isostatic or eustatic causes. Abandoned cliffs are not uncommon, and provide excellent opportunities to study the early stages of slope evolution. The classic study of this situation is by Savigear (1952). At Pendine, in South Wales, a sand spit appears to have grown eastwards from a headland, separating the cliffs to the north of it from the sea. If the assumption of eastward growth is correct, the date at which erosion of the cliff base ceased should become later towards the end of the spit, which is confirmed by an eastward transition from concave forms to vertical cliffs. On the basis of an inductive comparison of profile forms, Savigear found that steep rock cliffs evolve by slope replacement; the free face retreats parallel, and is replaced from below by a slope of about 32°, possibly with intermediate steeper replacement slopes (Fig. 48*A*). Below 32° a regolith cover

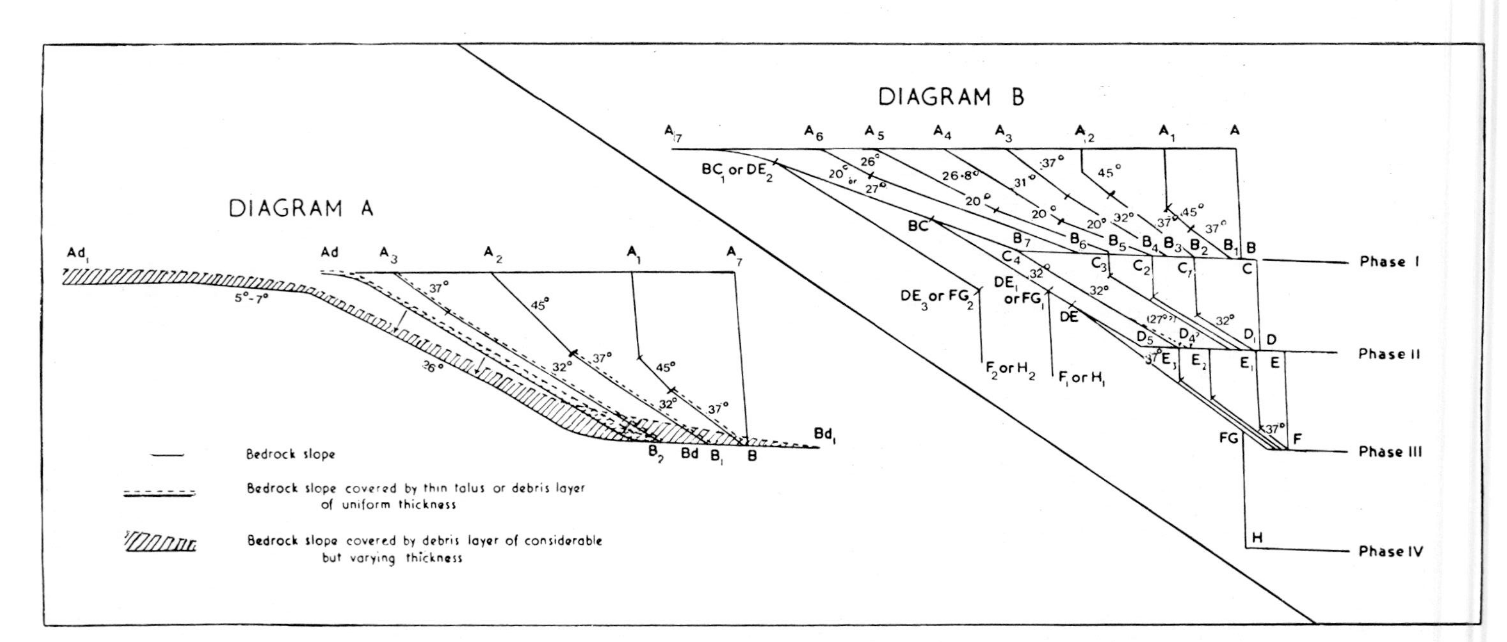

FIG. 48. Slope evolution on marine cliffs, after Savigear (1962). Diagram *A* = cliff abandoned by the sea, impeded basal removal. Diagram *B* = cliffing, and slope retreat during a series of sea-level changes. Successive periods of higher sea-levels, Phases I-IV, are separated by intervening falls to lower sea-levels.

is retained. Savigear suggested that regolith-covered slopes of 32° may retreat parallel with unimpeded basal removal, but with impeded basal removal they decline.*

Composite marine cliffs (also called slope-over-wall cliffs) possess two units of form: a lower, almost vertical, free face, and an upper, regolith-covered *seaward slope*. The terms *bevelled cliff* and *hog's-back cliff* have been applied to composite cliffs in which the seaward slope is respectively shorter and longer than the free face. One-cycle and two-cycle origins have been proposed for composite cliffs. In the one-cycle explanation (Fig. 49*A*), the seaward slope retreats by subaerial denudation contemporaneously with marine undercutting of the free face (Challinor, 1931; Arber, 1949); the form is thus time-independent. The two-cycle theory supposes the following sequence (Fig. 49*B*): (*1*) a simple cliff is cut; (*2-4*) sea-level falls, the cliff base is

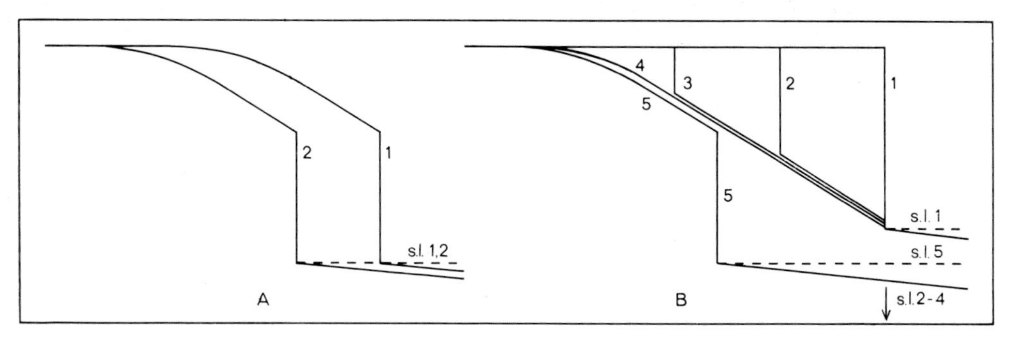

FIG. 49. Theories for the origin of composite marine cliffs: *A* = one-cycle; *B* = two-cycle.

abandoned, and the cliff evolves into a gentler slope; (*5*) sea-level rises, and renewed marine erosion cuts a further cliff, intersecting the gentler slope (Cotton, 1951; Savigear, 1962). In many temperate-latitude regions the slope reduction stage may have taken place under periglacial conditions, whilst the free face is likely to have been cut during and since the Flandrian transgression (*c.* 5000 B.C.). The chronology of composite cliff evolution is discussed by Orme (1962). The one-cycle theory requires parallel retreat of the whole cliff and slope, and it is unlikely that a regolith-covered slope of 40° and less can retreat as fast as a steep cliff. Consequently a two-cycle origin is more probable. Possible exceptions occur where the seaward slope is coincident with a bedding or fault plane (Wilson, 1952). With reference to cliffs of the southern hemisphere, Cotton (1967) suggested that composite forms on hard rocks had a two-cycle origin, with slope recession occurring in the later part of the last glacial period, and modern cliffs cut below since the Flandrian transgression, whereas on soft rocks, cliff recession has been contemporaneous with the transgression.

Investigating seaward slopes by profiling, Savigear (1956, 1962) made the discovery that they are not smoothly convex, but consist of a series of rectilinear segments. For a given area and rock type, the angles of these segments recur on several profiles. Savigear interpreted this by extending the cyclic theory, assuming each

* Using these terms as defined on p. 7.

segment-angle to have resulted from a phase of cliffing followed by a period of slope decline (Fig. 48*B*). This is analogous to the origin of polycyclic valleys. This intriguing suggestion awaits further investigation.

SCREES

Some of the earliest field and experimental studies specifically concerned with slopes were on screes (e.g. Leblanc, 1842; Davison, 1888; Piwowar, 1903). Screes differ qualitatively from most other slopes in having an external debris supply, the coarse-grained nature of which leads to an absence of soil and vegetation. Despite these substantial differences some central questions on slopes, for example process-form interactions and the balance of debris supply, transport and removal, can be investigated on screes, with the advantages of relatively rapid process rates and minimal disturbance by human agencies.

Scree slopes, unless undermined, are concave in profile. The angle of the upper, steeper part is normally 30°-38°; 35° is the most frequently reported value, and 42° the highest. This steep portion is commonly less than half the profile length of the scree and is succeeded by a longer, somewhat gentler, section, often between 33° and 25°. There is sometimes a concavity, which may attain quite low angles at the base (Fig. 50). Where undermined by a stream, glacier or the sea, the lowest part of the profile may steepen. From 56 surveyed profiles on limestone screes in North Wales, Tinkler (1966) found that the frequency distribution of angles was negatively skewed, resulting from the fact that the characteristic (most frequent) angle, 35°, lay close to the limiting (steepest) angle of 36°. Investigating a bedding sequence of 34 layers in a talus cone in pumice in New Zealand, Blong (1970) found a bimodal angle frequency distribution, with peaks at 26° and 30°. It is generally found that large rock fragments stand at steeper angles than smaller, but the reverse relation or an absence of correlation has also been reported. Platey fragments stand at gentler angles than polygonal blocks. The mean size of fragments usually increases downslope, but this may be mainly caused by the accumulation of relatively few very large boulders towards the base. There are also cases where fragment size shows no relation to profile position. Behre (1933) reports a downslope size decrease, which he attributes to progressive weathering during movement down the scree. In semi-arid Karroo (rainfall 230-380 mm) the size of the majority of fragments decreases downslope, and the scree is smoothly concave; a few large blocks lie near the base, but they play no part in determining the angle of slope. The maximum scree angle on sandstones in this area is 34°. On similar lithology in the sub-humid interior of Natal (rainfall 650-900 mm) it is only 35°, despite a soil and vegetation cover. In both areas, screes of similar fragment size stand at similar angles, demonstrating that particle size and not climate is the chief factor governing scree angles (Fair, 1948b).

The visual impression of great thickness of screes is deceptive. Many are less than 5 m thick (normal to the surface) and only a few instances exceeding 10 m have been reported. This may, however, be due to the difficulty of locating the rock surface beneath thick accumulations; McDougall and Green (1958) employed measurements

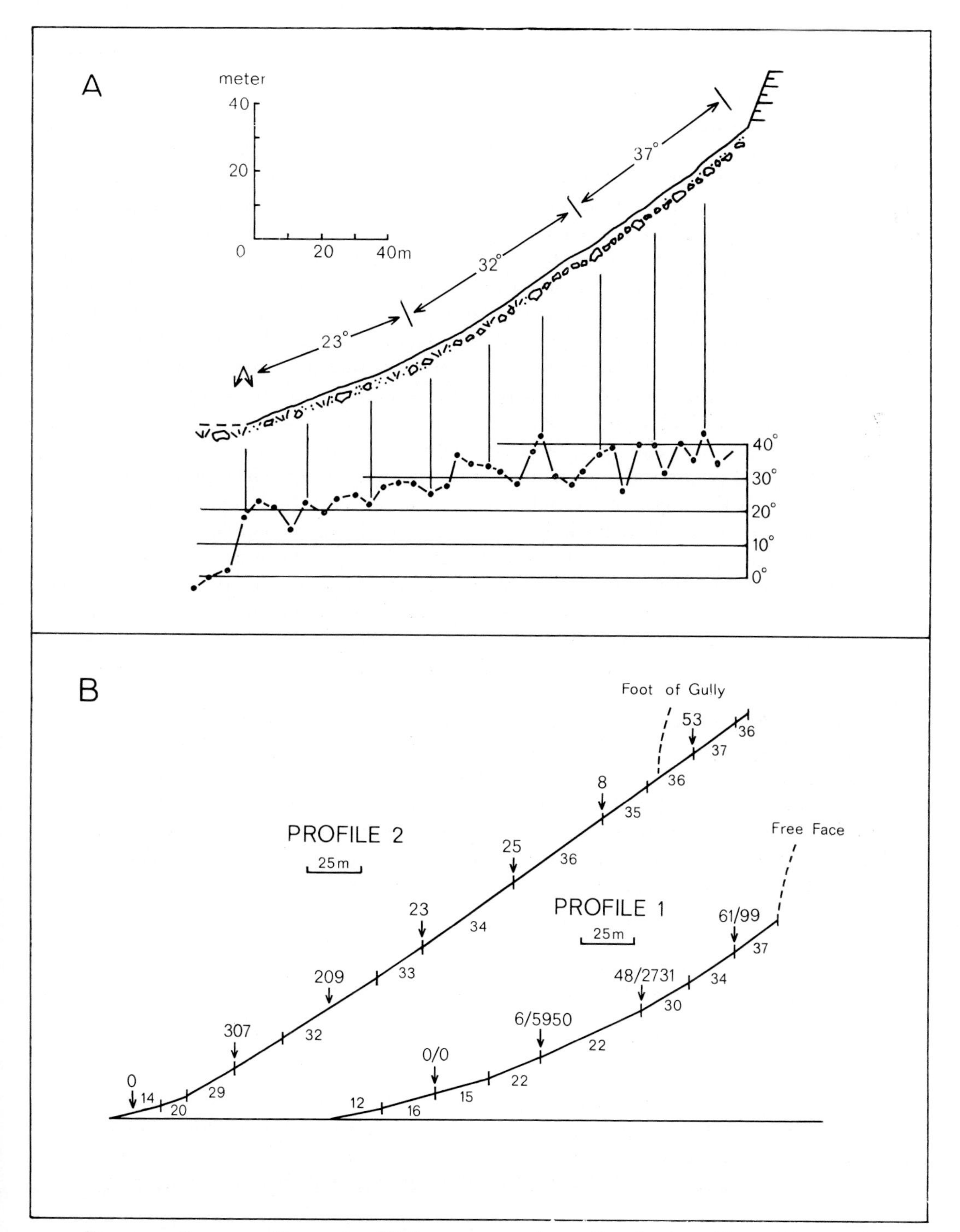

FIG. 50. Scree profiles. *A* = Kärkevagge, northern Lappland; with profile graph. After Rapp (1957). *B* = Southern Alps, New Zealand. Numbers above the profiles are the volumes of materials accumulated, on Profile 2 during a period of slush avalanching, and on Profile 1 during a period of rockfall followed by a period of slush avalanching. After Caine (1969).

of magnetic dip for this purpose; they found that on a 300 m high dolerite scree slope the scree thickness at first increased rapidly from the top downwards, below which it remained at a fairly constant thickness of about 90 m. Talus cones may be thicker. A 300 m long cone formed of Tertiary sandstone in Spitzbergen was 14-20 m thick 100 m from the top, and 14 m thick at the base (Rapp, 1957; the original gives horizontal thicknesses). Thin superficial layers of rock fragments, sufficiently thin for their surface form to be determined by that of the underlying material, are distinguished as *clitter* (Tinkler, 1966); angles of clitter slopes are gentler than screes in similar materials. The above summary of the form of screes is based on Behre (1933), Caine (1969), Hay (1942), C. A. M. King (1956), Koons (1955), McDougall and Green (1958), Morawetz (1932), Rapp (1957, 1960b), Robinson (1966), Tinkler (1966), Young (1956), and the summary of early work in Van Burkalow (1945).

The loose material of which screes are composed lacks cohesion; consequently the angle at which the material will stand is unaffected by the height of the slope. The *angle of repose* is the angle at which loose material will stand when piled. The *angle of sliding friction* is the angle of an inclined plane at which a body resting on the plane will first begin to slide (Van Burkalow, 1945). The former term should be confined to the angle attained following addition of material by fall from above. If a slope is undermined, mass movement of the whole scree may occur, and the resulting angle is not necessarily the same as the angle of repose. The further term *angle of settling* is suggested, defined as the angle at which loose material comes to rest following the removal of basal support or the steepening of the scree above the angle of stability by basal erosion.

A theoretical treatment of the packing and stability of spheroids of equal sizes is given by Allen (1969). Let a and b be the principal axes of a spheroid, c the fractional volume concentration (1—the fractional porosity), and ϕ the angle of initial yield. Two cases are considered: with the spheroids irregularly arranged, and with the major axes aligned parallel to the direction of sliding; in both cases, the maximum possible value of c for spheroids is 0·75. For each case, the angle varies with the shape of the spheroids and their packing. Among values of ϕ_i given are:

Spheres ($b/a = 1$)	$c = 0{\cdot}6,\ \phi_i = 15^\circ$
	$c = 0{\cdot}7,\ \phi_i = 32^\circ$
Flattish spheroids, ($b/a = 0{\cdot}2$)	
Axes irregularly arranged	$c = 0{\cdot}6,\ \phi_i = 17^\circ$
	$c = 0{\cdot}7,\ \phi_i = 21^\circ$
Axes aligned parallel	$c = 0{\cdot}6,\ \phi_i = 6{\cdot}5^\circ$
	$c = 0{\cdot}7,\ \phi_i = 7{\cdot}5^\circ$

Thus the angle of initial yield becomes less as the spheroids become flatter, more densely packed, and more regularly orientated. The low values obtained for the case of aligned axes indicates, however, by comparison with observed scree slopes, that it is not justifiable to assume that actual scree fragments will behave similarly to ideal spheroids.

Van Burkalow (1945) carried out scale model studies of factors affecting scree angles, using sand and small rock fragments. For material of uniform size, there is a slight tendency for the angle of repose to become gentler with increase in fragment size. But when coarse material is given an admixture of fine the angle becomes steeper, and in imperfectly sorted material, the larger the average size, the steeper the slope. Spherical fragments, or those with rounded edges, stand more steeply than angular material. Platey fragments were not tested. A smoother fragment surface and a higher density both produce gentler angles. Increasing the height of fall decreases the angle of repose, but if the scree is first formed by dropping material and subsequently undermined by removing a basal support, the angle of settling becomes greater as the height of the original fall is increased, presumably owing to firmer packing. Natural screes are invariably formed of imperfectly sorted material, therefore the normal occurrence of steeper slopes on coarser fragments agrees with the experimental results. The finding of an interaction between size and sorting may account for reported field instances where coarser fragments are not found on steeper slopes. Unless field observations are taken at a constant position on the scree profile, an apparent relation between fragment size and angle may be spurious, because of the strong correlation of both these last variables with distance down the scree. Natural screes have a finite, if small, margin of stability, and hence necessarily stand at an angle somewhat below the angle of repose.

Processes affecting screes fall into three groups: the accumulation of material on the surface, the downslope transport of material already present, and the removal of material. The main process of addition is rockfall. Rapp (1957, 1960, 1960b) found that on schists in northern Sweden infrequent but large rockfalls contributed most of the material, but on limestones and conglomerates in Spitzbergen small rockfalls were quantitatively more important. The formation of a scree by accumulation is a dynamic process, in which the kinetic energy of the falling rocks is expended on the scree on impact. Unless trapped in a cavity, the point at which a rock comes to rest depends on the scree angle. If the scree stands at less than a certain angle, the loss of energy at the average impact exceeds the interim gain in energy from gravity, therefore a certain number of impacts will bring the rock to rest. Above this critical angle, the rock will roll to the bottom (Jeffreys, 1932). Larger blocks travel further down the scree because they are less liable to be trapped in cavities, and also because volume, and therefore mass and momentum, increase as the cube of the diameter, whereas surface area and therefore friction only increase as its square. Melton (1965) measured whether boulder-covered slopes were stable by attempting to roll rocks down them, noting whether the rocks came to a halt.

Material from very large rockfalls is sometimes carried well beyond the base of the main scree slope. It has been suggested that fluidization, by the rock overriding and compressing entrapped air, is responsible (Kent, 1966; Shreve, 1968).

In cold climates, slush avalanching is both a further cause of accumulation and a major process of downslope transport of scree. Slush avalanching occurs with snow-melt, and is considerably more important than transport of rocks by snow avalanches (Rapp, 1960). Caine (1969) found that the amount of material deposited by slush avalanching showed a linear decrease with distance from the scree base, whereas the amount of accumulation from rockfall decreased linearly with distance from the foot of the free face. Other processes of transport on scree are talus shift (p. 57) and surface wash. Gardner (1969) measured the movements of stones on sandstone screes in the Canadian Rockies. Movements showed a very high variability, both spatial and temporal; thirteen transects showed mean movements of 0·5 m, eleven of 1-9 m, and the greatest individual stone movement was 71 m.

Fabric studies have potential applications to screes. Caine (1967) found a lack of preferred orientation of long axes on dolerite screes in Tasmania, and this was interpreted as indicating that the screes were largely accumulation forms directly produced by rockfall, and little modified by subsequent downslope transport. The same technique was applied to relict periglacial blockfields, for which it was suggested that those of allochthonous origin show preferred orientation whereas autochthonous blockfields, formed *in situ*, do not (Caine, 1968).

If a scree slope is undermined there may be substantial transport by debris sliding. The effects of this are illustrated in screes along the sides of the Austerdalsbre, Norway, a glaciated valley in which substantial glacial retreat has taken place since about 1800. Above the present glacier the screes are unstable, as indicated by lack of vegetation, and stand at 38°. Below the glacier snout, in the area abandoned by ice since 1935, the screes stand at 33°-35°, and have been colonized by herbaceous vegetation. Further down the valley 31°-33° segments occur, and there are scattered birch trees indicating stabilization. A possible explanation is as follows: 38° is the angle of settling, the result of undermining by the glacier; 33°-35° may be the angle of repose first produced by rocks falling from above; alternatively, it is possible that at 38° the slope possesses only short-term stability, and is transformed to one of 33°-35°, with long-term stability, by debris slides during snow-melt. The slope of 31°-33° is possibly a replacement slope, although intermediate stages, with two-segment profiles, were not located (Young, 1956).

Vegetation can be used as an index of scree stability. On non-calcareous British screes Leach (1930) found the stability succession: lichens and bryophytes→parsley fern (*Cryptogramma crispa*)→other ferns with grasses→heather (*Calluna*)→trees. Vegetation is probably a dependent factor, unable to exert an active stabilizing effect on scree.

Material is removed from screes by solution loss, surface and subsurface wash, and wind. Only fine particles and dissolved material are affected, hence removal from screes is subject to control by weathering. Whether the height and volume of a scree slope is time-independent is determined by the talus weathering ratio (p. 124).

Some screes may be largely relict, accumulation having taken place mainly during more intense frost weathering under periglacial conditions. This has been suggested for Wales (Ball, 1966), Cyrenaica (Hey, 1963), and is possibly the case in Sweden and Spitzbergen (Rapp, 1957). The most remarkable case described is in northern Chile, where at 800-1900 m altitude the majority of slopes over a substantial area are rectilinear, 28°-33°, stratified relict screes (Weischet, 1968). Stone stripes on screes are not necessarily periglacial in origin; contemporary re-formation, after disturbance, has been observed (Michaud, 1950; Miller *et al.*, 1954; Caine, 1963).

The convex, parabolic rock cores beneath screes, predicted by deductive models (p. 105), have not been observed in the field. Available evidence suggests that most screes are relatively thin veneers over rock surfaces not differing greatly in angle from that of the scree, and with little or no convexity. Rectilinear slopes at angles close to the angle of repose for screes, known as Richter denudation-slopes (p. 107), have been described by Bakker (1956) and Birot (1963), in both cases for compound scarps where marls underlie a limestone corniche. Tinkler (1966) found characteristic angles of 37°-39° for bedrock slopes in Wales. The existence of a slope with this form does not prove that it originated in the manner supposed by cliff recession models.

Low hills with one rectilinear slope at the scree angle, with or without a scree cover, are termed *flatirons*. They originate when part of the inclined rock slope beneath a scree becomes separated from the cliff by recession and dissection (Koons, 1955; Birot, 1963; Everard, 1964). Lange (1963) described planes of repose in caves, the suggested origins of which are analogous to the production of a Richter denudation-slope.

Screes are most frequent in hot, dry climates and in cold mountain regions, particularly where there has been glaciation. There are substantial differences in the processes operative in these two environments; in particular, slush avalanching and processes associated with frost action are absent from the former. This might be expected to produce form differences between the respective scree slopes, but there have been no comparative studies of this question.

BOULDER-CONTROLLED SLOPES AND DEBRIS SLOPES

The concept of *boulder-controlled slopes* was proposed by Bryan (1925). In deserts, well-jointed rocks first break up into boulders, which subsequently weather directly to fine material. The fine particles are removed by wash, whilst the boulders remain on the slope. If there is a layer several boulders thick, they will accumulate at the angle of repose; Fair (1948b) reports that in the semi-arid Karoo, boulders lie one or two deep. Frequently, however, there is only a single, discontinuous layer, with the boulders not touching, suggesting that it is the angle of sliding friction that is the upper limiting angle for boulder stability. The cover of boulders fills pre-existing hollows in the slope, protecting the subjacent rock, whereas initial rock projections are exposed to weathering. Consequently, by differential weathering, the bedrock comes to assume a rectilinear form at the angle of repose. Thinning of the protective boulder cover by weathering, or its occasional removal in catastrophic storms, can lead to parallel retreat of the rock slope by the same means of control.

This mechanism is not limited to screes beneath a cliff, but can take place on any slope in deserts, the requirements being direct weathering from boulders or stones to fine material, and removal of the latter before soil can accumulate. Steep rectilinear slopes with a discontinuous boulder cover are numerous on desert mountains in Egypt (Figs. 51, 52), and an origin by boulder-control of weathering is possible. In a study of boulder-mantled slopes in the type area of the Arizona desert, Melton (1965) found no correlation between fragment size and slope angle, and the majority of slopes were substantially below the angle of repose of the boulders. Melton argued from this that

Fig. 51. Desert slope in sandstone under an arid climate, Egypt. Intermediate between an irregular stepped form, caused by structural effects under control by weathering, and a boulder-controlled slope.

Fig. 52. Boulder-controlled slope in a semi-arid climate, Sudan. Bare rock faces occur at approximately the angle of repose of the boulders.

the concept of the boulder-controlled slope should be abandoned, but his evidence does not justify this conclusion.

Debris slopes occur beneath a free face, and have maximum angles of more than 25°, but the rocks fallen from the face above are partly buried by soil, and the slopes carry a vegetation cover. They are common in savanna climates, and in semi-arid to humid sub-tropical areas. They are concave in profile, and the size and proportion of large boulders projecting through the soil cover tends to increase towards the steep, upper part. They frequently occur beneath continuous cliffs formed by a caprock (Fair, 1947; Sparrow, 1966), but may also surround isolated inselbergs (Figs. 76-78, pp. 210-11). Their general similarity in form and occurrence to that of screes could indicate a similar origin, but this is not necessarily so. The formation of debris slopes has been the subject of some speculation, but there has been little study of their form or the processes active upon them.

XII PROFILE FORM

This and the two following chapters are concerned with slope form. Many studies include both a description of the observed slope forms and an attempt to explain these in terms of process or evolution. This need not always be the case. As in other environmental sciences, a foundation of descriptive material on form is essential for the discussion of origin; in geology and biology this basis was provided by the large volume of descriptive work carried out in the nineteenth century. Geomorphology lacks such a substantial foundation, and it is desirable that there should be more work of a purely descriptive nature, concerned with the survey of slopes and regolith characteristics. Both qualitative and quantitative descriptions are needed, and the methods used should be such as to permit comparison between studies from different areas. If the genesis of the forms is also considered, the initial description of the features to be accounted for must be in non-genetic terms. There is a need, however, for a branch of slope geomorphology concerned wholly with the description and comparison of form, without reference to their explanation; this would complement the existing independent subject of process study, the two branches together providing the basis for consideration of slope evolution.

The shape of the land surface can be considered as profile form and plan form. Profile form refers to the two-dimensional shape along a vertical plane, in particular the angle and shape of valley slopes and scarps along a plane that follows the direction of maximum slope. Plan form refers to the shape of the ground surface viewed vertically, as shown, for example, by the curvature of contours. A third major division of slope form *sensu lato* is concerned with the properties of the regolith, including micro-relief. Profile form is discussed in this chapter, preceded by an account of the measurement and description of slope angle. Aspects related to the occurrence and frequency of angles, although partly based on profile studies, are dealt with in Chapter XIII. Chapter XIV covers plan form, and Chapter XV regolith.

MEASUREMENT OF ANGLE

Slope may be expressed as degrees, percentage grade or unit rise. Percentage grade, equal to $100 . \tan \theta$, is used in soil conservation and engineering in the U.S. It has the advantage of giving twice the precision of angle in degrees without the use of decimals. Unit rise, or gradient, used in British engineering practice, only becomes the most convenient method at angles below *c*. 2°, and is therefore more suited to fluvial geomorphology than valley slopes. Height change in feet per mile is also sometimes used for gentle gradients. Angle is generally used for work on slopes; if measured in the field in degrees and minutes it is more convenient for calculation if converted into decimals of a degree. A difference of one degree becomes of decreasing significance as angle increases, in an approximately logarithmic manner (p. 172). The logarithm of angle in degrees may be used, but is mathematically less suitable than the logarithm of the tangent; the latter, however, is inconvenient in having negative values. If a logarithmic transformation is required, slope may be expressed as a unit termed *altan* (*a*ngle *l*og *tan*), where altan = $10(\log.\tan \theta + 3) = 10.\log(1000.\tan \theta)$; angles measured as 0° are given the altan value of 0·0 (strictly equivalent to 0·06°). Angles from 0° to 45° fall into the altan range 0·0-30·0; the midpoint of this range, altan 15·0, is equivalent to 1·82°. A conversion table between different systems is given as Table 9.

Slope angle is observed in the field over a distance measured along the ground surface, termed a *measured length.* A single observation consists of angle and ground surface distance. For most purposes it is sufficient to measure angles to the nearest half-degree and distance to 0·1 m. Local ground surface irregularities (micro-relief) amount to at least 10 cm; subtended over 10 m this is equal to $\frac{1}{2}$°, imparting a random error of this order to the angle measurements. In areas where slopes of less than 2° are extensive, and are surveyed over lengths of 20 m or more, measurement to the nearest 0°10′ is justified. The most convenient and widely-used instruments are an Abney level, linen tape and ranging poles; these give the required accuracy, are rapid to use, and need only one observer. It is preferable to take pairs of readings, upslope and downslope, repeating if they differ by more than $\frac{1}{2}$°; if this is not done, the instrument must be checked by paired readings at the beginning and end of every day, applying a correction if necessary. An alternative instrument, convenient if a series of angle measurements are to be taken from a single point, is a lightweight theodolite; with the use of a Ewing stadi-altimeter, horizontal distances and vertical height differences to over 30 stations per hour can be obtained (Pugh and Brunsden, 1965). Pitty (1968) has described a slope pantometer, suitable for obtaining large numbers of short (1-2 m) and equal measured lengths. A surveyor's level is accurate but slow; its use for slope, as opposed to altitude, measurement is undesirable in principle, in that the value required is obtained indirectly by trigonometric conversion. For special-purpose work requiring high accuracy of slope measurement, a theodolite is preferable. A recent development is the profile recorder, which is wheeled across the ground and reproduces a continuous scale record of the surface, with a local shape accuracy of 5 cm; its use will substantially augment the amount of slope data obtained.

True and apparent slope are analogous to true and apparent dip in geology. *True slope* is the angle of surface slope of the ground in the direction of maximum slope, i.e. perpendicular to the contour. *Apparent slope* is the angle of surface slope in any direction other than that of true slope. Angles measured by viewing in a downslope direction are recorded in the field as minus. The *aspect* is the horizontal direction faced by the true slope.

Profile curvature is the rate of change of angle with distance down the true slope, expressed in degrees per 100 m. Convex slopes have positive curvature, and concave slopes negative. To obtain the profile curvature at a point on the ground surface,

Table 9. Conversion between methods of expressing slope angle.
A. Based on degrees

Degrees and minutes	Degrees and decimals of a degree	Altan	Percentage Grade	Gradient, or unit rise. 1 in:
0°00′	0·00	0·00*	0·0	∞
0°10′	0·17	4·6	0·3	344
0°20′	0·33	7·6	0·6	172
0°30′	0·50	9·4	0·9	115
0°40′	0·67	10·7	1·2	85·9
0°50′	0·83	11·6	1·5	68·8
0°06′	0·10	2·4	0·2	573
0°12′	0·20	5·4	0·4	287
0°18′	0·30	7·2	0·5	191
0°24′	0·40	8·4	0·7	143
0°30′	0·50	9·4	0·9	115
0°36′	0·60	10·2	1·1	95·5
0°42′	0·70	10·9	1·2	81·9
0°48′	0·80	11·5	1·4	71·6
0°54′	0·90	12·0	1·6	63·7
1	1·00	12·4	1·8	57·3
2	2	15·4	3·5	28·6
3	3	17·2	5·2	19·1
4	4	18·4	7·0	14·3
5	5	19·4	8·8	11·4
6	6	20·2	10·5	9·5
7	7	20·9	12·3	8·1
8	8	21·5	14·1	7·1
9	9	22·0	15·8	6·3
10	10	22·5	17·6	5·7
11	11	22·9	19·4	5·1
12	12	23·3	21·3	4·7
13	13	23·6	23·1	4·3
14	14	24·0	24·9	4·0
15	15	24·3	26·8	3·7
16	16	24·6	28·7	3·5
17	17	24·9	30·6	3·3
18	18	25·1	32·5	3·1
19	19	25·4	34·4	2·9
20	20	25·6	36·4	2·8

Table 9A.—*continued*

Degrees and minutes	Degrees and decimals of a degree	Altan	Percentage Grade	Gradient, or unit rise. 1 in:
21	21	25·8	38·4	2·6
22	22	26·1	40·4	2·5
23	23	26·3	42·5	2·4
24	24	26·5	44·5	2·2
25	25	26·7	46·6	2·1
26	26	26·9	48·8	2·1
27	27	27·1	51·0	2·1
28	28	27·3	53·2	1·9
29	29	27·4	55·4	1·8
30	30	27·6	57·7	1·7
31	31	27·8	60·1	1·7
32	32	28·0	62·5	1·6
33	33	28·1	64·9	1·5
34	34	28·3	67·5	1·5
35	35	28·5	70·0	1·4
36	36	28·6	72·7	1·4
37	37	28·8	75·4	1·3
38	38	28·9	78·1	1·3
39	39	29·1	81·0	1·2
40	40	29·2	83·9	1·2
41	41	29·4	86·9	1·2
42	42	29·5	90·0	1·1
43	43	29·7	93·3	1·1
44	44	29·8	96·6	1·0
45	45	30·0	100·0	1·0
50	50	30·8	119·2	0·84
60	60	32·4	173·2	0·58
70	70	34·4	274·8	0·36
80	80	37·5	567·1	0·18
90	90	60·0*	—	0·00

* Values conventionally assigned.

Table 9—*continued*

B. Based on altan

Altan	Degrees and minutes	Degrees and decimals of a degree	Percentage grade	Gradient, or unit rise. 1 in:
0	0°00′–0°03′	0.00–0.06	0.0–0.1	∞–1146
1	0°04′	0.07	0.1	859
2	0°05′	0.09	0.2	688
3	0°07′	0.12	0.2	491
4	0°09′	0.15	0.3	382
5	0°11′	0.18	0.3	313
6	0°14′	0.23	0.4	246
7	0°17′	0.29	0.5	202
8	0°22′	0.37	0.6	156
9	0°27′	0.45	0.8	127
10	0°34′	0.57	1.0	101
11	0°43′	0.72	1.3	79.9
12	0°54′	0.90	1.6	63.7
13	1°09′	1.15	2.0	49.8
14	1°26′	1.43	2.5	40.0
15	1°49′	1.82	3.2	31.5
16	2°17′	2.28	4.0	25.1
17	2°52′	2.87	5.0	20.0
18	3°37′	3.62	6.3	15.8
19	4°33′	4.55	8.0	12.6
20	5°43′	5.72	10.0	10.0
21	7°11′	7.18	12.6	7.9
22	9°01′	9.02	15.9	6.3
23	11°18′	11.30	20.0	5.0
24	14°07′	14.12	25.1	4.0
25	17°33′	17.55	31.6	3.2
26	21°43′	21.72	39.8	2.5
27	26°38′	26.63	50.1	2.0
28	32°16′	32.27	63.1	1.6
29	38°28′	38.47	79.4	1.3
30	45°00′	45.00	100.0	1.0

Table 9—*continued*

C. Based on percentage grade

Percentage grade	Degrees and minutes	Degrees and decimals of a degree	Altan	Gradient, or unit rise. 1 in:
0·0	0°00′	0·00	0·00	∞
0·1	0°03′	0·05	0·0	1146
0·2	0°07′	0·12	3·1	491
0·3	0°10′	0·17	4·6	344
0·4	0°14′	0·23	6·1	246
0·5	0°17′	0·28	6·9	202
0·6	0°21′	0·35	7·9	164
0·7	0°24′	0·40	8·4	143
0·8	0°28′	0·47	9·1	123
0·9	0°31′	0·52	9·6	111
1	0°35′	0·58	10·8	98·2
2	1°09′	1·15	13·0	49·8
3	1°43′	1·72	14·8	33·4
4	2°18′	2·30	16·0	24·9
5	2°52′	2·87	17·0	20·0
6	3°26′	3·43	17·8	16·7
7	4°00′	4·00	18·4	14·3
8	4°35′	4·58	19·0	12·5
9	5°09′	5·15	19·5	11·1
10	5°43′	5·72	20·0	10·0
11	6°17′	6·28	20·4	9·1
12	6°51′	6·85	20·8	8·3
13	7°24′	7·40	21·1	7·7
14	7°58′	7·97	21·5	7·1
15	8°32′	8·53	21·8	6·7
16	9°06′	9·10	22·0	6·2
17	9°39′	9·65	22·3	5·9
18	10°12′	10·20	22·6	5·6
19	10°46′	10·77	22·8	5·3
20	11°19′	11·32	23·0	5·0
25	14°02′	14·03	24·0	4·0
30	16°42′	16·70	24·8	3·3
35	19°18′	19·30	25·4	2·9
40	21°48′	21·80	26·0	2·5
45	24°14′	24·23	26·5	2·2
50	26°34′	26·57	27·0	2·0

Table 9C.—*continued*

Percentage grade	Degrees and minutes	Degrees and decimals of a degree	Altan	Gradient, or unit rise. 1 in:
55	28°49′	28·82	27·4	1·8
60	30°58′	30·97	27·8	1·7
65	33°02′	33·03	28·1	1·5
70	35°00′	35·00	28·5	1·4
75	36°52′	36·87	28·8	1·3
80	38°40′	38·67	29·0	1·2
85	40°22′	40·37	29·3	1·2
90	42°00′	42·00	29·5	1·1
95	43°32′	43·53	29·8	1·1
100	45°00′	45·00	30·0	1·0

Table 9—*continued*

D. Based on gradient

Gradient, or unit rise. 1 in:	Degrees and minutes	Degrees and decimals of a degree	Altan	Percentage grade
1000	0°03′	0·05	0·0	0·1
900	0°04′	0·07	0·7	0·1
800	0°04′	0·07	0·7	0·1
700	0°05′	0·09	1·6	0·1
600	0°06′	0·10	2·4	0·2
500	0°07′	0·12	3·1	0·2
400	0°09′	0·15	4·2	0·3
300	0°11′	0·18	5·1	0·3
200	0°17′	0·29	6·9	0·5
100	0°34′	0·57	10·0	1·0
90	0°38′	0·63	10·4	1·1
80	0°43′	0·72	11·0	1·3
70	0°49′	0·82	11·5	1·4
60	0°57′	0·95	12·2	1·7
50	1°09′	1·15	13·0	2·0
40	1°26′	1·43	14·0	2·5
30	1°55′	1·92	15·2	3·3
20	2°52′	2·87	17·0	5·0

Table 9D.—*continued*

Gradient, or unit rise. 1 in:	Degrees and minutes	Degrees and decimals of a degree	Altan	Percentage grade
10	5°43′	5·72	20·0	10·0
9	6°20′	6·33	20·5	11·1
8	7°07′	7·12	21·0	12·5
7	8°08′	8·13	21·6	14·3
6	9°28′	9·47	22·2	16·7
5	11°19′	11·32	23·0	20·0
4	14°02′	14·03	24·0	25·0
3	18°26′	18·43	25·2	33·3
2	26°34′	26·57	27·0	50·0
1	45°00′	45·00	30·0	100·0

measured lengths *a* and *b* are set out respectively upslope and downslope from the point; their angles, θ_a and θ_b, and ground surface distances, D_a and D_b, are recorded. The profile curvature C_{ab} is given by

$$C_{ab} = \frac{\theta_a - \theta_b}{0{\cdot}5(D_a + D_b)} \times 100 = 200 . \frac{\theta_a - \theta_b}{D_a + D_b} \ {}^\circ/100 \text{ m} \qquad (12.1)$$

Thus if successive measured lengths, in a downslope direction, are 10 m at $-3°$ and 10 m at $-4{\cdot}5°$, the curvature at the point between them is $200(1{\cdot}5/20) = +15°/100$ m. On a slope profile an estimate of the curvature of the measured length itself may be required. This can be taken as the mean of the curvature of the two points bounding it. Alternatively, if *p*, *q* and *r* are successive measured lengths (in a downslope direction), the curvature of *q* is given by

$$C_q = 100 . \frac{\theta_p - \theta_r}{0{\cdot}5D_p + D_q + 0{\cdot}5D_r} \ {}^\circ/100 \text{ m} \qquad (12.2)$$

Profile curvature may alternatively be specified by radius of curvature, stating whether this is for a convex or concave slope. The vertical radius of curvature, R_v, in metres, is given by $R = 5730/C$, where C is curvature in degrees per 100 m.

SLOPE PROFILES

The earliest surveys of slope profiles for geomorphological purposes were by Tylor (1875) and Lake (1928). Their potentialities for the study of slope evolution were first fully demonstrated by Fair (1947, 1948, 1948b) and Savigear (1952); Strahler (1950) was the pioneer in the use of profiles in conjunction with statistical analysis. By 1956 profiles had become established as a standard descriptive technique in geomorphology.

Discussions of the methodology of profile survey and analysis are given by Savigear (1956, 1967), Young (1964, 1971) and Pitty (1966, 1967, 1968b, 1969). The main field studies employing the technique of profiling are by Fair (1947, 1948, 1948b), Savigear (1952, 1960, 1962), Pallister (1956, 1956b), Palmer (1956), de Béthune and Mammerickx (1960, summarizing previous work in Belgium), Fourneau (1960), Hack and Goodlett (1960), Young, (1963, 1970b), Mammerickx (1964), Clark (1965), Pissart (1966) and Lewin (1970).

PROFILE SURVEY

A *slope profile* is a line across the ground surface largely or entirely following the direction of true slope. It is normally surveyed by a series of measurements of angle and distance. The points between measurements taken are *profile stations* and the interval between stations is a *measured length.*

A *transect* is a line across the ground surface, often straight, that does not follow the true slope. Angle measurements taken along a transect show apparent slope. Whilst not normally employed in geomorphology, it is sometimes necessary to survey transects, for example in conjunction with work in plant ecology. In this case, additional measurements of the direction and amount of true slope should be taken at regular intervals along the transect, using a standard measured length, e.g. 5 m.

Most slope profiles extend from drainage divide to talweg. In field survey, the line of the profile should be extended beyond the divide and talweg to points where either a slope in the opposite direction to that surveyed has become clearly established (more than 2°), or a non-slope landform, such as a flood-plain, commences. In surveying a scarp or isolated hill it may be possible for the profile to follow the direction of true slope throughout. With profiles across a valley side, a difficulty arises close to the crest of a convex interfluve; a point is necessarily reached where the angle measured along the crest exceeds that measured perpendicularly to its axis. The same problem can arise close to the centre of a concave valley floor. Adherence to the rule of following the direction of true slope would cause the profile to turn at right angles, thence asymptotically approaching the crest or talweg. Such a profile ignores the geomorphological reality of the valley side. Pitty (1966) suggested that the profile should be terminated at the point where its angle equalled that measured longitudinally down the interfluve or talweg. This would confine the profile entirely to true slope, but gives an incomplete record of the valley side. In cases where the longitudinal slope is gentle a better solution is that at, or somewhat before, the above-defined point, the profile line should be continued perpendicularly to the line of the crest or talweg. These portions which record apparent slope can be indicated on the record of results, permitting them to be excluded from the analyses where appropriate.

In deciding the location of profiles, most studies have used purposive sampling, i.e. deliberate selection of the line of profile. Profiles are usually confined to slopes that are straight in plan, thus excluding valley-head and spur-end slopes. Purposive sampling has been criticized as subjective, and therefore open to the selection, conscious or otherwise, of profiles that will demonstrate some preconceived hypothesis. This depends on the use to be made of the results. For descriptive purposes, the selection of representative lines is frequently justifiable. Controlled sampling is desirable if the profiles are analyzed statistically as a sample drawn from the total ground surface of the area. Practical difficulties of access may render random sampling impossible; even in these circumstances, the careful selection of lines judged to be representative of the landforms of the area can be regarded as a simple form of stratified sampling, and does not necessarily preclude statistical treatment of the results.

If a random sample of profiles is required, some adaptation of the standard methods of point sampling is necessary, consequent upon the requirement that a profile follows the line of true slope. A possible method is to take a point sample (random or systematic) and extend profiles up and down the true slope from each point. However, lines of true slope diverge downslope on areas convex in plan, and converge downslope on those concave in plan. Consequently, a system based on the extension of profiles from points leads to under-representation of land that is convex in plan and lies on the lower parts of the area, and over-representation of low-lying slopes that are concave in plan. Close to the interfluves the reverse is true. In addition, the use of points as a basis produces a poorly stratified sample, sometimes with profile lines crossing. Bias caused by plan curvature also affects sampling systems which use either talwegs or drainage divides as baselines.

The following scheme for stratified sampling of profiles minimizes, although it does not eliminate, this source of sampling bias (Fig. 53). Using maps or air photographs, all talwegs and watersheds are marked. For each drainage basin, mid-slope lines are

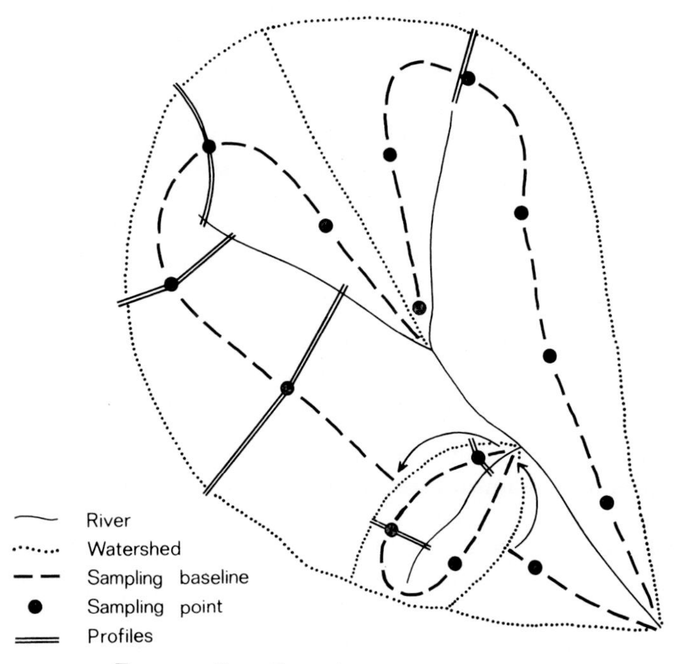

FIG. 53. Sampling scheme for slope profiles.

drawn, intermediate between the talweg and watershed. The mid-slope lines are arranged sequentially, using any convenient order, and this sequence is treated as one continuous line, the *profile sampling line.* Sampling points are marked at equal intervals along the profile sampling line. Profiles are extended upslope and downslope from the sampling points, terminating at talwegs, watersheds or non-slope areas. The procedure as described gives a non-stratified systematic sample. It may be stratified by taking one sampling point within each drainage basin; or by delimiting component slopes (p. 4), and either sampling one profile within each component slope, or confining sampling points to portions of the profile sampling line which lie on valley-side slopes. If such stratified procedures are used, adjustments are necessary to restore areal proportionality to the total sample.

In selecting the positions of profile stations during field survey, two different practices have been followed. In early work, stations were sited where the slope was judged visually to change, giving unequal measured lengths. The use of a standard unit, with consequently no choice in the siting of stations, has since been advocated. One criticism of the first method is that only those changes in angle that are seen in the field will be selected as profile stations; others that are equal in amount but are not seen may fall between stations, causing inequality in the record. The main disadvantage in using a unit length arises from the nature of slopes; most include a long section in which changes in angle are small and gradual, together with shorter sections of rapidly-changing angle. If a long unit length is used, breaks of slope in the latter sections, visibly apparent to the surveyor, are recorded with deliberate inaccuracy. If the unit length is made short enough to record such breaks, the time needed to survey the smooth sections is greatly increased. Uniform lengths are more convenient in certain types of analysis. The choice of the best method depends on the aims and circumstances of the survey. If methodological investigations into the nature of slope profile form are intended, then a standard length is preferable. If what is required is a record of the shape of a landform, for interpretation in geomorphological terms, the subjective choice of profile stations gives a more accurate record and is quicker.

The following working rules give a profile that records slope in greater detail where it becomes more irregular, is rapid to survey, and is subject to sufficient control of length to permit quantitative analysis:

(*i*) No measured length shall be more than 20 m or less than 2 m.
(*ii*) The difference between adjacent measured lengths shall not exceed 2° on slopes of less than 20°, or 4° on slopes exceeding 20°, unless both lengths are 2 m.

The suggested values of 20 m, 2 m and 2° are for topography of normal scale, and can be varied with unusually high or low drainage density, or with very gentle slopes. The field procedure may be illustrated by an example. Let P, Q and R be successive stations. Measured length PQ is measured as 10 m at 6°, and QR as 8 m at 9°. Since the difference exceeds 2°, R must be moved back until QR is 8° or less. If QR still exceeds 8° when reduced to 2 m, both previous measurements must be abandoned and Q moved back, re-locating P by backward measurement; new values are then obtained for the shortened PQ. With experience the need to move Q should rarely arise. The use of the profile recorder permits the selection and comparison of a range of alternative measured lengths subsequent to field survey.

The following supplementary measurements should be recorded:

(*i*) The longitudinal slope of the interfluve crest, or other crest of the profile.
(*ii*) The longitudinal slope of the talweg, or other base of the profile.
(*iii*) The plan curvature (p. 176) at the steepest point on the profile.
(*iv*) The aspect of the slope at the steepest point on the profile, and at other points if the direction changes substantially.

To plot the profile graphically, the angle and distance measurements are converted into rectangular co-ordinates and summed from one end, as illustrated in Table 10.

Table 10. Conversion of angle and distance measurements into rectangular co-ordinates

							Co-ordinates	
Measured length	D, distance, metres	θ Angle, degrees	cos θ	sin θ	Horizontal difference $= D.\cos\theta$	Vertical difference $= D.\sin\theta$	Horizontal metres	Vertical metres
1	10	−2·5	·999	·044	10·0	−0·4	10·0	−0·4
2	10	−2·0	·999	·035	10·0	−0·4	20·0	−0·8
3	10	−0·5	1·000	·009	10·0	−0·1	30·0	−0·9
4	10	0·0	1·000	·000	10·0	0·0	40·0	−0·9
5	10	+1·0	1·000	·018	10·0	+0·2	50·0	−0·7
6	10	+2·0	·999	·035	10·0	+0·4	60·0	−0·3
7	10	+4·0	·998	·070	10·0	+0·7	70·0	+0·4
8	20	+4·0	·998	·070	20·0	+1·4	90·0	+1·8
9	20	+6·0	·995	·105	19·9	+2·1	109·9	+3·9
10	20	+8·0	·990	·139	19·8	+2·8	139·7	+6·7
31	10	+24·0	·914	·407	9·1	+4·1	420·1	+80·2
32	10	+26·0	·899	·438	9·0	+4·4	429·1	+84·6
33	10	+27·5	·887	·462	8·9	+4·6	438·0	+89·2
34	10	+29·5	·870	·492	8·7	+4·9	446·7	+94·1
35	10	+30·0	·866	·500	8·7	+5·0	455·5	+99·1

The profile is plotted without vertical exaggeration in the first instance. The scale selected will depend on the measured lengths employed and the total length of the profile; under normal conditions 1:500 is suitable for the initial plot, for inspection and graphical analysis. The points are joined by straight lines, and the angle written above each length.

Accurate, closely-contoured maps would in theory provide the necessary information from which to construct slope profiles, but comparisons of profiles drawn from maps with those surveyed in the field have repeatedly shown that the former are very inaccurate. Contours can be used to obtain the mean angle of a slope more than 100 m in height, but are unsatisfactory for obtaining the local slope angle at a point. In the case of the maximum steepness of a slope, the estimate from contours tends to be substantially too low but by a variable amount, so that the value obtained by applying a correction factor is subject to a large error.

Slope angle can be obtained from air photographs by a modification of the method for measuring height differences. The parallax difference between two points on a stereoscopic pair of photographs is measured by stereometer, and slope angle obtained from the formula:

$$\tan \theta = \frac{fp}{CW} \qquad (12.3)$$

where f = focal length of camera lens, in mm, p = difference in parallax between the two points, in mm, C = mean horizontal distance between the two points, as measured on the photographs, in mm, and W = mean of the distances between the principal point and the transferred principal point, in mm.

This formula is only approximate (Allum, 1966). More refined methods for slope measurement are described by Meckel *et al.* (1964). Heighting from air photographs can, with refined instruments and substantial time, be carried out to the nearest 1 m; but even with this accuracy, profiles using measured lengths of 20 m would be subject to an error in angle of 3°. Profiles that are to be used to study the detailed form of slopes should therefore always be obtained by field survey.

PROFILE ANALYSIS

Profile analysis is the division of a profile into a number of parts, each of which possesses certain properties of form. The definition of the parts must be based entirely on the form of the profile, and not be in any way dependent on process or evolution. The basis of such a system was established by Savigear (1952, 1956), who noticed that on certain surveyed profiles there were parts in which the angle remained approximately constant; these he termed rectilinear segments. Smoothly curved portions of the profile were termed convex or concave elements. This system has been extended and quantified by Young (1964, 1971). The division into rectilinear, convex, and concave portions has remained the principal basis of the analysis of profile form.

The terminology and definitions given here follow those of Young (1964, 1971). Savigear's original terms segment and element are retained in preference to the more logical but less convenient 'rectilinear profile unit' and 'convex or concave profile unit'. The previously suggested term 'phase of development' should not be used in profile analysis, as it carries genetic implications. The terms are defined in Fig. 54.

These definitions of segment and element are qualitative. Consequently, there is a measure of subjectivity involved in the analysis. As an example, consider a portion of a profile with the succession of 10 m measured lengths 7°, 6°, 7°, 6°, 7°, 8°, 7°, 8°; this may be interpreted either as one 80 m segment at 7° or as two 40 m segments at $6\frac{1}{2}$° and $7\frac{1}{2}$°, and observers with different predilections will make different decisions in

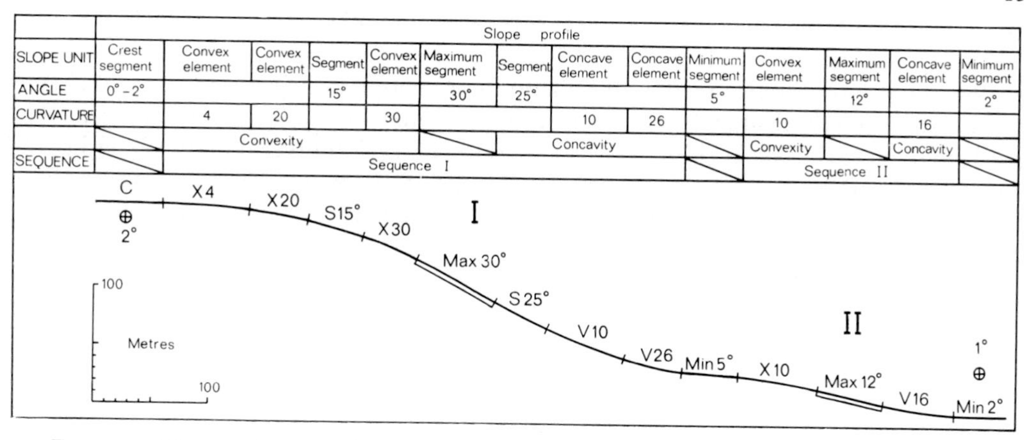

FIG. 54. The terminology of slope profile analysis. Based on Young (1964), modified. Values of curvature are given in degrees per 100 m.

Slope unit — a segment or an element.
Segment — a portion of a slope profile on which the angle remains approximately constant.
Element — a portion of a slope profile on which the curvature remains approximately constant.
Convex element — an element with a downslope increase in angle (i.e. with positive curvature).
Concave element — an element with a downslope decrease in angle (i.e. with negative curvature).
Maximum segment — a segment which is steeper in angle than the slope units above and below it; it may also form the lowest unit on a profile, having a gentler unit above it.
Minimum segment — a segment which is gentler in angle than the slope units above and below it.
Crest segment — a segment bounded by downward slopes in opposite directions.
Basal segment — a segment bounded by upward slopes in opposite directions.
Irregular unit — a portion of a slope profile within which there are frequent changes of both angle and curvature.
Convexity — all parts of a slope profile on which there is no decrease in angle downslope, but excluding maximum, minimum, and crest segments.
Concavity — all parts of a slope profile on which there is no increase in angle downslope, but excluding maximum, minimum, and crest segments.
Profile sequence — a portion of a slope profile consisting successively of a convexity, a maximum segment, and a concavity.

such a case. Where the question is to decide whether a convexity is a smoothly-curved element or a series of rectilinear segments, as in a succession such as $-16°$, $-16°$, $-17°$, $-17\frac{1}{2}°$, $-19°$, $-19\frac{1}{2}°$, $-20\frac{1}{2}°$, $-20\frac{1}{2}°$, the opportunity for differences of interpretation are considerable. Where subjectivity is present in the analysis of field data, the results obtained by different observers, and even by the same observers working in different areas, are not quantitatively comparable.

To achieve objectivity in the identification of a slope unit it is necessary to specify the maximum permissible level of variability of angle or curvature within it. The coefficient of variation is suitable as a measure of variability. Let the lengths and angles of the measured lengths be denoted by l_i and θ_i respectively. Using standard statistical formulae, the mean angle, $\bar{\theta}$, and the coefficient of variation of angle, V_a, are given by:

$$\theta = \frac{\Sigma l\theta}{\Sigma l} \text{ degrees} \tag{12.4}$$

$$V_a = 100 \times \frac{\sqrt{\dfrac{\Sigma l\theta^2}{\Sigma l} - \bar{\theta}^2}}{\bar{\theta}} \text{ per cent} \tag{12.5}$$

The coefficient of variation is used in preference to the standard deviation because the significance of a variation of one degree is greater on a gentle than a steep segment (p. 172). Thus the succession 32°, 34°, 30°, 32° is acceptable as a 32° segment, whereas 2°, 4°, 0°, 2° would not be classed as a segment; the coefficients of variation are 4% and 71% respectively. As the mean approaches zero the coefficient of variation becomes infinite, therefore where $\bar{\theta}$ is less than 2° it is replaced by 2 in the denominator of equation *12.5*.

The mean curvature, $\bar{C}$, and its coefficient of variation, V_c, are obtained by substituting curvature for angle in equations *12.4* and *12.5*. For example, consider successive 10 m measured lengths at −3°, −4°, −5°, −6½°, −7°, −8°, −9°; the curvatures of the middle five lengths are +10, +12½, +10, +7½, +10°/100 m, giving a convex element with $\bar{C}$ = 10°/100 m and V_c = 15·8%.

The meaning of 'approximately' in the above definitions may therefore be stated quantitatively. A segment is a portion of a slope profile within which the coefficient of variation of angle does not exceed a specified value, V_{amax}; it is described by its length, L, $\bar{\theta}$, and V_a. An element is a portion of a slope profile within which the coefficient of variation of angle does not exceed a specified value, V_{cmax}; it is described by L, $\bar{C}$, V_c and the angles of its terminal measured lengths. Thus Profile *H7* in Fig. 56 (p. 154) includes a segment of 92 m at 11·6° (V_a = 9%), and a 0°-8° convex element of 241 m at 0·9°/100 m (V_c = 24%).

The values selected for V_{amax} and V_{cmax} depend on the purpose for which the profile is being analyzed. If the same value is selected for both, most profiles fall almost entirely into segments; this is possibly a consequence of the fact that angle is measured directly and curvature indirectly. The following values are suggested:

(*i*) For general purposes of profile analysis: V_{amax} = 10%, V_{cmax} = 25%. Slope units identified on this basis have a good standard of internal uniformity, but the profile is not fragmented into an excessive number of short units.

(*ii*) To identify selected portions of the profile having a high standard of internal homogeneity: $V_{amax} = 5\%$, $V_{cmax} = 10\%$. This is unsuitable if the entire profile is to be apportioned into units.

(*iii*) To divide the profile into relatively few units, giving a schematic impression of the landscape: $V_{amax} = 25\%$, $V_{cmax} = 50\%$. The individual units identified by these values do not have a subjectively acceptable standard of homogeneity.

The simplest method for the identification of slope units is by graphical inspection. The profile is plotted, and tested by laying a transparent rule along successive parts; those parts in which the profile stations lie nearly along a straight line are marked as segments. The intervening parts are inspected for portions where the curvature remains relatively constant, provisionally identified as elements. Alternatively, the profile is plotted as a graph of angle (ordinate) against distance (Fig. 55), and provisional segments selected where the plotted line is nearly horizontal; elements are similarly

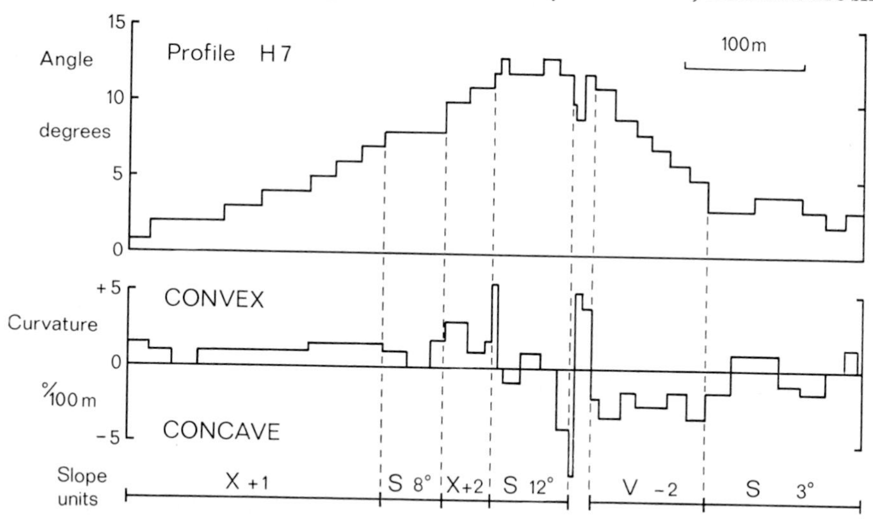

FIG. 55. Profile graphs of angle and curvature. The profile is shown in Fig. 56. Division into slope units is by graphical inspection.

identified by graphing curvature against distance. For the units provisionally identified in this way, the coefficients of variation are calculated, and where these coefficients depart substantially from the values of V_{amax} and V_{cmax} adopted, the units are lengthened, shortened or rejected as appropriate. This method of graphical analysis does not, without excessive calculation, identify the best possible slope units, but gives results of sufficient precision for many purposes.

If an entire profile is to be divided into slope units in an optimal manner, computer calculation is necessary. Each measured length on a profile belongs not to one but to several overlapping slope units; thus in the succession 10°, 10°, 10°, 9°, 8°, 7°, 7°, 7°, the length at 8° falls into two alternative segments ($V_{amax} = 10\%$) as well as into an element which overlaps both. This problem can be overcome by allocating each measured length to the longest slope unit into which it falls. This procedure is termed the *system of best units* (Young, 1971). *Best units analysis* is defined as the division of a slope profile into segments and elements such that the coefficients of variation, of angle or curvature respectively, do not exceed specified values. Where two or more segments overlap, the overlapping portion is allocated to the longest unit. *Best segments analysis* and *best elements analysis* are the division in a similar manner into segments only and elements only respectively. A computer program has been written for these analyses (Young, 1971). Fig. 56 shows the effects of varying the values of V_{amax} and V_{cmax} for two profiles of contrasting form. The units obtained using $V_{amax} = 10\%$ and $V_{cmax} = 25\%$ correspond fairly well to the results likely to be obtained by subjective analysis.

One of the critical questions concerning slope form is whether a slope is smoothly curved or consists of rectilinear portions separated by breaks of slope. By defining slope units as having a specified maximum value of variability, it is shown that there is no single answer to this question. Consider a convexity with the succession of 10 m measured lengths: $-16°$, $-16°$, $-17°$, $-17\frac{1}{2}°$, $-19°$, $-19\frac{1}{2}°$, $-20\frac{1}{2}°$, $-20\frac{1}{2}°$. There are segments of 40 m at 16·6° and 19·9°, with $V_a = 8\%$ and 7% respectively. The curvatures of the middle six lengths are 10, 15, 20, 20, 15, 10°/100 m, giving an element with $\bar{C} = 15°/100$ m and $V_c = 27\%$. Thus at $V_{amax} = 10\%$ and $V_{cmax} = 25\%$, this part of the profile would be classified as two segments; by raising the value of V_{cmax} it falls into one convex element. Which of these solutions is significant in relation to slope evolution involves considerations beyond those of the analysis of profile form.

The system of best units analysis can be applied irrespective of the lengths over which the angles are recorded. The results of the analysis, however, are substantially affected by the lengths used. It is therefore essential that, in giving the results of profile analysis, the working rules for measured lengths adopted in field survey should be stated.

A different system of analysis was proposed by Savigear in 1967. Adjacent measured lengths with the same angle are combined as intercepts, and the angle differences between adjacent intercepts are termed *profile discontinuities*. Using the intercepts and discontinuities, the profile is divided into *neutral*, *positive* and *negative* units (corresponding respectively to segments, convex elements and concave elements on the earlier system), separated by *neutral*, *positive* and *negative discontinuities*. Thus what on the earlier system would be described as successive segments of 10° (upper) and 15° followed by a concave element commencing at 12° would on the 1967 system be termed a three-unit profile of the type 'neutral unit-positive discontinuity-neutral unit-negative discontinuity-negative unit'. The succession of units and discontinuities are then classified according to one of two systems, profile unit relationships or profile unit groups.

Apart from the terminology, the units are similar on the two systems; the main difference is that in the 1967 system angular discontinuities between adjacent units, or *breaks of slope*, are given equal importance to that of the units. Criticisms of the 1967 system are that the higher orders of classification are excessively complex for practical use, and that the exact procedure for the determination of units and discontinuities is not specified. The emphasis given to breaks of slope deserves further investigation; it is essential, however, that the initial stages of analysis should not contain any tendency for preferential identification of abrupt, as compared with gradual, changes in angle.

If the term *break of slope* is to be employed in profile analysis it must be defined quantitatively. Suitable parameters are the curvature or, alternatively, the maximum ground surface distances over which a change in angle takes place. For slopes of average dimensions a minimum curvature (C_{min}) of 100°/100 m is suitable; this has the advantage of being identifiable using the '2 m-2°' rule for profile survey (p. 146). However, a curvature that would subjectively be regarded as an abrupt break of slope amid chalk or granite topography would appear to be a smoothly-curved convexity in badland relief, and hence the value of C_{min} used to define a break of slope should be stated. In order to identify breaks of slope by profile analysis, obtain the curvatures of each measured length, select all those that exceed C_{min}, and combine any two or more such curvatures. In Fig. 56, using C_{min} = 100°/100 m, profile *H5* does not contain any breaks of slope, but profile *MT41* includes convex and concave breaks.

CURVE FITTING

An alternative method for the analysis of convexities or concavities is to obtain from the rectangular co-ordinates best-fitting polynomial curves; the deviation of the surveyed points from the curves can then be calculated. This method has been used by Savigear (1956; 1962, footnote 12), Hack and Goodlett (1960), Lewin (1969) and Young (1970b). Curves can also be fitted to entire profiles, but the value of doing so is doubtful. Curve fitting, for which standard computer programs are available, is fully objective and potentially useful. It is desirable, however, to limit the curves to quadratic and cubic forms, otherwise curve-fitting becomes a mathematical exercise with no interpretative value. Among possible uses are the indication of whether the curvature of an element remains constant, or changes with distance from the crest. It is possible that a type of slope unit additional to segments and elements might be identified, having a relatively constant rate of change of curvature; some pediments may be of this form. Such units can also be identified by calculating the rate of change of curvature, C', for each measured length, obtained from the formula:

$$C'_q = \frac{C_{pq} - C_{qr}}{l_q} \times 100 \quad {}^{\circ}/100\ \text{m}/100\ \text{m} \qquad (12.6)$$

The values of C' are graphed against distance, or employed in best units analysis in the same way as those of angle or curvature.

Curve fitting is particularly applicable to the question of whether convexities and concavities are smoothly curved or contain rectilinear segments. The method is to take the segments indicated by best units analysis, fit second-order polynomial curves ($y = a+bx$) to them, obtain the residual sum of squares and the number of degrees of freedom for each segment, and sum these. A single third-order polynomial ($y = a+bx+cx^2$) is then fitted, and the residual sum of squares and degrees of freedom obtained. The variance ratios obtained from these two sources are compared, and the

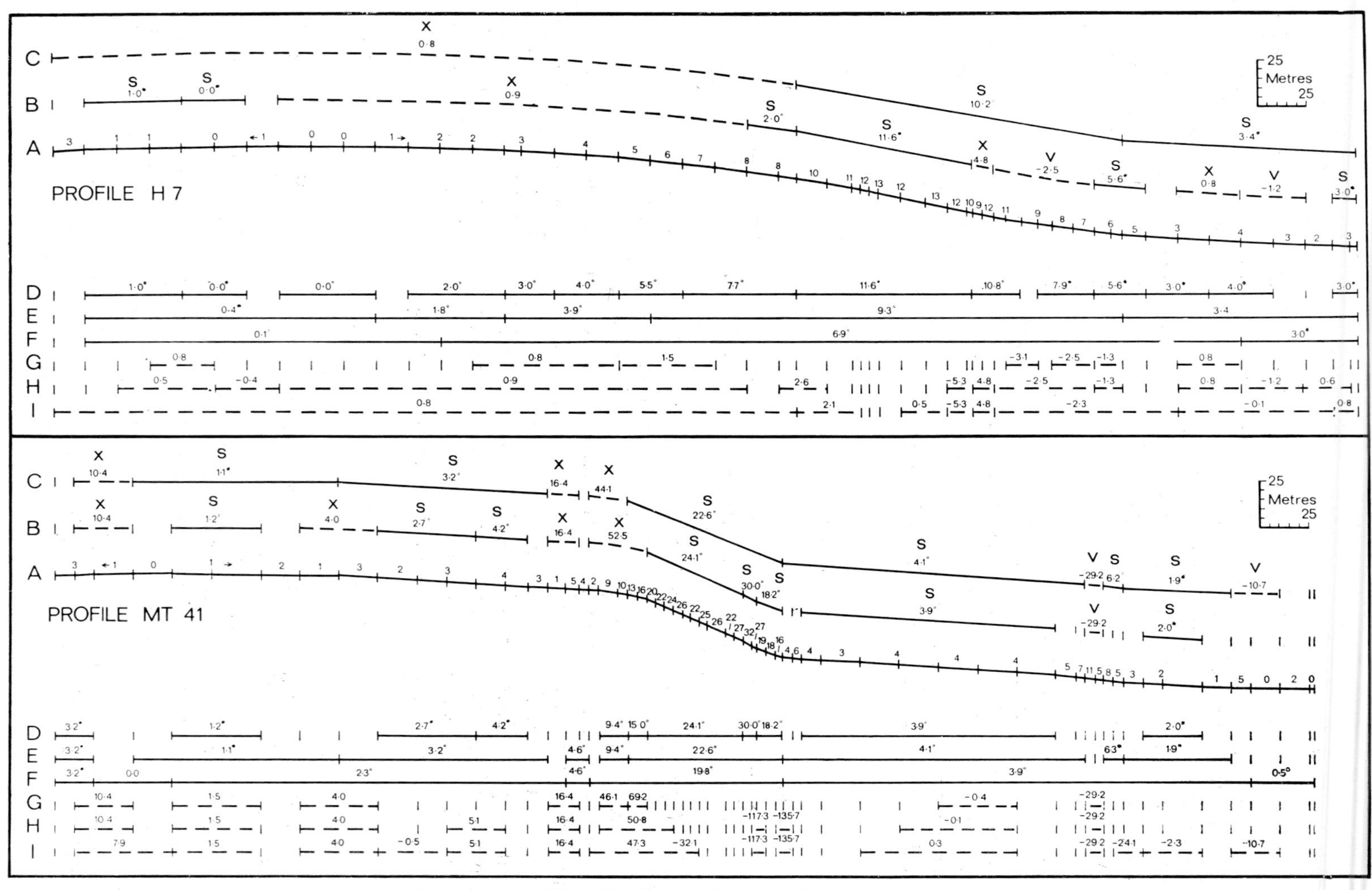

Fig. 56. See caption opposite.

significance of the difference between them may be examined by means of an F-test, although there are reservations concerning the appropriateness of this test for data with some measure of spatial auto-correlation. An example is shown in Table 11. The

Table 11. Comparison of the fit of four 2nd-order polynomial curves with one 3rd-order polynomial curve to a convexity. After Young (1970b)

Measured lengths, nos.	Mean angle of segment, degrees	Order of polynomial	Residual sum of squares	Degrees of freedom	Variance ratio	F	p
4–23	1·1	2	·536	18	·011		
23–30	1·7	2	·010	6	·002		
30–40	2·4	2	·083	9	·009		
40–47	4·7	2	·198	6	·033		
4–47		(Sum)	·828	39	·021	20·5	< ·01
4–47		3	16·951	41	·413		

results in this case show that the fit of four segments is better, by a factor of about 20, than that of one convex element, indicating clearly that the convexity is segmented. This result was found to be generally true for convexities on sandstones under a moist savanna climate in part of the Mato Grosso, Brazil (Young, 1970b). For seaward slopes in south-west England, Savigear (1962) found differences between the fit of segments and of smooth curves to be non-significant.

FORM IN PROFILE

Most slopes appear on casual inspection to be smoothly curved. Rectilinear forms strike the eye only in the special cases of cliffs and screes. Early writers, from Gilbert White in the eighteenth century, described slopes as rounded, and noted the widespread occurrence of two curved components, an upper convexity and a lower concavity. It therefore came as a challenge to established ideas when, in the 1950s, the existence of rectilinear segments on regolith-covered slopes was demonstrated.

FIG. 56. Two slope profiles with analyses into best segments, elements and units, illustrating the effects of varying the maximum coefficients of variation of angle, V_{amax}, and of curvature, V_{cmax}. A = measured lengths, with angles to the nearest degree; B and C = best units; D, E and F = best segments; G, H and I = best elements; S = segment (angle in degrees); X = convex element; V = concave element (curvatures in degrees per 100 m); segments consisting of one measured length are left blank. Maximum coefficients of variation:

Curve:	*B*	*C*	*D*	*E*	*F*	*G*	*H*	*I*
V_{amax}, per cent:	10	25	10	25	50	—	—	—
V_{cmax}, per cent:	25	50	—	—	—	10	25	50

Fig. 57. An usually long rectilinear slope, in arid mountain relief north of Kabul, Afghanistan.

Fig. 58. A convex crest over a rectilinear slope, steepening to vertical cliffs near the base. Limestone hill in rainforest climate, Malaya.

On a perfect convex-concave slope there is a point of inflection at the junction of the two elements, and, using profile survey of measured lengths, it is not unlikely that, due to micro-relief, an apparent short maximum segment would be recorded on such a slope. The lengths of many recorded maximum segments are such, however, that they cannot be caused by this source of observational error. On various rock types, in humid climates, slopes with a rectilinear maximum segment occupying at least 10% of the total ground length of the profile are substantially more common than slopes on which the maximum segment is short or absent. This applies even in the classic instance of supposedly rounded slopes, the chalk, both in England (Clark, 1965) and Belgium (Fourneau, 1960). Very long maximum segments, occupying 30-50% of the slope, are particularly common where their angle is steep, 30°-35°, but can also occur at all angles down to 2° or less. Richter denudation slopes (p. 107) are a special case of long segments of *c.* 35°.

Where the maximum segment is steep, over 28°, adjacent segments only a few degrees less steep sometimes occur, either above or below the maximum segment (Young, 1963; Clark, 1965). The possible origin of this feature is discussed below (p. 171).

It has not been determined whether most convexities and concavities are smoothly curved or consist of series of segments with relatively angular intersections. Examples of both may be cited. The following succession of 26 measured lengths, each of 20 m, was surveyed on a granite slope on the Nyika Plateau, Malawi: (cf. Fig. 2, p. 2): 0°, $-\frac{1}{2}$°, -1°, -2°, $-2\frac{1}{2}$°, $-3\frac{1}{2}$°, -4°, $-4\frac{1}{2}$°, -5°, -5°, -6°, $-6\frac{1}{2}$°, -7°, $-7\frac{1}{2}$°, $-7\frac{1}{2}$°, $-7\frac{1}{2}$°, -8°, $-8\frac{1}{2}$°, $-9\frac{1}{2}$°, $-9\frac{1}{2}$°, $-9\frac{1}{2}$°, -10°, -10°, $-10\frac{1}{2}$°, $-11\frac{1}{2}$°, $-12\frac{1}{2}$°. If equal increments of $\frac{1}{2}$° are interpolated between 0° and $12\frac{1}{2}$° the difference between the theoretical smooth curve and the surveyed profile nowhere exceeds 1° (non-significant at nearly 50% on a χ^2-test). In contrast, Savigear (1962) gives the following succession of 6 m lengths on an apparent convexity: -17°, $-16\frac{1}{2}$°, -17°, $-20\frac{1}{2}$°, -21°, -20°, $-18\frac{1}{2}$°, -20°, -24°, $-26\frac{1}{2}$°, $-26\frac{1}{2}$°, -28°, $-24\frac{1}{2}$°, $-27\frac{1}{2}$°, $-26\frac{1}{2}$°, $-25\frac{1}{2}$°, -27°; this falls clearly into segments of 17°, 20°, and 26°. Savigear has claimed that on seaward slopes surveyed in south-west England most of the convexities are formed by successive segments of increasing angle, although a statistical comparison between the fit with polynomial curves and with subjectively-determined segments gave inconclusive results. For valley slopes of Famenne, Belgium, Seret (1963) states that truly convex or concave parts of slopes rarely exceed 5-7 m in length, and that apparently curved parts 'are generally formed by a series of rectilinear portions'. On valley slopes surveyed in Exmoor and Wales, smooth convexities and concavities, with ten or more measured lengths showing successive increases in angle, are common (e.g. Profile *H7*, Fig. 56, p. 154). It is unnecessary for there to be further debate on this point in qualitative terms; it can be answered with precision, for given profiles, by statistically-based statements using variability parameters for segments and elements.

Where smoothly-curved elements occur, it is theoretically possible for the curvature to remain constant over the element, or to increase or decrease downslope. Most examples described fail to disprove the null hypothesis that there is no variation in curvature with distance. Hack and Goodlett (1960) found that convexities on sandstones in Virginia approximated to power functions, of the form $H = cL^f$, where H = fall in height below the crest, L = horizontal distance from the crest, and c and f are constants. The value of f varied from 1 to 2; for a typical example, $H = 0{\cdot}016L^{1{\cdot}7}$. On pediments, most descriptions indicate that curvature, as well as angle, decreases downslope, giving a curve that is possibly of exponential form.

Smoothly curved slope elements are most frequent on, but not confined to, homogeneous rocks, such as granite and chalk. Lewin (1970) found that on chalk, 4th-degree polynomial curves could be fitted to convexities with errors of less than 1·2 m. Segments are particularly common on alternating rocks of differing lithology, for example the sandstones and shales of the Coal Measures and Millstone Grit (Carboniferous). Fourneau (1960) compared the frequency of convex-concave slopes and slopes with an intervening maximum segment for parts of Belgium; a maximum segment was more commonly present on six of the seven rock types surveyed, including chalk and clay, the exception being limestone. The convexity was proportionally longest on sandstones and shortest on shales.

Most surveys of profile form discussed above refer to temperate climates. It is commonly supposed that slopes in arid regions are segmented; this has not been demonstrated, and there are certainly exceptions. Under polar conditions benched slopes on which segments are frequent have been reported (Pissart, 1966). Smooth convex elements are common under both rainforest and savanna climates (Figs. 2, 59, 90). Residual hills in West Africa include both segmented and smoothly-curved forms (Savigear, 1960).

FIG. 59. A convexity and a rectilinear slope separated by a break of slope. Formed on granite under a high-altitude savanna climate, the eastern margin of the Nyika Plateau, northern Malawi.

The proportions of the total ground length of the profile occupied by the convexity, maximum segment and concavity differ substantially according to whether a free face occurs. On slopes without a free face there is a wide variation, but it is unusual for the concavity to occupy more than 50%. In steep-sided valleys subject to rapid stream erosion the maximum segments occupy over half the slope, and the concavity is short or absent. Clark (1965) describes a group of profiles from Devil's Dyke, a steep valley in chalk, in which the proportion occupied by composite 26°-32° maximum segments increases downvalley with increase in valley depth. In Britain the convexity frequently occupies over 40% and sometimes over 80% of the slope (Young, 1963, Table II; Clark, 1965). On sandstones in Virginia, profile form is convex-over-rectilinear, with only small areas of concavity (Hack and Goodlett, 1960). On gently-undulating plains with a savanna climate, in Malawi, the proportions of convexity, maximum segment and concavity are typically about equal. An extreme case of a multi-convex landscape occurs on sandstones in part of the Mato Grosso, Brazil, where most slopes are 95-99% convex (Young, 1970b). Valleys in which the concavity is entirely absent also occur under rainforest in central Malaya.

On slopes that include a free face the convexity is short. The free face forms the maximum segment, and normally is succeeded by a steep, 25°-35°, rectilinear or slightly concave unit, termed the debris slope in King's model of slope form (p. 37). A basal concavity frequently occupies a high proportion of the total slope length. Where the caprock that forms the free face is removed, slopes formed in the underlying weaker rocks invariably have a much longer convexity (e.g. Fair, 1947, 1948). The presence on the slope crest of a steep-sided inselberg, even if quite small, can substantially alter the slope form; on adjacent interfluves in the same rock, without and with an inselberg, the ratio of convexity, maximum segment and concavity may be changed from about 50:20:30 to 5:5:90. Another common form in slopes with an inselberg at the crest is a two-sequence slope, the sequences being separated by a long minimum segment (Young, 1970b).

For the explanation of features of slope profile form, three extreme viewpoints may be distinguished. The first is that the features are functions of external variables, such as climate, geology and valley depth. Examples of this type of explanation are Louis' (1961) hypothesis that different climates are associated with different valley forms (p. 235); Baulig's (1940) conjecture that the convexity is relatively longer on rocks providing permeable debris, which has neither been confirmed nor refuted by field studies; and Clark's (1965) attempt, for an area with uniform rock and climate, to correlate form parameters with valley depth, which achieved only limited success in terms of the percentage variance accounted for. This approach leads to the use of multivariate analysis (e.g. Melton, 1957, see p. 214). The second type of explanation is that form features are the product of the local denudational history. Thus Young (1963) concluded that many of the features of slopes in three upland areas of Britain were associated with successive stream rejuvenations, with intervening periods of stillstand. Savigear's (1962) explanation of the segmented convexities of seaward slopes is in similar terms. These two kinds of explanation are in time-independent and time-dependent terms respectively. The third approach is the contention that there are certain basic features common to all slopes, regardless of rock, climate or local morphological evolution. On the level of broad generalization, the slope evolution systems of Penck and King are of this type. An example of a more specific nature is the view that all apparently-curved parts of slopes consist of segments and breaks of slope.

Many of the results at present available concerning profile form can be stated only as loosely-defined generalizations, because the source material is not quantitatively comparable. *A fortiori*, explanations of the observed features can only be tentative. For progress to be made, quantitative comparability of results from studies of different areas is necessary, and this requires greater rigour in methods of survey and analysis. Profiling has already been the means of achieving some of the most substantial results in slope geomorphology; but the studies so far carried out can be regarded only as the early, trial stages in the use of a technique with considerable potential.

XIII ANGLE

THE frequency distribution of slope angle within a given area can be estimated by measurements of true slope at individual points. The points may be randomly sampled or, more conveniently, a systematic sample may be obtained by measurement at regular intervals along transects. Except for small areas, however, much of the field survey time in such a method is spent in locating the points. For most purposes, angle frequency is better obtained from slope profiles. To obtain the best estimate of angle frequency within an area the profile lines should be obtained using a profile sampling design (p. 145). Most published frequency distributions are based on non-randomly sited profiles; the results, therefore, strictly refer only to the surveyed profiles, and are not necessarily applicable to the area as a whole. It is also possible to obtain angle frequency by constructing a morphological map and measuring the areas of each facet (Gregory and Brown, 1966). This avoids sampling error by giving complete areal coverage, but is subject to the substantial limitations of accuracy present in morphological mapping (p. 179).

To plot the frequency distribution of angle from a profile sample, class limits for angle are selected, and the measured lengths within each angle class summed and converted to percentages. Three-degree classes are suitable for representing the overall distribution, and one-degree classes for indicating frequency peaks and troughs. In most areas the frequency distribution displays marked positive skewness. Normalization can often be achieved by transformation of the original angle data to altan form (p. 137); in most areas, transformation to $\sqrt{\theta}$ or log. θ is less successful in approximating the data to a normal curve (Speight, 1971). Similar angle frequency distributions can occur in areas having substantially different profile forms (Pitty, 1968b). Thus, if a profile is rotated about a horizontal axis, the angle frequency is unaltered but convex and concave portions are interchanged. The information on form that is conveyed by angle frequency distributions is substantially increased if obtained separately for slopes with positive, zero and negative curvature (Fig. 60, *A-C*).

Most angle frequency distributions show that gentle slopes, *c*. 1°-4°, occupy a greater area than almost level, 0°-1°, land (Young, 1961, 1970b; Seret, 1963; Molchanov, 1965; Macar and Pissart, 1966; Gregory and Brown, 1966; Speight, 1970). The maximum frequency occurs in this gently sloping range. The frequency then decreases, often in a manner which approximates to a negative exponential decline (Figs. 60*A*, 63). From the limited data available it appears that a high proportion of all slopes on erosional topography are less than 10°. Expressed in terms of altan, frequency distributions for areas of varied relief usually show standard deviations of 0·3-0·65 altan,

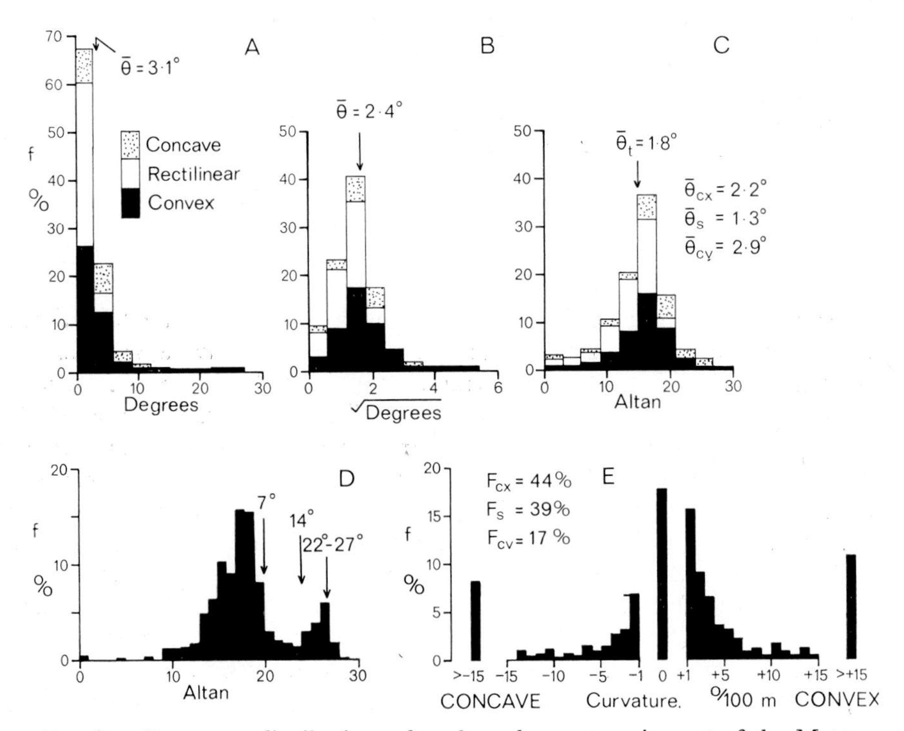

FIG. 60. Frequency distributions of angle and curvature in part of the Mato Grosso, Brazil. *A*, *B* and *C* show the same data, *A* = by angle (marked positive skewness); *B* = by square root of angle (smaller positive skewness); *C* = by altan (slight negative skewness). Rectilinear slopes show a sharp cut-off above 5°, altan = 20, and concave slopes are concentrated towards the steeper angles. *D* = a sub-region of the basin, showing a bimodal angle frequency distribution. *E* = frequency distribution of curvature. For meaning of symbols see p. 163.

irrespective of the mean. If the landscape is partitioned into relatively homogeneous units and the altan frequencies of these summed separately, the standard deviation is reduced to 0·2-0·4 altan (Speight, 1971). The mean angles for convex, rectilinear and concave parts of profiles may differ from each other substantially (Fig. 60*C*). Whether the approximate log-normality of angle frequency distributions is caused by the existence of polycyclic landscapes or is a feature also of relief of monocyclic origin has not been investigated.

Curvature can similarly be represented as a frequency distribution, separating convex (positive) from concave (negative) curvatures (Fig. 60*E*). The apparent frequency of zero curvature, i.e. rectilinearity, is greatly influenced by the minimum angular unit employed in the original observations, and there are theoretical arguments for partitioning the zero class between the two lowest classes of convex and concave curvatures, either equally or proportionally. Since, however, the extent of almost rectilinear slopes is of considerable geomorphological interest, it may be standardized as a class of near-zero curvature, defined as having curvatures in the range plus to minus 1°/100 m. The following classes of profile curvature are suggested:

$C = > +100°/100$ m	Markedly convex
$C = +10$ to $+100°/100$ m	Moderately convex
$C = +1$ to $+10°/100$ m	Slightly convex
$C = +1$ to $-1°/100$ m	Near-zero curvature
$C = -1$ to $-10°/100$ m	Slightly concave
$C = -10$ to $-100°/100$ m	Moderately concave
$C = > -100°/100$ m	Markedly concave

Frequency distributions of curvature show high values at low curvatures 1°-5°/100 m, a rapid decline in frequency with increase in curvature, and the occurrence of occasional slopes of very high curvatures, over 100°/100 m. The relative proportions of the ground surface having convex, near-zero and concave curvatures is an important descriptive parameter in characterizing a landscape.

The following are the main parameters for use in describing the angle and curvature frequency characteristics of a landscape. Means and standard deviations should be based on distributions that have been as far as possible normalized. These parameters may be used in combination with others describing the vertical dimensions of a landscape (e.g. relative relief) and its horizontal recurrence-interval of elements (e.g. drainage density or valley width). In each case, σ is the standard deviation for the corresponding mean.

$\bar{\theta}_t, \sigma_t$	Mean angle, all slopes
$\bar{\theta}_{cx}, \sigma_{cx}$	Mean angle, convex slopes (i.e. slopes with positive curvature)
$\bar{\theta}_s, \sigma_s$	Mean angle, slopes with near-zero curvature
$\bar{\theta}_{cv}, \sigma_{cv}$	Mean angle, concave slopes
F_{cx}, F_s, F_{cv}	Percentage frequencies of slopes with positive near-zero and negative curvature
Convex : concave ratio	Ratio $F_{cx}:F_{cv}$
$\bar{C}_t, \sigma_{c,t}$	Mean curvature, all slopes
$\bar{C}_{cx}, \sigma_{c,cx}$	Mean curvature, convex slopes
$\bar{C}_{cv}, \sigma_{c,cv}$	Mean curvature, concave slopes

CHARACTERISTIC AND LIMITING ANGLES

CHARACTERISTIC ANGLES

Characteristic angles are those which most frequently occur, either on all slopes, under particular conditions of rock or climate, or in a local area (Young, 1961). They appear as modes on a graph of angle frequency distribution. The class with the maximum frequency is the *primary characteristic angle*. Characteristic angles are a property wholly of form, carrying no implications as to their cause.

Table 12 lists reported instances of characteristic angles. Every whole-degree angle from 0° to 40° is represented. The data is on too variable a basis for quantitative comparison, but there are certain recurrent groups:

(*i*) Over 45°. There may be marked relative peaks within the angle range represented by cliffs, but frequencies relative to the total land surface area are low.

(*ii*) 30°-38°, typically 33°-35°. The absolute area occupied by such slopes is again small, but a marked peak, interrupting the decline in frequency with increase in angle, is common.

(*iii*) 25°-29°, typically 25°-26°. This peak is common to many areas. Despite the adjacent total range in angle with group (*ii*), for individual regions it is separated from the latter by a clear frequency minimum.

Local areas may possess characteristic angles within the range 10°-24°, but there is no indication of significant groupings.

(*iv*) 5°-9°. In some instances this peak represents concavities or footslopes lying below steeper slopes.

(*v*) 1°-4°. In many areas the primary characteristic angle lies in this range.

Table 12. Characteristic slope angles

Source	Location	Climate	Rock	Characteristic angles (degrees)
Fair, 1947, 1948	Natal	Sub-tropical	Dolerite, sandstone	4, 12, 18, 23, 27, 33, 36, 42, 45
Strahler, 1950	U.S.	Temperate	Various	15, 26, 33, 35, 38, 42, 45
Savigear, 1956, 1962	S.W. England	Temperate	Various	2, 5, 11, 15, 26-9, 32, 37, 40, 45
Young, 1961	U.K.	Temperate	Sandstone, shale	4, 9, 15, 24-6, 33-5
Seret, 1963	Belgium	Temperate	Shale, limestone	3, 5-7, 13, 18, 26, 31
Macar, 1963	Belgium	Temperate	Sedimentaries	5-7, 30
Johnson, 1965	Pennines, U.K.	Temperate	Sandstone, shale	6, 11, 19, 26, 37
Clark, 1965	U.K.	Temperate	Chalk	8-9, 26-8, 32
Gregory and Brown, 1966	U.K.	Temperate	Various	7-9
Macar and Pissart, 1966	Belgium	Temperate	Various	6, 18-21
Pissart, 1967	N. Canada	Polar	Various	5-7, 32
Molchanov, 1967	Urals	Temperate	Various	1, 3-4, 9-12, 16-18, 31-3
Lewin, 1969	Yorks Wolds, U.K.	Temperate	Chalk	4-6, 22-9
Carson and Petley, 1970	Pennines, U.K.	Temperate	Sandstone, slate	20, 25-6, 33
Blong, 1970	N. Zealand	Temperate	Pumice	27, 34

Although attention is usually concentrated on frequency peaks it would be of equal or greater significance with respect to slope evolution if frequency troughs, angles which are under-represented in the landscape, were to be demonstrated.

LIMITING ANGLES

Limiting angles are those which describe the range of angle within which given forms occur or given processes operate, either on all slopes, under particular conditions of rock or climate, or in a local area (Young, 1961). They include *upper* (maximum) and *lower* (minimum) *limiting angles.* Lower limiting angles for processes were first defined, as '*seuils de fonctionnement*', by Tricart (1957).

Under humid climates, the upper limiting angle for the presence of a regolith cover, or the lower limit for a free face, is normally 40°-45°. For a continuous vegetation and soil cover, without rock outcrops or scars in the root mat, the upper limit is typically 30°-36°. The lower limit for the occurrence of well-developed terracettes (p. 202) is normally 32°-33° (Young, 1961; Clark, 1965), but has been recorded as low as 22° (Macar and Pissart, 1964). On sandstones and shales of the Pennines the following limits occur under grassland: discontinuous regolith with projecting rock outcrops, 41°-49°; continuous regolith but with scars in the turf cover, 36°-40°; terracettes, 33°-36°; smooth regolith and grass cover without substantial microrelief, $<33°$ (Young, 1961); occasional soil slips occur on angles down to 20° (Carson and Petley, 1970). Exceptionally steep regolith-covered slopes, of 70°-80°, have been reported from rainforest climates. These have short-term stability, but are subject to recurrent landsliding; it has been suggested that the regolith is held in place by the binding action of tree roots, but this is unproven (Wentworth, 1943; White, 1949). These applications of the concept of limiting angles involve associations between properties of form, namely angle in relation to ground surface features (microrelief, regolith and vegetation). If comparative work on this aspect is to be extended, it will be necessary to develop greater precision in the description of ground surface features. The interpretation of these form associations involves considerations of process.

A stable slope, in the engineering sense of the term, is a slope below the limiting angle for rapid mass movements; on the time scale of landform evolution it is the angle of long-term stability that is relevant. In the case of deep landslides on slopes possessing appreciable cohesion there is no single angle of stability, but a height-angle relation, as in the upper curve of Fig. 61.

For given rock and climatic conditions, surface landslides (p. 79) occur on gentler slopes than deep landslides. On boulder clay in Durham, the lower limiting angle for surface slides is 29° (Fig 61). On London Clay, deep rotational slips occur on 13°-20°, whereas the lower limit for shallow mass movements is *c.* 10° (Skempton, 1964; Hutchinson, 1967). In Southern California, shallow soil slips are restricted to parts of slopes that are locally steeper than 38½° (Bailey and Rice, 1967; Rice *et al.*, 1969). Sandstones in the Pennines show no signs of rapid mass movements below 20° (Carson and Petley, 1970). On glacial drift in Ontario, abandoned meander bluffs cease to

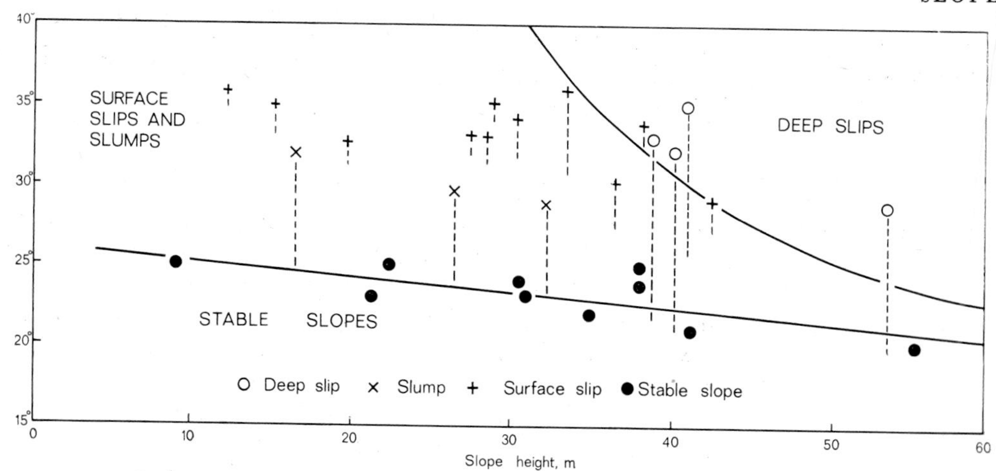

FIG. 61. Deep and shallow landslides in relation to angle and slope height, on boulder clay in Durham. Based on Skempton (1953).

be affected by surface slides below 24° on sandy deposits, and 20° on clays (Packer, 1964).

Two different explanations of the lower limiting angles for surface landslides have been proposed. In the case of London Clay, the observed limiting angle corresponds with stability conditions calculated by using the residual shearing strength, whereas deep slides are related to the peak strength (p. 81) (Skempton, 1964; Hutchinson, 1967). This explanation, however, is not applicable to regolith material in which substantial amounts of sand, gravel or stones are present, since materials of this nature show only small differences between peak and residual shearing strengths. For a stability analysis of slips confined to a shallow regolith, conditions approximate to a rectilinear slope of infinite length with the angle of the failure plane equal to the slope of the ground surface. The stability equation for such a slope is

$$\gamma . z . \cos \beta . \sin \beta = c' + (\gamma . x . \cos^2 \beta - u) . \tan \phi' \quad (13.1)$$

where γ = bulk density of soil, z = depth of failure plane, c = effective cohesion, ϕ' = effective angle of shearing resistance, u = pore water pressure along the failure plane, and β = slope angle = failure plane angle (Skempton and Delory, 1957). The strength reaches a minimum when the regolith becomes fully saturated, giving a water table at the ground surface and groundwater flow parallel to the surface. Under such conditions $u = z . \cos^2 \beta$, and the maximum stable angle is given approximately by:

$$\tan \beta \simeq \tfrac{1}{2} \tan \phi' \quad (13.2)$$

(Carson and Petley, 1970). For typical regolith material on non-clay soils $\phi' = 43°$, therefore $\beta = 25°$. Further tests on a range of natural regolith materials, and if possible under field conditions, are necessary before this suggested mechanism can be taken as being of general application.

The existence of lower limiting angles for the occurrence of slow mass movements has also been suggested. Tricart (1957) held that solifluction ceased to occur below a certain critical angle, which would vary with regolith characteristics and climate. The presence in part of Arctic Canada of long rectilinear slopes at 6°-8° has been interpreted as corresponding to the lower limit for relatively rapid solifluction (Pissart, 1967). For a material that follows the laws of plastic flow there is a critical stress below which no strain occurs (Fig. 17, p. 44); it follows that for such material resting on a slope, acted upon by the stress arising from its weight, there is a minimum angle for flow. Souchez (1966) assumed that soil creep followed the laws of plastic flow, and developed a model of regolith movement and slope evolution, but this assumption is probably incorrect (p. 111). In the mechanisms of both solifluction and soil creep there is a substantial component of movement associated with expansion and contraction of the regolith, which occurs equally on level ground; if the downslope movement associated with a given amount of expansion and contraction varies with sin θ, then there is no sharp lower limiting angle but a gradual decrease in rate with decrease in angle. Lower limiting angles for slow mass movements could be substantiated only if it could be shown experimentally that under certain conditions regolith follows the laws of plastic flow.

THE INTERPRETATION OF ANGLE FREQUENCY

The main feature of overall angle frequency distributions is the fact that gentle slopes are considerably more extensive than steep. There are two possible explanations of this. It may represent the relative survival time of slopes of differing angles. If steep slopes are more rapidly altered or destroyed by denudation than gentle slopes, whatever may be the manner of evolution, the latter will survive in the landscape for longer than the former; it follows that at a given moment of time, gentle slopes will occupy a larger area than steep. On the basis of this argument, the existence of a negative exponential decline in frequency with increase in angle suggests that the rate of destruction of a slope increases exponentially with increase in angle. The alternative explanation is that angle frequency distributions reflect morphological history, the predominant gentle slopes representing extensive survival of late-stage landforms and the low-frequency steep slopes corresponding to recent rejuvenation.

In many tropical landscapes, both arid and humid, there is a marked visual distinction between gentle and steep slopes, with slopes of intermediate angle, about 8°-20°, relatively uncommon. If this subjective impression is correct, angle frequency distributions from such landscapes will be bimodal; Fig. 60*D* shows an example. On sandstones under savanna in Northern Australia, M. A. J. Williams (1969) reports a trimodal distribution, with peaks at 10°-11° and 4°-5° representing hillslopes and pediments respectively, and a peak at 0°-1° formed by alluvial flats. It is possible that angle frequency distributions are representative of a fundamental morpho-climatic distinction: in the absence of special structural conditions, landscapes of temperate and rainforest regions tend towards unimodal distributions, whereas those of arid, semi-arid and savanna regions tend towards bimodal distributions.

Five reasons for the occurrence of characteristic angles have been proposed, of which the first three are only explanations in a limited sense:

(*i*) *Angle frequency peaks result from random variation.* Speight (1971) claims that many supposed characteristic angles are statistically non-significant, lying within the confidence limits for a normally distributed population (as log.tan θ). His method, however, involves the prior grouping into classes which for steeper angles are relatively wide (a $\frac{1}{4}$-$\frac{1}{2}$-$\frac{1}{4}$ smoothing function applied to classes of 0·1 log.tan units), which necessarily eliminates frequency variations of a finer order. Statistical testing of the significance of frequency peaks is desirable, but there are difficulties in that the precise nature of the distribution to be taken as the null hypothesis is not known.

(*ii*) *Particular conditions of structure produce characteristic angles.* This is a statement of an association between features of form, and not in itself an explanation in terms of process. Gregory and Brown (1966) showed that for part of the North Yorkshire Moors the total angle frequency shows a relatively smooth exponential decline from a 1° maximum to 35°, but for each of 13 geological formations, marked characteristic angles occur (Fig. 62). Strata of similar lithology do not necessarily give similar angles, the superposition of differing lithologies being important. Similar associations between angle and lithology have been demonstrated in Belgium (de Béthune and Mammerickx, 1960; Macar and Fourneau, 1960; Macar and Lambert, 1960; Macar, 1963). In the Famenne, a comparison of shales and limestones showed that whilst some characteristic angles were confined to one rock type only, others were present on both (Seret, 1963).

(*iii*) *Certain angles are intrinsic features of slope evolution, irrespective of rock type and climate.* This hypothesis is in itself a statement about form, although carrying implications with respect to process. It can only be supported by convergence of information from different environments. It is true of the 33°-35° group of characteristic angles, but has not been demonstrated for gentler slopes.

(*iv*) *Characteristic angles reflect local morphological evolution.* This hypothesis requires the assumption that slopes become gentler with time, and may be outlined using Davisian terminology. Assume that an area has been peneplaned, giving extensive gentle slopes. It is uplifted, and a partial cycle proceeds to a stage at which the valleys have moderate slopes, part of the relict peneplain remaining. A second rejuvenation produces steep inner valleys. Individual valleys will now have polycyclic forms, and the three partial cycles will be represented by angle frequency maxima. The values of the characteristic angles correspond to the stages of cyclic development reached, so there is no reason why certain angles should be preferentially represented. An example of this occurs in Exmoor (Fig. 63); the main support for a cyclic explanation is the form of individual valleys, but additional evidence comes from a comparison of the characteristic angles on two rock types:

Shales	3°-4°	9°-10°	23°-25°
Sandstones	4°-5°	13°-15°	28°-29°

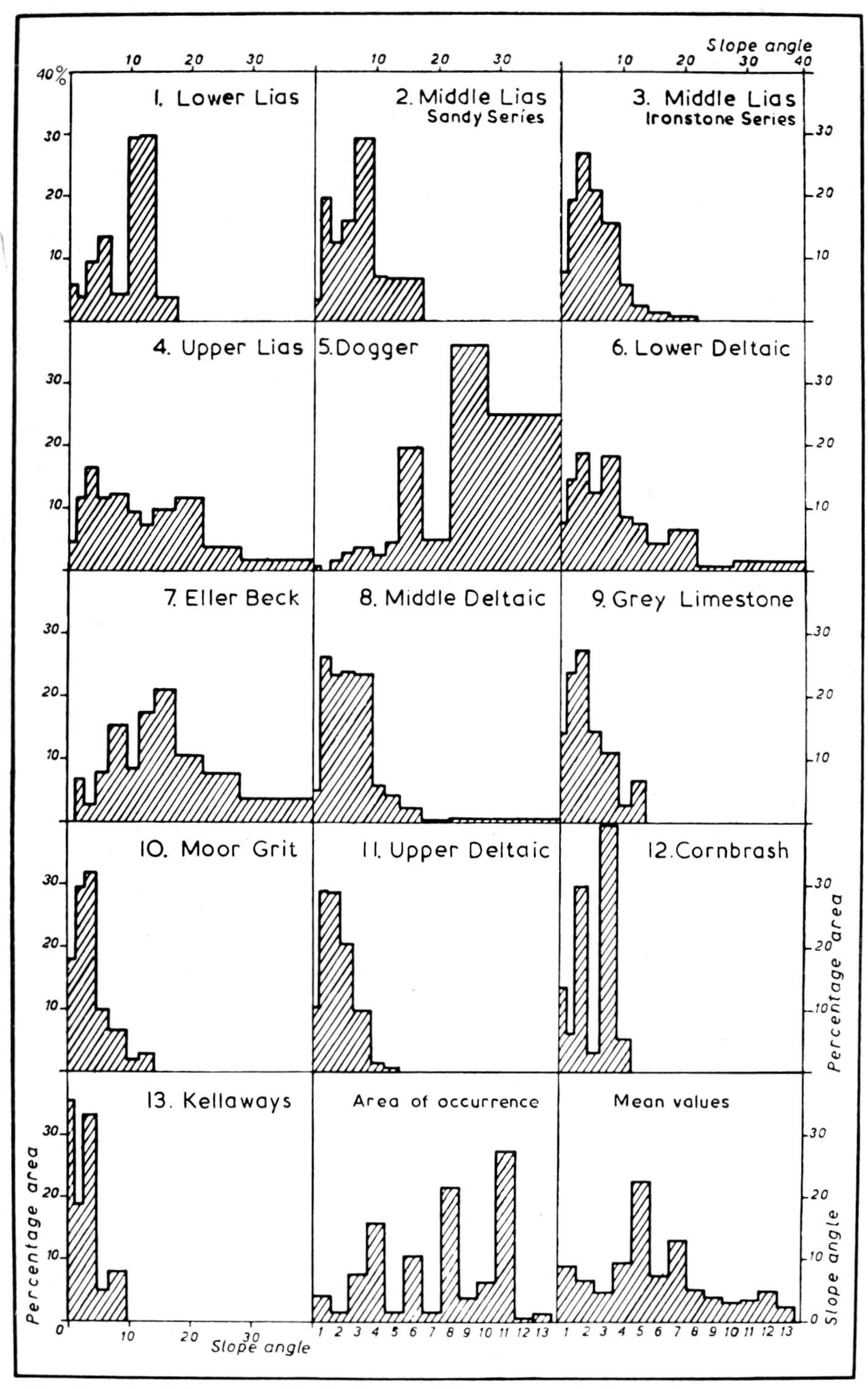

FIG. 62. Angle frequency distributions on different geological strata, in part of northern England. After Gregory and Brown (1966).

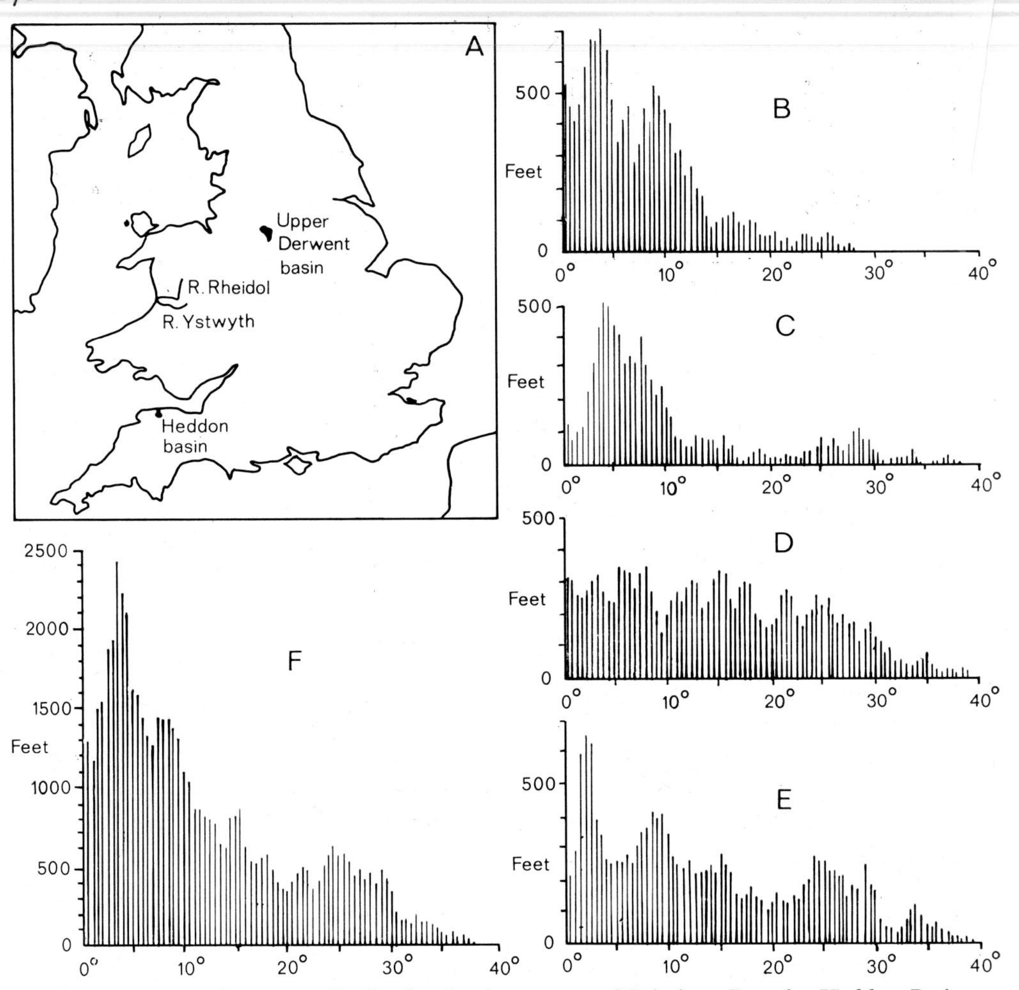

FIG. 63. Angle frequency distribution in three parts of Britain. *B* = the Heddon Basin; shales. *C* = the Heddon Basin; sandstones. *D* = central Wales, between the Rivers Rheidol and Ystwyth; shales. *E* = the Upper Derwent Basin; sandstones and shales. *F* = all areas combined. After Young (1961).

These suggest that the three pairs correspond to three epicycles of erosion, each being represented by a slightly higher angle on the more resistant rock (Young, 1961).

Ruhe (1950) found that the angle frequency distributions on glacial drift differed according to the age of the drift. The peak frequencies, which refer to mean (crest to base) angle were: Kansan (oldest), 9°-22°; Iowan and Tazewell, 2°-3°; and Mankato and Cary (youngest), 0°-2°. The increase in angle with age was attributed to the fact that integrated drainage is not yet established on the younger drifts.

(*v*) *Characteristic angles are related to limiting angles for processes.* Angles in the >45° group may represent the angle of long-term stability, although for hard-rock cliffs this is difficult to verify experimentally (p. 120). On slopes of unconsolidated loess in Iowa and Tennessee it has been demonstrated on the basis of shear strength tests that two observed characteristic angles, 77° and 51°, correspond to the expected residual slopes following vertical cleavage failure and oblique shear failure respectively (Lohnes and Handy, 1968). The 33°-35° group of characteristic angles lies close to both the limiting angle for a continuous regolith cover and the angle of repose of scree. The formation of rectilinear 33°-35° slopes from initial steeper irregular slopes can be explained on the assumption that rock outcrops undergo more rapid retreat than regolith or scree covered slopes.

The 25°-29° group frequently occurs as rectilinear segments, having relatively angular intersections with the steeper maximum segments above or below them (Savigear, 1956, 1962; Young, 1961, 1963; Clark, 1965; Carson and Petley, 1970). This may be explained on the assumption that microrelief, turf scars and soil slips indicate relatively rapid slope retreat; slopes of *c.* 35° are short-lived features, and in the absence of rapid basal erosion they are transformed by surface mass movements into segments of less than 30°. The angular segment intersections suggest that this transformation is by replacement rather than decline. Thus 25°-29° slopes are not necessarily the first to be produced following stream dissection of an area, but they are the first to be developed that are not relatively quickly reduced to a gentler angle.

The above explanation was originally proposed on evidence of limiting angles for ground surface forms (Young, 1961). Carson and Petley (1970) have subsequently attempted to explain segments at *c.* 25° as representing the angle of long-term stability for non-clay regolith material. They measured the peak and residual shearing strength of the regolith from two areas in Britain in which there occurred segments at 33° and *c.* 25°, separated by angular intersections. Values of effective cohesion were found to be zero. The stability angle of the slope is related to the reduction of the regolith. With a regolith of stones only, pore water pressure never rises above zero owing to the large interstices, and the slope stands at 35°, the angle of repose for coarse fragments. As soon as fine material is present, pore water pressure occurs. For fine material with some stones, the effective angle of shearing resistance ϕ' was found to be 42°-44°; therefore, from equation *13.2*, the maximum stable angle β is 24°-36°. For soil without stones, $\phi' = 35°\text{-}36°$, therefore $\beta = 19°\text{-}20°$. As reduction progresses, the maximum stable angle will decrease, not continuously, but in steps; this explains the occurrence of segments at these angles, and the angular breaks of slope between them.

The evidence that the frequency distribution of segments is polymodal is stated by Carson and Petley to be statistically non-significant, and in matching the experimental results to the field data they employ special pleading. This hypothesis deserves further investigation, but the evidence available is insufficient for it to be regarded as established.

The possibility that characteristic angles in the 5°-9° group may be associated with the limiting angle for relatively rapid solifluction has been noted above. Such slopes occur amid concavities both in polar climates and in regions having relict periglacial features, but the mechanism has not been demonstrated.

The two main hypotheses to account for characteristic angles are thus that they are time-dependent features, representing stages of morphological evolution, and that they are time-independent forms related to the mechanism of process. The former explanation would be supported by correspondence with evidence from local denudation chronology, the latter by the repeated occurrence of the same angles in areas with similar present-day environmental conditions.

ANGLE CLASSIFICATION

The division of a continuous variable such as angle into discrete classes is inevitably to some extent arbitrary, but for some purposes it is convenient to treat a range of angle values as belonging to a single class. The object of classification is simplification, but this object is defeated both if there are too many classes and if too wide a range of values is included in each class. There is some consensus of opinion that for a single variable, seven is the optimum number of classes; this may be related to the capacity of the English language to denote about seven measures of quantity or frequency. In selecting class limits, the first principle is that for normal conditions the frequencies in each class should not be very unequal. Secondly, it should be possible to make general statements that apply to the whole of each class but not to values lying outside it. Such statements may be of either a scientific or practical nature. Other things being equal, round numbers are desirable. In classifications designed for a specific practical purpose the class limits can be entirely determined by user-requirements, to the sacrifice of other principles.

A comparison of existing classifications shows there is widespread agreement that the class limits, expressed in angle or percentage units, should follow an approximately geometric progression. For example, the classifications of Gellert (1967), based mainly on surface processes, and Macgregor (1957), based on the practical significance of angle with respect to land use, are both approximately logarithmic. In Fig. 64 the vertical scale is logarithmic, and column *A* shows an eclectic system based on 17 previous classifications (devised mostly for use in soil survey and land classification). Column *B* shows the FAO soil survey system, itself based in part on the most widely used soil survey scheme, that of the U.S. Department of Agriculture. Numerically simple schemes are obtained by making each upper class limit twice the angle value of the lower, based either on 1° (1°-2°-4°-8°-16°-32°-64°) or 10° (1¼°-2½°-5°-10°-20°-40°). Speight carried the logarithmic principle to its logical extent in a scheme based on equal increments of log. tan. (Fig. 64, column *C*), which he extended far into the under 2°, or 'gradient', range.

In designing a classification based on geomorphological significance, there is again a clear indication that the steeper classes should cover a wider angle range than the gentler. It is desirable that common characteristic angles, in particular 35°, should lie centrally within a class, whereas limiting angles should form class limits. The limits should be such that it is possible to estimate visually the proportions of a landscape lying within each class. The following angle classification designed for slope geomorphology has seven classes, with subdivisions of the end members which

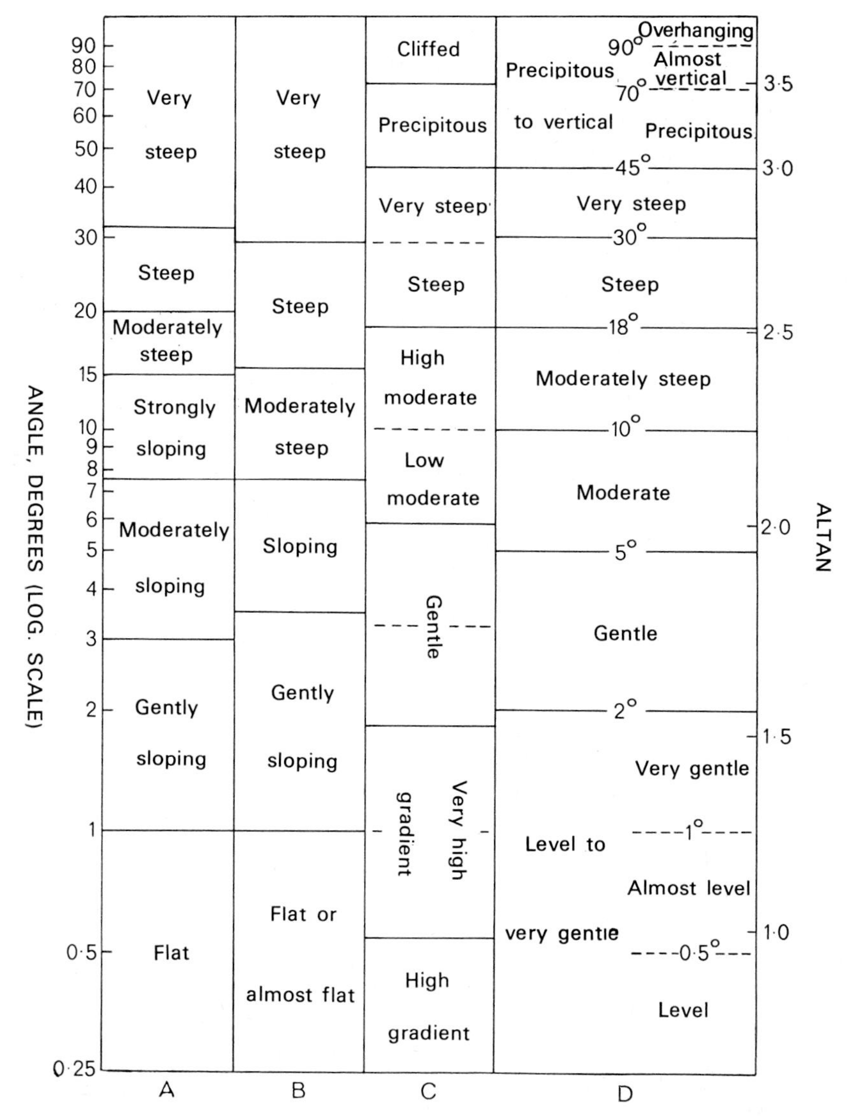

FIG. 64. Angle classifications: A = eclectic, based on 17 previous classifications (Curtis *et al.*, 1965); B = classes used in the FAO Soil Survey Manual; C = system based on equal increments of altan; D = classification used here.

may be employed where either of the latter are especially frequent or significant (Fig. 64, column *D*):

0°-2° *Level to very gentle.** This corresponds to what are thought of as 'flat' areas in typical erosional landscapes; erosion surface remnants commonly include land of

* 0°-2° is taken to mean 0·00°-1·99°, 2·00° being placed in the 2°-4° class; i.e. for all classes, values recorded as lying on the limit are put into the higher class.

up to $1\frac{1}{2}°$. For agriculture and most other practical purposes such land may be put into the 'no-limitation' category. This class may be subdivided into:

$0°$-$\frac{1}{2}°$ *Level.* Direction of slope often indeterminate.
$\frac{1}{2}°$-$1°$ *Almost level.* Sufficient to define direction of slope.
$1°$-$2°$ *Very gently sloping.* Whilst appearing nearly level in a normal land-form context, in some erosional plains the maximum segments of valley sides may fall within this class.

2°-5° *Gentle.* On what is generally considered to be a peneplain, 5° is the steepest acceptable slope (cf. Cotton, 1941, p. 186). Most pediments are included. In many areas this class contains the primary characteristic angle. Practical limitations are relatively slight.

5°-10° *Moderate.* The range in which, in many regions, angle frequency declines steeply.

10°-18° *Moderately steep.* Landscapes that are neither plains nor steeply dissected contain substantial areas of this and the preceding class. Practical limitations are substantial, but such land may still be of moderate economic value.

18°-30° *Steep.* This includes the 25°-29° group of characteristic angles. Above 18° arable cultivation requires special measures, and many forms of land use become uneconomic.

FIG. 65. A landscape dominated by very gentle slopes. The mean angle is 0·9°, and 93% of the slopes are between 0·3° and 3·0°. Formed on sandstones under semi-deciduous rain-forest, the Mato Grosso, Brazil.

30°-40° *Very steep.* This class contains the 30°-38° group of characteristic angles; its upper and lower limits are close to the limiting angles for a regolith cover and for a smooth and continuous regolith cover respectively. Such land has little economic value.

Over 45° *Precipitous to vertical.* Normally formed by cliffs. This class may be subdivided into:

45°-70° *Precipitous.* Usually irregular or stepped cliffs.
70°-90° *Almost vertical.* Usually sheer rock faces.
Over 90° *Overhanging.*

XIV PLAN FORM AND SLOPE CLASSIFICATION

PLAN FORM

PLAN form refers to the shape of the ground surface along a horizontal plane. A slope may be *plan-convex*, as is the case for spur-end slopes, *plan-concave*, as for valley-head slopes, or *straight in plan*. Plan curvature is specified by the horizontal radius of curvature, R_h; values of R_h are given as positive to indicate plan-convexity and negative to indicate plan-concavity, in conformity with the convention for profile curvature.* The change in aspect over a distance of 25 m is 29° for $R_h = 50$, and 2·9° for $R_h = 500$. These values are taken as class limits in the following classification of plan curvature:

$R_h = < +50$	Notably convex in plan
$R_h = +50$ to $+500$	Slightly convex in plan
$R_h = +500$ to -500	Almost straight in plan
$R_h = -500$ to -50	Slightly concave in plan
$R_h = < -50$	Notably concave in plan

To measure plan curvature in the field, set out ranging poles at the observation point and at a standard distance, L, on either side of it at the same height; the three poles thus stand along a contour. A suitable distance for normal relief is $L = 25$ m. Measure the horizontal angle, γ, subtended at the central pole by the two lateral poles. Measurement to the nearest degree is sufficient, using a prismatic compass. For plan-convex slopes, γ exceeds 180°. L is measured as a chord, not a radius, therefore the radius of curvature is given by

$$R_h = \frac{\frac{1}{2}L}{\sin\left(\frac{\gamma - 180}{2}\right)} \qquad (14.1)$$

Plan curvature may also be obtained from contoured maps, using a transparent template overlay.

Flow-lines are lines following the direction of true slope (Fig. 66). Flow-lines converge downslope on plan-concave slopes, and diverge downslope on plan-convexities. The *slope-catchment* for a given point is the area from which surface flow, following the direction of true slope, would cross the point; it is expressed in square metres of catchment area per metre of contour length. To obtain the slope catchment for a given point from contoured maps, draw a short baseline through the point and parallel to the contours. Extend flow-lines from both ends of the baseline upslope to the water-

* The opposite convention is employed by Troeh (1964, 1965) and Speight (1968).

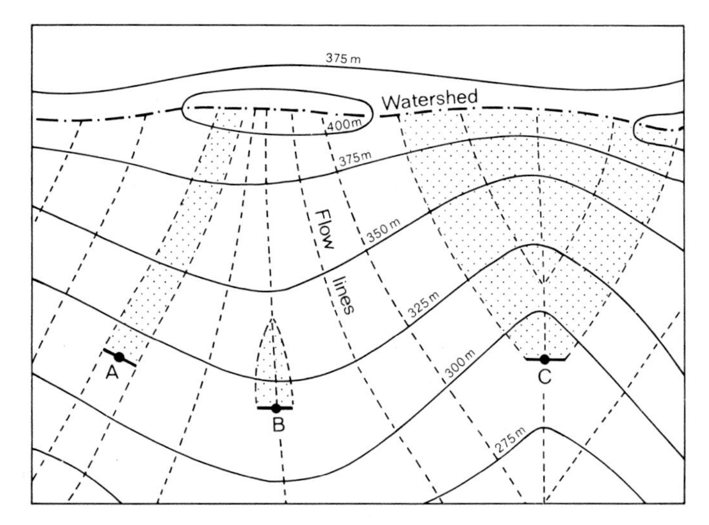

FIG. 66. Flow-lines and slope-catchments. The slope-catchments of points *A*, *B* and *C* are shaded.

shed, and mark in the latter; measure the area enclosed by the flow-lines and the watershed, and divide by the length of the baseline. An alternative procedure is given by Speight (1968). Values of slope catchment increase downslope, linearly on straight slopes, and faster or slower than linearly on plan-concave and plan-convex slopes respectively. Equations showing the effects of plan curvature on water discharge are given by Kirkby and Chorley (1967, p. 16).

Convergence and divergence of flow-lines affects downslope regolith movement. If it is assumed that denudation is entirely by soil creep, and that the rate of creep varies only with the slope angle, then net ground loss will be less on straight slopes than on plan-convex slopes, and less than either on plan-concave slopes (Louis, 1935; Baulig, 1940). With respect to plan-convex slopes this is possibly true, but on plan-concave slopes the assumptions are unrealistic; both surface flow and throughflow will also converge, therefore the soil will be moister and soil creep more rapid. If surface wash also occurs, the volumes of both wash and debris to be transported will be affected by curvature to the same extent, but as the transporting power of wash increases considerably more than linearly with its volume, the net effect will be that plan-concave slopes are subject to faster denudation than straight slopes. These arguments are on *a priori* grounds, as no measurements of the relation between plan curvature and the rates of surface processes are available.

The methods of form description so far described employ slope angle as a basis. It is alternatively possible to describe the shape of the ground surface taking elevation as the basic unit. For profile form alone, if a slope is described by an equation $z = f(x)$, where z = elevation and x = horizontal distance, then dz/dx gives the slope angle, as a tangent, and d^2z/dx^2 the curvature. Combining profile and plan form, any part of a slope can be described by a trend surface, defined by an equation of the form:

$$z = a+bx+cy+dx^2+exy+fy^2+gx^2+hx^2y \ldots \quad (14.2)$$

where z = elevation and x and y are mutually perpendicular horizontal coordinates.

This approach has been applied by Troeh (1964, 1965) to describe a special class of slopes on which: (*i*) the contours approximate to segments of concentric circles, and (*ii*) slope gradient varies linearly with distance from the origin of these circles. Such slopes have the form of a paraboloid of revolution, and may be expressed in cylindrical coordinates by the equations:

$$\begin{aligned} Z &= P+SR+LR^2 \\ G &= dZ/dR = S+2LR \\ L &= \tfrac{1}{2}(d^2Z/dR^2) \end{aligned} \quad (14.3)$$

where Z = elevation of a point, P = elevation of the point of origin, R = radial distance from the origin, S = gradient (as a percentage) at the origin, G = gradient at a distance R from the origin, and L = half the rate of change of gradient with distance from the origin (as per cent per unit distance). Positive and negative values of G indicate concave and convex profile curvature respectively; positive and negative values of L indicate respectively plan-concavity and plan-convexity. These equations provide a means of describing pediments, alluvial fans and other landforms with an approximately conical form. They have also been applied to randomly sampled points, giving good correlations between the parameters G, L and R and the independently-observed soil drainage class (Troeh, 1964).

Greysukh (1967) has proposed a method for defining the shape of the ground surface at a point. Eight equally-spaced radii are extended from the point, the elevations of the centre and of each radial point are obtained, the slopes of the radii calculated, and these slopes plotted as a function of the angle of rotation. Each shape of the ground surface yields a characteristic pattern.

If profile and plan form are to be considered jointly, the trend surface is an appropriate method. Given lengthy computer calculations, any slope can be described by a trend surface, and the residuals from this surface defined. It is possible to show that certain classes of slopes, or parts of slopes, conform to given orders and types of trend surface; for example, convexities on chalk might lend themselves to this approach. The disadvantage of this method lies in the interpretation in geomorphological terms of the results obtained. For most purposes it is better to treat profile and plan form as independent properties of a slope.

SLOPE MAPPING

Geomorphological mapping has become a substantial branch of the subject; the following is not intended as a comprehensive treatment, but is concerned mainly with those aspects which are related to slopes. Four types of geomorphological maps are considered: morphological maps, slope angle maps, maps based on combined morphological parameters, and maps constructed on a genetic basis.

MORPHOLOGICAL MAPS

The term *morphological map* refers specifically to one type of geomorphological map, that based on the recognition of areal slope units. Morphological maps are compiled on relatively large scales, typically 1 : 10 000. The technique was developed by Savigear (1956, 1963, 1965) and Waters (1958); modifications have been suggested by Macar (1963), Curtis *et al.* (1965) and Gregory and Brown (1966).

The technique is based on the assumption that the landscape can be divided into uniform morphological units, bounded by morphological discontinuities (Fig. 67).

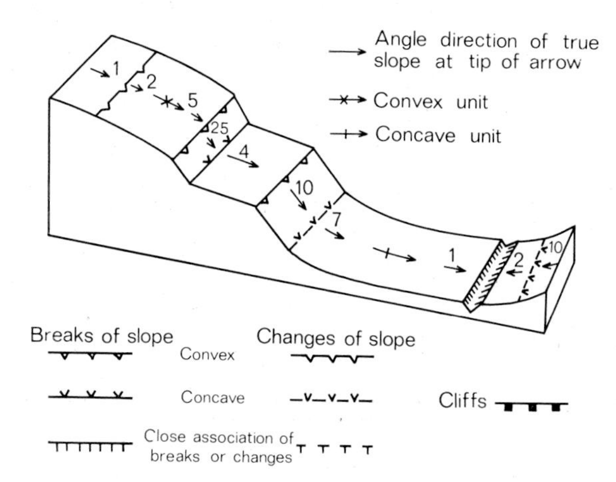

FIG. 67. The symbols used in morphological mapping. After Curtis *et al* (1965).

Originally, all units were assumed to be rectilinear, corresponding to segments on a slope profile, but it has since been found necessary to recognize curved units, the areal equivalent of slope elements. The definitions are as follows:

Morphological units

Slope facet (or *plane morphological unit*) — sloping plane area of the ground surface.

Flat — level area of the ground surface.

Curved morphological unit — area of the ground surface which has approximately constant profile curvature.

Convex morphological unit — curved morphological unit with positive profile curvature.

Concave morphological unit — curved morphological unit with negative profile curvature.

Cliff — slope of more than 40° formed of bare rock.

Irregular morphological unit — a unit which possesses surface irregularities too small to be represented on the scale of the map.

Slope facets and curved morphological units may be subdivided according to whether they are plan-convex, straight, or plan-concave.

Morphological discontinuities

Break of slope — angular discontinuity of the ground surface separating morphological units.

Change of slope — gradual change of angle of the ground surface separating morphological units.

Inflection — line or zone separating adjacent convex and concave morphological units.

Morphological units may also be enclosed by boundary lines which are not morphological discontinuities.

The symbols employed are illustrated in Figs. 67 and 68. The '*V*' symbols on discontinuities point downslope, and are drawn on the steeper side of the line. This is the original and most widely-used convention; the altered symbols later used by Savigear (1960, 1965, 1967) are visually less satisfactory. Juxtaposition of these basic signs on convex crests and concave valley floors yields diamond-shaped and *X*-shaped symbols respectively. Adjacent convex and concave breaks that are too close together to be mapped separately are shown by a joint symbol; to indicate narrow incised gullies, having two convex breaks at the upper limit of the incision and a *V*-shaped floor, two such joint symbols are used, with the shorter lines pointing slightly downstream. A continuous line indicates breaks of slope, and a broken line changes of slope. Inflections and other boundaries that are not discontinuities may be distinguished as dotted lines.

Morphological maps are difficult to interpret visually. It is of some help if a light shading is added, increasing in density as the slope becomes steeper and leaving nearly level ground unshaded. The problem of distinguishing cartographically between facets of similar slope but at different relative heights has not so far been solved. It is equally true, however, that monochrome contour maps are hard to interpret. If morphological maps were to be produced in colour, possibly cartographic conventions could be found that would give the plastic appearance necessary for rapid visual appreciation.

In early examples of morphological maps discontinuity lines were allowed to end in mid-slope, as in Fig. 69*A*. This is unacceptable; by the logic of the method, any given point on the ground surface must be allocated to only one morphological unit. The reason for the earlier practice was that only discontinuities were recognized as boundaries between units. Such boundaries may, however, be transitional, with no more rapid change of angle at the boundary than within the adjacent units. One such case is the inflection (Fig. 69*B*). Another is where a facet steepens gradually and evenly in a lateral direction along the slope; if the facet is first defined by the angle near to its gentler end, the range of angle that it is permissible to include within the facet will be exceeded at some point along the slope, and a boundary line must be drawn in this position (Fig. 69*D*). Thus Fig. 69*A* should correctly be shown as Fig. 69*C*. Whatever the shape of the ground, morphological units should be completely enclosed by boundaries, as in the example shown in Fig. 68.

The emphasis which this method places on facets and breaks of slope is partly due to historical circumstances. It was developed at Sheffield, where much of the surround-

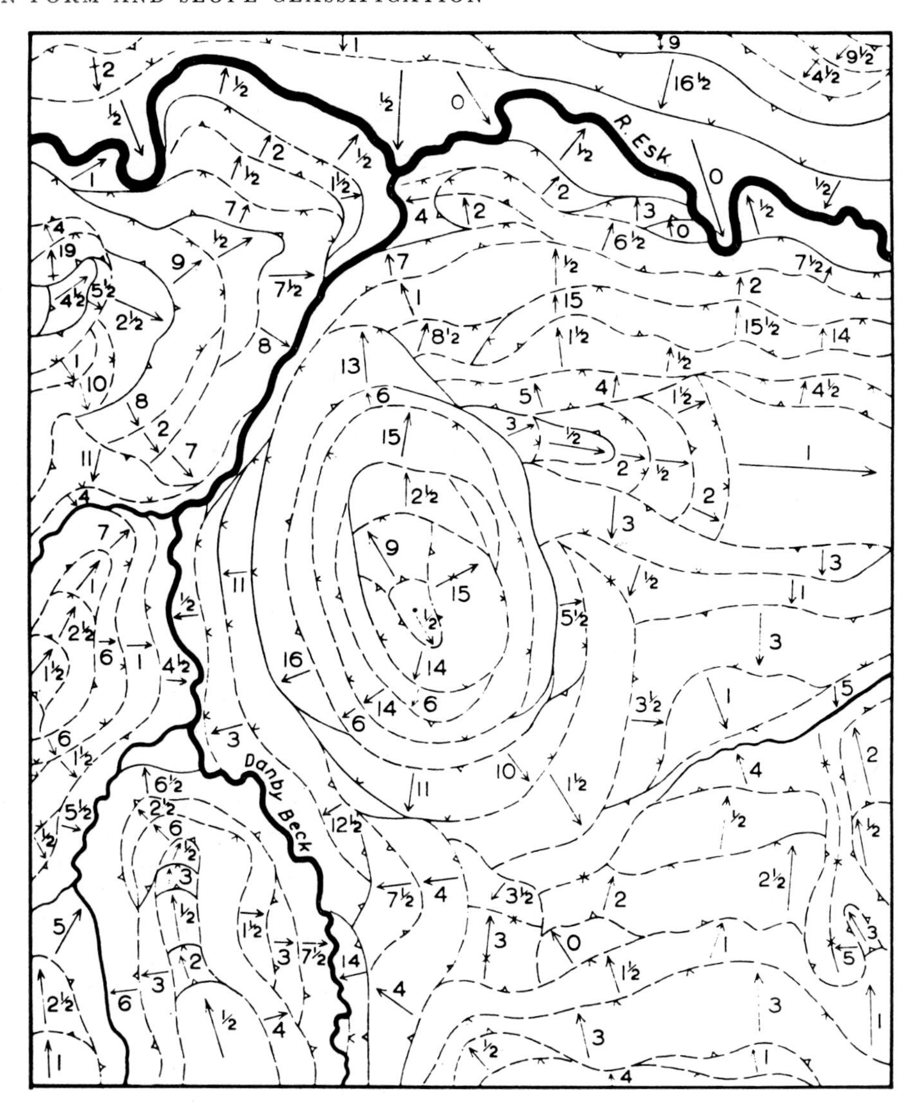

Fig. 68. A morphological map of part of Eskdale, Yorkshire, England. After Gregory and Brown (1966).

ing country is a cuesta topography formed from alternating sandstones and shales, and most of the land surface appears to fall into facets, distinctly separated from each other. It is in country of this type that morphological mapping is most easily carried out. Even with the addition of curved areal units, it is difficult in practice to apply the method to landscapes in which smoothly-curved slopes are extensive, as, for example, Chalk Downlands.

The absence of precision in the definitions of both morphological units and

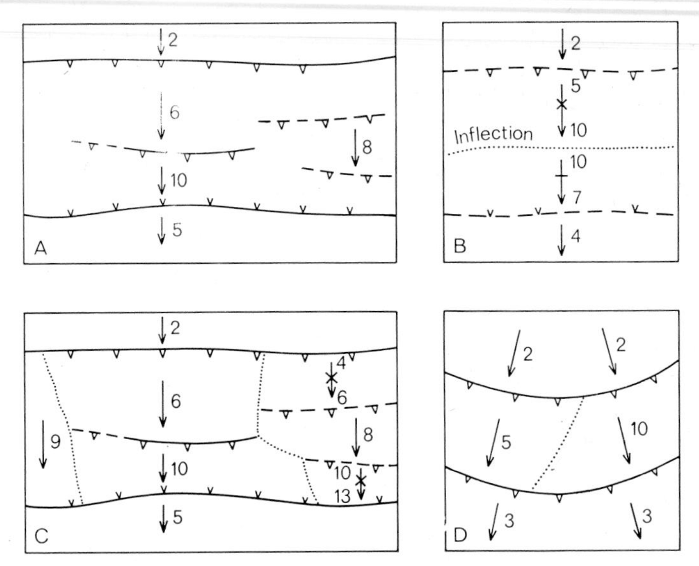

FIG. 69. The use of boundaries which are not morphological discontinuities, but which separate morphological units.

discontinuities leads to a lack of rigour in the system. It is not sufficient to define a facet by its mean angle alone; the permissible departure from this mean must be given, and a procedure specified for determining whether a given part of a slope belongs to a particular facet. It might be specified, for example, that a facet of angle θ° may include any area of the slope which, when observed over a 10 m length, has an angle of $0{\cdot}9\theta$-$1{\cdot}1\theta$, i.e. a permitted range of $\pm 10\%$ of the mean angle. In the corresponding specification for curved areal units, a range of $\pm 25\%$ of the mean curvature is appropriate. Breaks and changes of slope are similarly defined only in qualitative terms in the existing system. There are two alternative means of quantitative definition: by maximum ground surface distances over which the transition between land included in the two adjoining units takes place, or by maximum values of profile curvature. For example, a break of slope might be defined either as a zone in which the transition between adjoining units occupies less than 5 m, or as a zone with a profile curvature, measured over two 5 m lengths, exceeding 50°/100 m. Corresponding values for a change of slope could be 20 m and 10°/100 m respectively; where the latter values are exceeded a separate curved unit, and not a discontinuity, should be mapped.

The accuracy of maps produced by existing methods could be investigated by means of point-sampling, although there are no records of any such tests. Trials carried out by the British Geomorphological Research Group have shown that different observers independently mapping the same area produce different results, i.e. the method is not replicable. It is doubtful if accurate maps, based on quantitatively defined morphological units and discontinuities, can be produced without surveying sub-parallel profiles at frequent intervals down all slopes. Precision of definition and accuracy of survey are necessary if the maps are to be used as a basis for quantitative analysis and comparisons between areas, but to adopt such standards would greatly increase the time taken in field survey. The main value of morphological mapping is as a field shorthand; it is a simple means of recording the approximate areal relations between units of slope form, and is employed in this way both in general geomorphological description and in soil survey. A morphological map is a useful reconnaissance document, but in its present form it has serious limitations as a basis for slope studies.

SLOPE ANGLE MAPS

There are two kinds of map based primarily on slope angle: isoclinal maps and average slope maps. *Isoclinal maps* show the slope angle at a point, and are normally obtained by transformation of contour maps. Classes of angle are selected. For each class limit, the distance apart of contours on the base map is determined from the formula

$$H = \frac{1000\,V.\cot\theta}{S} \qquad (14.4)$$

where H = contour separation in mm, V = vertical contour interval in m, and S = map scale. (If V is in feet, 305 is substituted for 1000 in equation *14.4*.) For example, on a 1 : 25 000 map with a vertical interval of 10 m, a 5° slope is represented by contour lines $1000 \times 10 \times 11{\cdot}43/25\,000 = 4{\cdot}6$ mm apart. Dividers are set to this contour spacing, or a template is made, and isoclines are drawn separating areas with steeper and gentler slopes than the class limit. The methods to be followed on crests, valley floors and slopes curved in plan are given by Blenk (1963). The resulting map of isoclines is shaded, with increasing density for steeper slopes. With monochrome representation the difference between the gentle slopes of interfluves and of valley floors is not always clear, but the addition of prominent blue stream lines overcomes this difficulty. By reducing the scale of the finished map a fine, plastic visual effect can be achieved (Miller and Summerson, 1960). It is alternatively possible to map, for standard areal units, the percentages of steep slopes and gentle slopes, above and below selected angles (Thrower, 1960).

Isoclinal maps have the merit of simplicity, showing the distribution of a single but important parameter of landforms. Their main limitation arises from the known inaccuracy of angles derived from contours. For this reason they can be used to give a generalized impression of the distribution of steep and gentle slopes, but cannot provide reliable estimates of the proportions of the ground surface lying within different angle classes. An isoclinal map based on sub-parallel profiles, a method practicable for limited areas using a profile recorder, would be considerably more accurate, and of value both in slope studies and for applied purposes.

Average slope maps show the average slope over distances of a mile or kilometre, as derived from contours. Straight lines are drawn across the base map and the number of contour crossings is counted. The formula for average slope, in which the correction for obliqueness of crossings assumes random contour directions, is

$$\tan\theta = \frac{V.N}{0{\cdot}6366\,K} \qquad (14.5)$$

where V = vertical contour interval (m or ft), N = number of contour crossings per km or per mile, and K = 1000 for metric units or 5280 for feet and miles. For example with a vertical interval of 10 m, 23 crossings per km gives

$$\tan \theta = 230/636 \cdot 6 = 0 \cdot 362, \text{ or } \theta = 20°.$$

This is called the *Wentworth method* (Wentworth, 1930; Zakrzewska, 1967). Bonniard (1929) gives a method of obtaining a *clinographic curve*, which represents graphically the mean slope between each pair of contours.

The property represented on average slope maps is not in fact the value that would be obtained by taking the average of field measurements of angle, but is an artificial parameter. Average slope maps can only be produced at scales smaller than those of isoclinal maps. They are a useful means of regional geomorphological description (Raisz and Henry, 1937; Trewartha and Smith, 1941; Calef and Newcomb, 1953).

COMBINED MORPHOLOGICAL PARAMETERS

A classification of the shapes of parts of the ground surface is obtained by combining profile and plan curvature. If profile curvature is divided into convex, rectilinear and concave, and plan curvature into plan-convex, straight and plan-concave, a qualitative grouping into nine classes is obtained. If seven profile curvature divisions (p. 163) and five plan curvature divisions (p. 176) are used, the number of classes is increased to 35; further subdivision on the basis of angle classes is possible. Schemes constructed on this basis have been given by Ruhe and Walker (1968) and Ahnert (1970). Maps based on such classes would be a means of showing the nature of the landforms, but would be difficult to produce accurately. For the purpose of quantitative characterization of an area, class frequencies obtained from a systematic point sample are preferable. The class to which a given point on a slope belongs may be determined by three measurements: the angle over a standard distance in an upslope direction, the corresponding downslope angle, and the plan curvature; the profile curvature and estimated angle at the point are obtained from the two angle values.

An objective means of mapping landforms is obtained by combining slope with other morphological properties. The parameters used vary according to the scale of mapping. For initial mapping at a large scale, Speight (1968) devised a system based on four parameters, and using nine classes of slope angle, four of profile curvature, four of plan curvature and three of unit catchment (Fig. 70). Many of the theoretically possible 432 classes did not occur. Those that were present were grouped subjectively into seven land-form elements: crests, hill slopes, concave foot slopes, convex foot slopes, swales, plains, and water courses. The resulting map of these elements was compared with a map of land systems obtained by conventional air photograph interpretation methods, and substantial differences were found in the percentages of each land system occupied by the landform elements. This is a potentially valuable approach to landform mapping.

If the map scale is progressively made smaller, a point is reached at which it is no

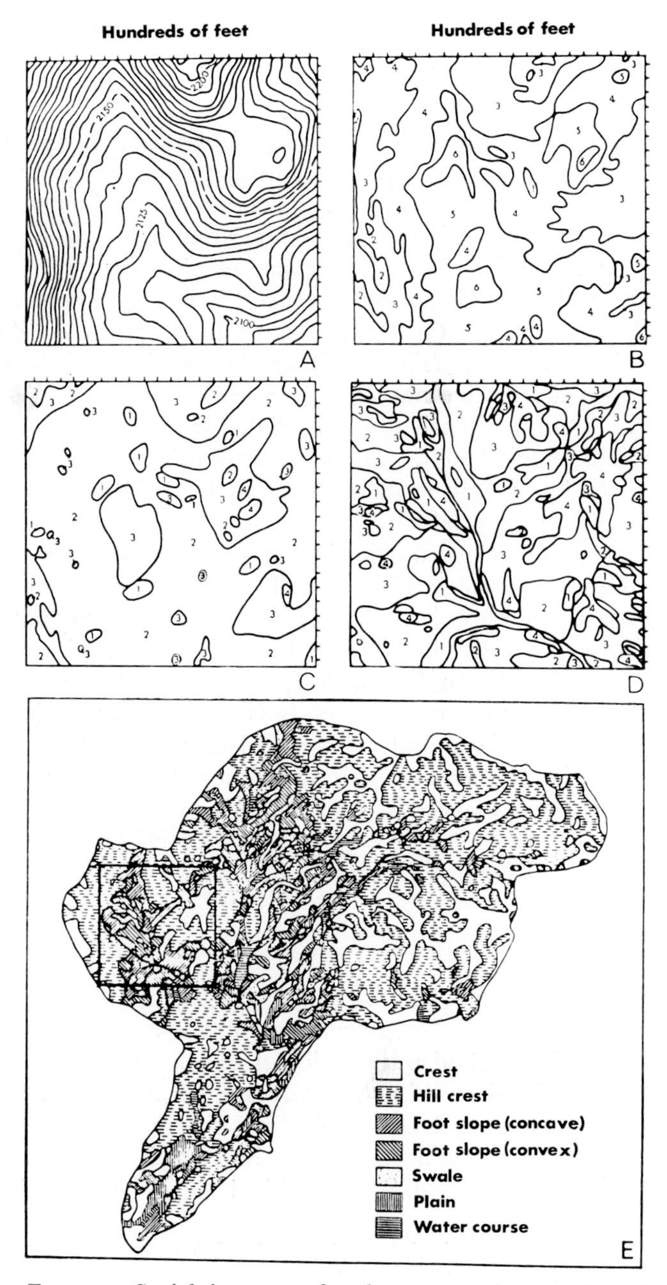

FIG. 70. Speight's system for the parametric mapping of land-form elements. *A* = contour map, vertical interval 5 ft (1·52 m). *B* = slope angle. *C* = profile curvature. *D* = plan curvature. *E* = land-form elements; the square encloses the area shown in *A-D*. The terminology used for *C* and *D* differs from that of the original. After Speight (1968).

longer practicable to map properties of individual points of the ground surface. The unit of mapping then becomes the valley as a whole, and the necessary descriptive parameters change. Actual slope angle is replaced by mean slope, or by the percentages of defined gentle and steep slope categories; profile curvature is replaced by terms describing whether interfluves and valley floors respectively are angular, rounded or flat (cf. Ollier, 1967); unit catchment is no longer relevant, and is replaced by mean valley spacing or drainage density, and in place of plan curvature, other means of describing plan form characteristics become necessary, e.g. whether linear or non-linear, orientated or randomly directed. In addition, relative relief becomes an important descriptive attribute. The resulting classes of combined parameters can be grouped and given descriptive names, e.g. rugged mountains, dissected plateaux, undulating plains. A comprehensive system of this kind is described by Van Lopik and Kolb (1959; cf. Speight, 1967). Methods for obtaining quantitative parameters of plan form are given by Speight (1968). A simple but effective use of the method is in a landform map of Cyprus produced by Thrower (1960).

The approach of combined parameters is a methodologically sound means of landform mapping. The initial stage of data collection is objective and quantitative. The values of each parameter are then grouped according to some numerically rational system, e.g. equal division of the value or its logarithm. Combination of the single-parameter classes yields multi-attribute classes. Subjectivity occurs only at the final stage, in which the number of mapping units is reduced by deciding that certain bounding values of the multi-attribute classes are unimportant, and combining the classes separated by such values. The main criticism of work so far carried out is that it has been based on contour maps. Detailed mapping of combined parameters based on field survey, using a systematic grid of points, would give a landform map equal or superior to that produced by morphological mapping.

GENETIC GEOMORPHOLOGICAL MAPS

In this method of geomorphological mapping, landforms are shown according to their presumed origin; past and present processes, together with the 'age' of the landforms, are also given. The method was developed mainly in continental Europe, each country employing modifications of a basically similar approach. Discussions are given by Klimazewski (1963) and Gellert (1967); Gilewska (1967) shows the same area of Silesian scarpland mapped according to the French, Hungarian, Russian and Polish methods. The International Geographical Union (1968) has published a unified key.

The system of Verstappen and Van Zuidam (1968) may be taken for illustration. The eight genetic groups of landforms distinguished are forms of structural, volcanic, denudational, fluvial, marine, glacial and periglacial, aeolian and solutional (karst) origin. Within each group, between 20 and 53 forms are allotted mapping symbols; thus the forms of denudational origin include planation surface, sediment, solifluction, earthflow, badlands, peat and ferruginous duricrust. There are additional groups of symbols for 'morphometry' (e.g. sharp crestline, asymmetrical rounded valley) and for lithology. In all, 326 landform symbols are given, to which are added letter symbols for 'age', e.g. Holocene, Pleistocene. In addition to this scheme for the general geomorphological map, there are two special-purpose maps, a morpho-conservation map and a hydro-morphological map, each with a different set of symbols. The morpho-conservation map includes colour shading for slope angle classes, together with signs for breaks of slope.

'The distinct characterization of an area after its morphogenesis [is] the foremost aim of geomorphological mapping' (Gellert, 1967). Insofar as such maps are intended to present the final results of research, two disadvantages from the point of view of the map user are the multiplicity of symbols and the very disparate nature of the information represented. It has also been suggested that these maps may be used as a basis for further research (e.g. Ottmann and Tricart, 1964). But for investigating the origin of landforms they are fundamentally unsuited, owing to the high degree of subjective interpretation involved in compilation; examples are the mapping of a fault-line scarp, creep, an inselberg, and a nivation hollow, and in all attributions of age to forms. Such interpretation begs the question of the origin of such forms, committing the same error that is present in Davisian landform terminology.

A further defect of the continental systems is the unsatisfactory representation of valley slopes. Slopes with special properties, such as scarp slopes or screes, are allotted symbols, but normal valley slopes may be covered by shading or symbols for bedrock type, presumed surface processes, micro-relief, superficial deposits or supposed 'age'. With such diversity of treatment, it is impossible to use the map for purposes other than as an excursion guide. If it is to be used as a basis for any kind of study, uniformity of the attributes represented is necessary. If, for example, terracettes are shown, then it is necessary that a category of micro-relief be recognized, classes defined within it, and these classes systematically surveyed for the whole area of the map. Although much ingenuity and effort has gone into the construction of maps of this type, they contribute little to the sum of knowledge. For both pure and applied slope studies they are inferior to maps of slope angle, or maps of a small number of landform properties classified and represented in a systematic manner.

SLOPE CLASSIFICATION

The aim of slope classification is similar to that of the classification of other environmental features: to produce sub-groups of individuals about which more meaningful statements can be made than about the populations as a whole. Systems for classifying slopes are either classifications of whole slopes or classifications of parts of slopes. In the first of these groups the individuals are entire slopes, from crest to base and in some cases with a finite lateral extent. In the second group the individuals are essentially points on the ground surface, although in practice small areas are involved. Some whole-slope classifications are based upon the presence, absence or frequency of properties belonging to parts of slopes. The validity of any whole-slope classification of this nature depends on the rigour with which the component units are determined. Classifications of parts of slopes can be used to produce large-scale maps of the properties classified; isoclinal maps are an example. Whole-slope classifications can be represented only on medium to small-scale maps. The following discussion is concerned with principles; no classification system, however, can be assessed on theoretical grounds alone, and there is a need to test the various schemes that have been proposed by more extensive comparisons with observed data.

CLASSIFICATIONS OF WHOLE SLOPES

(*i*) *By origin.* In this approach, the slopes are grouped according to some factor assumed to be causally related to the nature of the slope. The factors that may be employed are the *agency responsible*, e.g. fluvial valley slopes, glaciated valley slopes, marine cliffs, and glacial depositional slopes; the *climate*, e.g. slopes of polar, temperate, arid, and humid tropical climates; *vegetation*, e.g. slopes carrying rainforest, savanna, or steppe; and *rock type* or *structure*, e.g. granite or limestone slopes, scarp and dip slopes. Where this basis is used, the criteria employed in allocating a slope to a class should be the genetic factors alone, independently observed, and not properties of the slopes themselves. Provided that the principle is followed, it is then possible to ask whether slopes associated with particular genetic factors possess properties in common, for example whether slopes formed on granites are predominantly convex, concave or neither. This method is similar to the genetic approach employed in the Russian approach to soil classification.

One structure-based classification, first used for soil catenas and subsequently for scarps (p. 119), may usefully be extended to all slopes. *Simple slopes* are formed of one rock type. *Compound slopes* are formed of two rock types, commonly a more resistant overlying a less resistant stratum, e.g. sandstone over shale. *Complex slopes* are those formed of more than two rock types, or more than one succession of two types, e.g. sandstone/shale/sandstone.

The validity of this approach depends on whether the factors can be independently observed. For rock type this is so, and such classifications can safely be used as a basis for further investigations. Climate, however, may have changed within the time-scale necessary to cause appreciable form change, and hence the slopes may possess relict features. It must therefore be made clear, for example, whether a class of 'arid climate slopes' refers to those actually found in arid regions at the present day, whatever may have been their origin, or to a hypothetical class of slopes formed wholly under arid conditions, which may not necessarily be represented at all in the present landscape. These two classes are distinct from each other; the extent to which the former corresponds to the latter is a matter for investigation. The danger of circular argument is apparent. The same reservations apply to classifications based on vegetation. Relict features also complicate the use of classifications based on the agency responsible. It cannot be assumed that the agent currently active at the base of the slope, e.g. a river, glacier or the sea, has been active for the entire period during which the present slope form has evolved.

(*ii*) *By the presence or absence of given features.* Examples are the divisions between slopes *with* and *without a free face*; *with* and *without a regolith cover*; and into those with *impeded*, *partly impeded* and *unimpeded basal removal* (p. 7). As a means of classification this approach is legitimate provided that the property concerned can be directly observed. This is true of the first two examples cited above, but is in some cases difficult to establish in the third. All three involve distinctions that are of basic importance in relation to surface processes and slope evolution.

(*iii*) *By surface form*. A division of valley slopes into *valley-head slopes*, *spur-end slopes* and *valley-side slopes* has been described above (p. 4 and Fig. 4). It is based partly on plan form and partly on the position of slope in relation to drainage lines.

The classification of whole slopes in terms of their profile form requires the prior determination of this form for the component parts. This may be obtained by analysis into slope units (p. 148). In order to avoid excessive complexity the values used for maximum permissible coefficients of variation should be those that give the generalised form, namely $V_{amax} = 25\%$ and $V_{cmax} = 50\%$. The next stage is to identify convexities, concavities, and maximum and minimum segments, from which the profile is divided into slope sequences (p. 149). Slopes may then be classified on a hierarchical system with three categories. The uppermost category is based on the number of sequences. The primary division is into *single-sequence* and *multi-sequence* slopes, and the latter are subdivided according to the number of sequences present. The unit for classification in the second and lowest categories is the sequence. The sequences are grouped according to whether they are *convex-concave*, *convex-rectilinear-concave*, or combinations lacking both or either a convexity and or concavity; cases of the latter type are of infrequent occurrence.

There are two alternative methods that may be used in the lowest category. The first, the order of slope unit method, is based on the succession of convex, rectilinear and concave slope units; schemes of this type have been put forward by Savigear (1967) and Ahnert (1970b). The second, the proportional method, is based on the relative lengths of the convexity, maximum segment and concavity.

In the *order of slope unit method*, the units presented are listed from the crest downwards, using the symbols X = convex element, V = concave element, M = maximum segment, N = minimum segment, S = segment (other), I = irregular unit, and oblique strokes to indicate divisions between sequences. For example the simplest type of single sequence, convex-rectilinear-concave slope is that consisting of three units only, *XMV*; if the convexity consisted of two elements separated by a segment the slope class would become *XSXMV*. The method of profile analysis employed here does not identify breaks of slope as such, but represents them as slope elements of high curvature. If, however, a method of analysis that distinguishes breaks of slope is employed, these may be included in the classification, representing convex and concave breaks by x and v respectively. For example, a profile on which the convexity includes adjacent segments separated by angular breaks could be represented as *XSxSxMV*. The incorporation of the lengths, absolute or relative, of slope units into this method requires the addition of a fourth category to the classification.

The *proportional method* does not employ slope units as such in the classification, except insofar as they are used to determine the sequences. The basis is the relative extent of the convexity, maximum segment and concavity, expressed as percentages of the total ground surface length of the sequence. The following adjectival terms are employed:

Predominant:	occupying > 80% of the sequence
High:	occupying 50–80% of the sequence
(unqualified):	occupying 10–50% of the sequence
Low:	occupying < 10% of the sequence

The possible classes of sequence, identified on this basis, are given in Table 13.

Table 13. Classification of slope sequences according to the proportions occupied by the convexity, maximum segment and concavity

	Percentage of sequence occupied by:		
Class of sequence	Convexity	Maximum segment	Concavity
1. Predominantly convex	>80	< 10	< 10
2. High-convex	50–80	10–40	10–40
3. High-convex, low-rectilinear	50–80	< 10	10–50
4. High-convex, low-concave	50–80	10–50	< 10
5. Predominantly rectilinear	< 10	>80	< 10
6. High-rectilinear	10–40	50–80	10–40
7. High-rectilinear, low-convex	< 10	50–80	10–50
8. High-rectilinear, low-concave	10–50	50–80	< 10
9. Predominantly concave	< 10	< 10	>80
10. High-concave	10–40	10–40	50–80
11. High-concave, low-convex	< 10	10–50	50–80
12. High-concave, low-rectilinear	10–50	< 10	50–80
13. Low-convex	< 10	40–50	40–50
14. Low-rectilinear	40–50	< 10	40–50
15. Low-concave	40–50	40–50	< 10
16. Equally proportioned	10–50	10–50	10–50

The usefulness of classification systems tends to vary inversely with their complexity. The order of slope unit method has a large, and theoretically infinite, number of classes, and for this reason the proportional method is to be preferred. Its 16 classes of sequence provide a basis for the quantitative comparison of slopes from different rock types, climatic regions, and other environmental circumstances.

CLASSIFICATIONS OF PARTS OF SLOPES

(i) By surface form. Points on, or parts of, the ground surface may be classified on the basis of angle, profile form, plan form, or a combination of these properties. These methods have been described above (pp. 163, 173, 176).

(ii) By regolith, ground surface features, or processes. Bryan (1925) grouped desert relief of Arizona into cliffy, debris-mantled (or boulder-controlled) and rain-washed slopes;

Strahler (1950) distinguished high-cohesion slopes at >40°, repose slopes at 30°-35°, and 'slopes reduced by wash and creep' and gentler angles; Jahn (1960) divided slopes of Spitzbergen into rock walls, screes and scree cones, and solifluction terraces. These classifications are based partly on the nature of the regolith and ground surface, and partly on the processes assumed to be operative. These types of criteria should not be confused. The nature of the ground surface is a directly observable property, and is therefore a legitimate basis for a classification. The following are among the types of ground surface: free face (cliff), rock slope, scree, debris slope (p. 135), clitter (p. 130), and regolith-covered slope. In some cases the processes active can be inferred from the ground surface, but such inferences should not be included in the definition of a class. It is not legitimate, for example, to describe regolith-covered slopes with a continuous vegetation cover as 'creep slopes', since this begs the question as to whether creep, wash, solution, solifluction or some other process is the predominant transporting agency upon them. Slopes can only be classified on the basis of dominant process if this has been established by direct measurement.

(iii) By combined properties. The hillslope elements of Wood (1942), as modified by King, have been described above (Fig. 12, p. 37). Where all four elements are present the slope is termed a 'standard hillslope'. The elements are defined mainly on the basis of profile form, but criteria involving the ground surface, regolith, and processes are also given. Slope models based on similar component units have been proposed by Ruhe (1960) and Savigear (1960, footnote 2).

The nine-unit landsurface model proposed by Dalrymple (1968) is of a similar nature. It originated from observations of slopes and soils in the North Island of New Zealand. The model is 'based both on form and contemporary geomorphic and pedogenetic processes'. The units that make up the model are shown in Fig. 71. The upper part of the profile is divided into units 1-3, the interfluve, seepage slope and convex creep slope. These are nearly level, gently sloping and markedly convex respectively; the interfluve has a smooth soil cover, the seepage slope has shallow surface depressions and percolines, and the convex creep slope has terracettes on its steeper portions. Unit 4, the fall face, consists of bare rock. Unit 5, the transportation mid-slope, is characterized by the occurrence of recent and old mass movement scars, as well as terracettes. The colluvial footslope, unit 6, is rectilinear to concave, with a smooth regolith cover. Units 7-9 are associated with the river. The alluvial toeslope represents the flood-plain or river terrace, whilst the channel wall and channel bed are parts of the river. Unit 1 occurs once only on every profile. The other eight units need not all be present. The units that occur on a given slope are listed in order from the crest to base; where the same unit occurs more than once it is followed by a bracketed numeral on the second and subsequent occurrences (Fig. 71 (*a*) and (*b*)).

The processes stated to occur on the nine units are for the most part inferred from the ground surface and regolith composition; they have not been established by measurement. The main process on the colluvial footslope is stated to be deposition; the impossibility of the continuance of net accumulation under steady state conditions has been noted (p. 197). The inclusion of presumed processes in the descriptions of unit

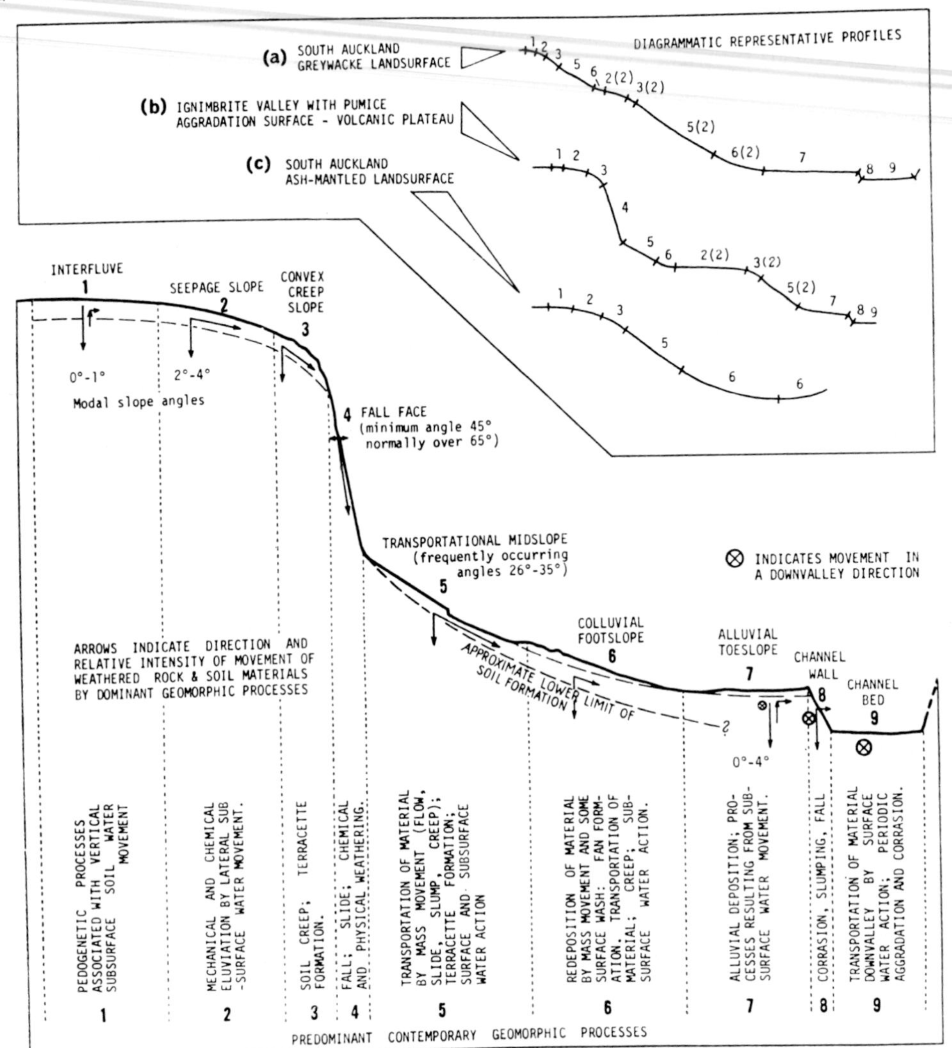

FIG. 71. Diagrammatic representation of the hypothetical nine unit landsurface model. After Dalrymple *et al.* (1968).

could, if the purpose of the model is misunderstood, inhibit the objective study of whether such processes are in fact predominant.

There is a clear correspondence between the hillslope elements of the Wood-King model and the units of the nine unit landsurface model. The 'crest' of the former is subdivided into units 1-3; the 'scarp' and the fall face are expressly equivalent, whilst the transportational midslope, an irregular facet at 26°-35°, clearly corresponds to the 'debris slope'. The colluvial footslope resembles the 'pediment' in respect of surface form, except that it is more steeply sloping. Units 7-9 are not part of the slope proper.

The main differences between the models are associated with the fact that King based his descriptions on semi-arid conditions, whereas the nine unit model is based on a region with a humid temperate climate. This contrast between the type-climates accounts for differences between the ground surface features of the debris slope and the transportational midslope, and also for the greater steepness of the colluvial footslope as compared with the pediment.

Both models were devised to apply primarily to slopes that include a free face. Although claimed to be applicable to slopes without one (Fig. 71 (*c*)), the definitions of units become forced in such cases; in particular, a regolith-covered maximum segment does not correspond, either in surface form, regolith characteristics or mode of origin, to the debris slope beneath a free face. This restriction greatly limits the extent of applicability of both models, since slopes lacking a free face are of substantially greater world extent than those possessing one.

Apart from the disadvantage of restricted applicability, both models have the same major limitation. This arises from the definitions of the component units, which are based on combinations of properties (profile form and the surface features and composition of the regolith) and processes. Such definitions assist in identifying the corresponding units in different areas, but they leave unanswered the question of how the model is to be applied if these various criteria do not all occur in conjunction. Suppose, for example, that beneath the debris slope there occurs a part of the profile that has low angles but is gently convex. On the standard hillslope model this cannot be classified as a pediment, owing to its convexity. On the nine unit model it could be placed on the basis of profile form as a recurrence of the seepage slope, but if, however, it has recent and old mass movement scars, it then becomes uncertain whether it should be allocated to the transportational midslope on account of the signs of mass movement, or to the colluvial footslope on the basis of surface form. These models can be used as a static basis of comparison, for the identification of areas in which units similar in all aspects are found. They lack the inbuilt flexibility, and capacity for the incorporation of variations, that is necessary in models if they are to take account of the variety of forms to be found in the natural environment.

XV REGOLITH AND MICRO-RELIEF

THE dynamics of the regolith can be divided into formation, movement, and destruction or removal. Formation is usually by rock weathering. Destruction and removal are by solution loss and basal removal respectively. Between these stages the regolith is present as a layer on the slope surface, partly *in situ* and partly moving downslope. That portion of the bedrock which does not go directly into solution passes into the regolith as a stage prior to its removal from the slope. A quantitative knowledge of regolith dynamics, i.e. the manner and rate of its formation, movement and removal, would solve most problems of slope evolution, but we are far from possessing such knowledge.

An interrelationship exists between slope form and regolith. The regolith cannot be considered wholly as an environmental factor, or independent variable, as can structure and climate, nor is it wholly a dependent variable. The first part of this chapter is concerned with the regolith as a property of form. In the second section, the effects of differences in regolith properties upon surface processes, slope form and evolution are discussed. These two sections thus refer to the influence of slope upon regolith and the influence of regolith upon slope respectively. The final section covers micro-relief, a topic which involves both ground surface form and regolith. Slope-soil relations from a pedological aspect are discussed in Chapter XX.

REGOLITH PROPERTIES

The form characteristics of a slope include the properties of the regolith. Two properties which are of particular relevance to slope evolution are the thickness of the regolith and its degree of reduction, of which representative parameters include the percentage coarse fraction (>2 mm) and the percentage clay ($<2\mu$). In theoretical discussions of slope evolution assumptions are frequently made about the properties of the regolith for which there has, until recently, been little observational basis. A further application is in deductive work using process-response models. When comparing the results of such models with actual slope form, the difficulties caused by equifinality (p. 20) are lessened if, in addition to the shape of the ground surface, observations of the regolith are available. Less work has been done on regolith characteristics than on surface form, and the results quoted may prove to be of local application only.

Variations in regolith properties are conveniently discussed in terms of a slope comprising a convexity, a maximum segment and a concavity. On three areas of Palaeozoic sedimentaries in Britain, Young (1963) found that across the length of the convexity the regolith thickness either remained constant or increased slightly downslope. Thickness usually increased downslope on maximum segments, the amount of increase being greater the steeper the segment. Over both the convexity and the maximum segment there was no evidence of systematic change in the degree of reduction, except that where the slope exceeded 25°, the proportion of large stones increased. Other reported results accord with these trends, except that Clay-with-flints overlying Chalk thins downslope across the convexity (Ollier and Thomasson, 1957). This, however, is a drift deposit originating near the crest, not material derived from the bedrock.

A variety of conditions, with respect to both thickness and reduction, occurs on concavities. Young (1963) found that the thickness usually showed a further downslope increase, which was either slight or considerable, although cases in which it remained unchanged also occurred. On the lower part of the concavity the degree of reduction increased considerably, the soil becoming almost stoneless; the angle at which this change commenced varied between 9° and 20°. On chalk an abrupt increase in the clay and silt fractions at the upper limit of the concavity was noted by Furley (1968). It is not always the case, however, that the soil becomes finer textured on concavities. On a forested area of sandstones in Virginia, Hack and Goodlett (1960) found the debris was much coarser in valley-head slopes than on convexities. On Mesozoic sandstones in Britain, Furley found no difference in clay percentages between convexity and concavity, whilst some pedological studies have reported a coarser regolith on the concavity. In temperate latitudes it has been demonstrated that the regolith on some concavities originated as a periglacial deposit (p. 241).

In terms of single variables, Ahnert (1970b) found that on gneissic schists in North Carolina the regolith thickness varied inversely with slope angle and directly with distance from the crest. Degree of reduction showed no significant correlations.

Under humid tropical conditions the thickness of the weathered mantle commonly exceeds 10 m, and sometimes 100 m. It may thin abruptly close to interfluve crests. Most of this thick weathered layer is *in situ*, as demonstrated by undisturbed quartz veins. The moving layer of the regolith is only of the order of 1 m thick. Stone lines have been held to represent the boundary between material *in situ* and that which has moved downslope, but this is not established (p. 52). The degree of reduction remains constant over the convexity and maximum segment. Under savanna environments a common feature is that the upper slope regolith gives place to more sandy material on the upper and middle parts of the concavity, which is in turn replaced by clay on the lower part (p. 256). Under rainforest conditions in Malaya, Swan (1970) found that the silt-plus-clay fraction was higher on concavities than on upper slope sites.

A different approach, which treats slope deposits in a stratigraphic manner, has been developed in Australia. It is based on evidence of buried soil profiles, which are interpreted in terms of periodic phenomena or *K* cycles (Butler, 1959, 1967). The theory assumes that, as a result of climatic changes, slopes have undergone alternating unstable and stable phases. During an unstable phase (*Ku*) downslope regolith transport occurs, with denudation on the upper part of the slope and accumulation on the lower part. During the succeeding stable phase (*Ks*) such transport ceases, and mature soil profiles are developed. The next unstable phase produces further erosion and deposition, either truncating or burying these soils. The regolith is consequently interpreted as consisting of successive depositional layers, the youngest uppermost, each having been laid down during an unstable phase. Walker (1962) distinguished three *K* cycles in New South Wales, equating stability with humid climatic phases and instability with increased surface transport under drier conditions (Fig. 72). In Papua,

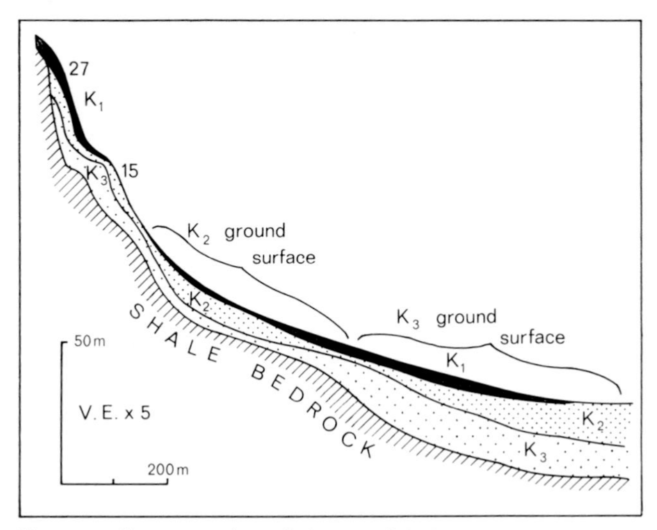

FIG. 72. Interpretation of the regolith in terms of *K*-cycles, the K_3 layer being the oldest deposit, and K_1 the youngest. New South Wales. After Walker (1962).

Mabbutt and Scott (1966) distinguished eight layers in the regolith, using pedological and mineralogical evidence; on the basis of these layers they constructed a complex chronology of morphogenesis, involving phases of weathering, stripping, aggradation and dissection. Whilst potentially a powerful means of elucidating slope evolution, this approach needs to be applied with caution, and its fundamental assumptions cannot be regarded as proven.

In addition to systematic trends, the regolith shows considerable random variation in properties. For samples drawn from maximum segments, Carson (1967) found the following coefficients of variation: soil depth (as estimated by auger penetration) 34%, percentage stones 41%, average stone size 34%, and silt-plus-clay 3%. The coefficient of variation of clay content for apparently uniform soil mapping areas has been recorded as 26% (Webster and Beckett, 1968). The magnitude of these variations is such that in statistically-based studies of regolith characteristics the number of samples should be at least ten.

Comparing these results with the assumptions made in theoretical studies, the most realistic generalizations for observed slopes are that the regolith remains unchanged in thickness and reduction across a convexity, and thickens down a rectilinear segment.

The common assumption that the degree of reduction increases downslope on the concavity is usually confirmed with respect to a decrease in stoniness, but with respect to an increase in clay content it has not been substantiated. Moreover, it may be that the change to a finer regolith, where it occurs, takes place not gradually and progressively but abruptly, over a short distance on the upper concavity. The regolith in temperate latitudes shows one feature that is possibly of greater significance in relation to slope evolution than the variations in its properties with slope form; this is its thinness, when compared with the dimensions of the slope profile. Except where periglacial or other drift deposits are present, the regolith on all parts of the slope is usually less than 1-2 m, which if plotted to scale on most slope profiles would be little more than the thickness of the line representing the ground surface. Diagrams accompanying theoretical discussions of slope evolution invariably show a regolith very much thicker than this in relation to the dimensions of the slope. Whilst vertical exaggeration may be necessary for clarity, arguments based on such diagrams may not be valid in terms of a very much thinner regolith.

A common fallacy is that the concavity of slopes is an accumulation zone. It is explicitly made by Furley (1968), and is implicit in the *K*-cycle approach. An accumulation zone must refer to a part of the slope which, over a period of time, undergoes a net gain of material (the passive balance of denudation of Jahn, 1954, 1968; see p. 101). The ground surface will be progressively raised. Such a condition, however, can only be temporary. It can occur where a steep slope has been formed by undercutting, and the undercutting agency has then ceased. Examples are former river bluffs over a flood plain, and abandoned sea cliffs. A concavity is then formed by accumulation. Most concavities, however, are found to be features of the bedrock form, and are not infillings of a former angular rock junction (Peel and Palmer, 1956; Young, 1963). If the concavities are assumed to be zones of accumulation, they cannot have come into existence under present conditions, for if an initial rectilinear slope developed a slight concavity, the latter would be quasi-simultaneously eliminated by accumulation. There remains the possibility that all present-day concavities are relict landforms, the product of conditions no longer operating, and are currently being eliminated. Whilst not impossible this is a drastic hypothesis, requiring stronger evidence than has yet been put forward before it can be substantiated (Young, 1969b).

INFLUENCE OF THE REGOLITH ON SLOPES

Insofar as the retreat of a given slope is subject to control by removal, the properties that influence its resistance are in part not those belonging to the rock but to the regolith derived from it. The permeability of the underlying rock is also significant. This point was first emphasized by Wolfe (1943), who showed how in an area of gently-dipping sedimentaries, subsequent ridge-and-vale topography had developed, despite the fact that the bedrock itself was rarely exposed. Differential lowering of the ground surface was in this instance attributed mainly to differences in the permeability of soils derived from the scarp-forming and vale-forming rocks, both shales.

It is a matter of importance, on which relatively little work has been done, to determine which properties of the regolith affect its resistance. These properties will differ according to the type of surface process which is dominant. Where landsliding is involved, shear-strength is appropriate. The major difference is between clays, which are unstable above *c.* 10°, and most other rock and regolith types for which the angle of long-term stability exceeds 25°. Chorley (1959, 1964) attempted to apply tests derived from soil mechanics to slopes in general, and obtained some statistically significant correlations. There is, however, no reason why shear-strength or penetrometer resistance should be applicable to the denudation of stable slopes, and unless such causal links can be demonstrated, it remains a possibility that the observed correlations are coincidental. The attempt by Carson and Petley (1970) to extend shear-strength analysis to a shallow slope regolith has been noted (p. 171).

The main soil variables affecting denudation by wash are probably infiltration capacity, aggregation, and stability. Jungerius (1965) describes a case where permeable sands, believed to have been laid down as river terraces, proved so much more resistant than the surrounding shale-derived clay soils as to become interfluve cappings, causing inversion of relief. Accelerated erosion experiments with bare soil demonstrate the importance of aggregate stability in controlling resistance to wash. Experiments on soils from northern England showed that brown earths possessed more resistance to wash than podzolic soils, and limestone soils more resistance than those derived from sandstones, the difference in both cases being caused by greater aggregate stability in the more resistant soils (Bryan, 1969). A criticism of this work is that sieved material was used in the wash tests. Soil texture is the main determinant of infiltration capacity, and also affects degree of aggregation. Aggregate stability is influenced by many factors, particularly organic matter content. Andre and Anderson (1961) found that soils derived from acid igneous rocks were two and a half times more erodable than those derived from basalt. Schumm and Lusby (1963) found that seasonal changes in the condition of the soil caused a corresponding change in its infiltration capacity.

With respect to removal by creep, it would be expected on *a priori* grounds that fine-textured soils would move more rapidly than coarse, since they undergo greater expansion and contraction for a given percentage moisture change. For the same reason, soils with a high content of expanding-lattice clay minerals probably creep more rapidly than those containing largely fixed-lattice minerals.

The above aspects taken together demonstrate the overall importance of permeability, either of the regolith or of the underlying rock, in determining the rate of ground loss from a slope. This is in accordance with the observed high resistance of limestones, and greater resistance of sandstones than shales.

DURICRUSTS

Duricrusts are hard, massive layers in the regolith formed by material of secondary, pedological, origin. On the basis of composition they are divided into *silcrete*, *calcrete*, *indurated bauxite* and *laterite* (= ferricrete, indurated plinthite), which are respectively siliceous, calcareous, aluminous and ferruginous. The term laterite is used here to refer only to ferruginous materials which occur naturally in an indurated condition. The geomorphological effects of duricrusts, of which laterite is much the most common, were first recognized by Woolnough (1918, 1930).

The effects of duricrusts on slope form and evolution are comparable to those of horizontally-bedded resistant rock strata. The slope consists of a free face, debris slope and pediment, and evolves by parallel retreat. The upper part of the free face is formed by the indurated material. The lower part, together with all or part of the debris slope and pediment, is formed from the deep layer of highly-weathered, crushable material that underlies the duricrust. In the case of laterites this may comprise the mottled and pallid zones, the latter kaolinized (Walther, 1915) (Fig. 73). Slope retreat is caused by

FIG. 73. A 60 m high laterite-capped cliff, in the Dallol Maouri, Niger, W. Africa. The plateau is underlain by 3 m of ferruginized sandstone overlying a 5 m mottled zone, and the remainder of the cliff consists of the former pallid zone, a kaolinized sandstone. On the right, where the capping is removed, the crushable kaolinized material is being rapidly dissected. The debris slope is partly vegetated. The present climate is semi-arid, and the laterite relict from more humid conditions.

removal of the soft underlying material, followed by vertical fissuring of the duricrust; blocks of the latter break off, falling into and subsequently becoming embedded in the debris slope. Processes causing undermining include surface wash, slumping, and the eluviation of clays by subsurface seepage. Positive evidence of the latter process is found in the case of some ferricretes which show cambering close to the scarp (Moss, 1965). Slopes showing evidence of retreat by such undermining processes are aptly termed *breakaways*.

Laterite breakaways occur most widely in savannas, but are found also in tropical climates ranging from rainforest (Hunter, 1961) to semi-arid (Brammer, 1956, 1956b). In southern Uganda, where an extensive sheet of laterite is in course of dissection, the slopes of an entire area are dominated by breakaway forms with underlying pediments (Pallister, 1956, 1956b), and a similar landscape has been described in northern Nigeria (Dowling, 1968). Parallel retreat of calcrete-capped slopes occurs in Cyprus (Everard, 1963, 1964). Where a duricrust capping is eliminated by intersection of opposing slopes, rapid decline or gully-dissection of the underlying material occurs. Thus slopes with a duricrust layer are effectively compound slopes, with respect both to form and to manner of retreat.

The general question of laterite formation will not be discussed; a review is given by Maignen (1966). But it is appropriate to mention the hypothesis of Trendall (1962), since this calls substantially upon mechanisms of slope evolution. The orthodox theory for the origin of extensive, fairly horizontal laterite duricrusts is that they originated as groundwater laterites, formed by the taking of iron into solution and its irreversible precipitation, in ferric forms, within the zone of seasonal fluctuation of the water table. A very gently undulating erosion surface is assumed. Trendall found that to obtain the iron held in a 10-m thick laterite overlying granites in Uganda would require a 200-m thickness of rock. He consequently challenged the view that the present altitude of such a layer corresponds to a long period of stable base-level. His hypothesis is that horizontal laterite sheets develop on interfluves during a prolonged period of vertical stream erosion. It is assumed that the two main denudational processes operating are solution and surface wash. The critical assumption is that 'the fastest surface lowering will take place where the opportunity for solution between the upper and lower limits of the water table is greatest; that is . . . at the crests of the ridges'. Surface wash will, however, cause greater lowering further from the crest. The net effect will be that somewhere between the crest and the stream there will be a point of minimum lowering, which will lead to the production of a convex break of slope. This break will be self-intensifying, through its influence on the runoff/infiltration ratio.

This ingenious theory is unacceptable on several counts. There is no evidence that solution lowering is greatest near interfluves; the formation of an angular break of slope as a result of continuous variation in the intensity of processes has never been demonstrated; and where the water table is sloping, iron in solution will migrate laterally towards the valleys. More generally, it calls for too fine a balance to be maintained between stream erosion and two different denudational processes. A simpler explanation of the progressive concentration of iron derived from a great thickness of rock is that the laterite, with the pallid zone beneath it, has subsided progressively downwards as the rock beneath is lowered by solution, during the later stages of a prolonged period of planation under a stable base-level (de Swart, 1964).

In a part of the Mato Grosso, Brazil, there appears to be contemporary and continuous formation of a laterite sill during slope retreat. Ferricrete, 30 cm thick, caps steep ledges which separate gently-sloping, broad, convex interfluve areas from steeper lower valley sides. Abundant seepage occurs from the ledges. Immediately above the top of the ledge, the ferricrete lies at a depth of less than 1 m. Some 50 m upslope it gives place to a soft, iron-mottled horizon, whilst beyond that, and beneath most of the crest area, there is no identifiable iron-rich layer. The implication is that the laterite is simultaneously being formed (by lateral seepage) and destroyed (by slope retreat), whilst producing an effect on the slope form comparable to that of a caprock (Townshend, 1970; Young, 1970b).

MICRO-RELIEF

Micro-relief refers to irregularities of the ground surface with dimensions of an order smaller than that of a slope sequence. Examples are terracettes, solifluction terraces, rill channels, gilgai phenomena and termite mounds. Most micro-relief features are developed entirely in the regolith; the bedrock/regolith junction also has considerable irregularities but these are not (or have not been shown to be) related to the surface micro-relief. A definition in terms of dimensions is unsatisfactory, since there is an overlap in scale with some valley slopes, e.g. in badland relief. Most micro-relief features, however, have a 'height' range (i.e. perpendicular to the slope as a whole) of between 10 cm and 5 m, and 'horizontal' dimensions (i.e. parallel to the slope) of 20 cm to 20 m. Surface form features of a smaller order of magnitude may be termed *nano-relief*. It has not been determined whether there is a continuous gradation between surface irregularities of different magnitudes or whether discontinuities can be distinguished. There are possibly four distinct orders of scale, exemplified by the valley-interfluve, the terracette, the faces of soil aggregates and the faces of textural particles (cf. Savigear, 1967); this question awaits quantitative investigation.

The following approach to micro-relief, based in part on Stone and Dugundji (1965), treats ground surface irregularities as wave phenomena. Consider first the case where recurrent features such as terracettes are apparent. The features are surveyed, either by levelling or with the profile recorder, and plotted. The mean profile of the slope, i.e. excluding micro-relief, is obtained, either by graphical estimation or by fitting a third-order polynomial curve, and plotted, and is termed the *macro-slope* profile. The crest and trough of each micro-relief feature are identified as the points farthest from the macro-slope. The following parameters are obtained for each individual feature, from which are derived their mean values (Fig. 74*A*):

Upper semi-wavelength	L_a
Lower semi-wavelength	L_b
Wavelength	L
Upper semi-wave height	H_u
Lower semi-wave height	H_l
Wave height	H
Parallel symmetry index	$I_h = L_a/L$
Perpendicular symmetry index	$I_v = H_u/H$
Mean upper-half angle	ϕ_a
Mean lower-half angle	ϕ_b
Roughness index	$R = \bar{H}/\bar{L}$

The mean half-slope angles could theoretically be derived from L, A and θ (the angle of the macro-slope), but are better obtained directly from the co-ordinates of the crests and troughs. The roughness index is independent of θ; it is zero on a perfectly

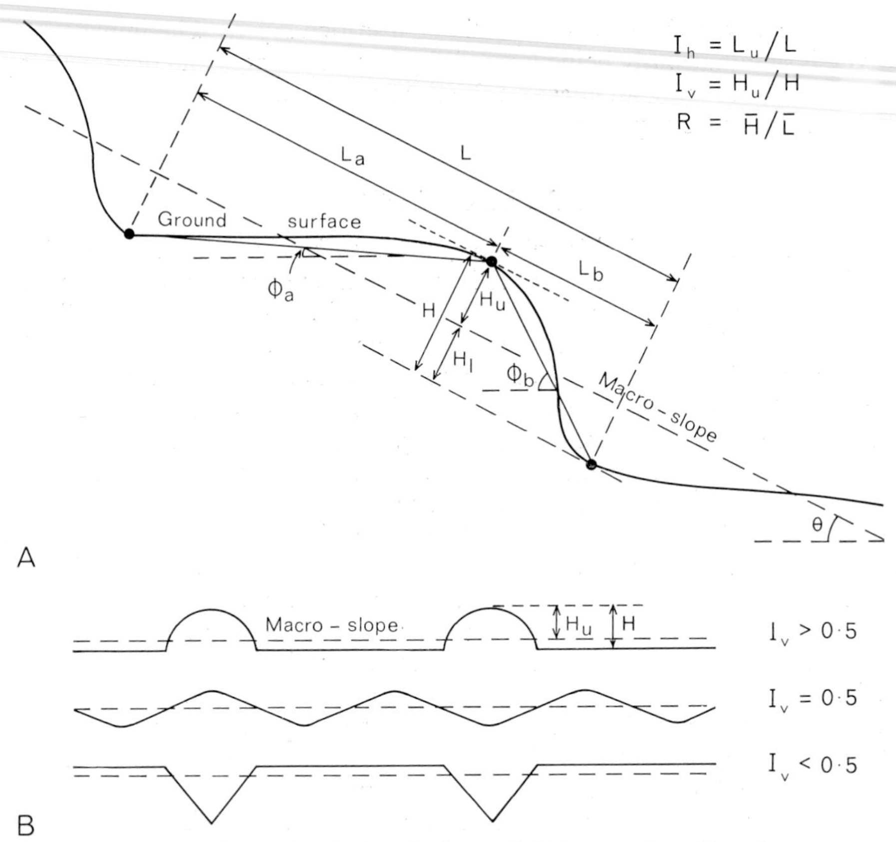

FIG. 74. Parameters for the description of micro-relief. For meaning of lettering, see text.

smooth slope, becoming 1·0 where wave height equals wavelength. Parallel and perpendicular symmetry denote directions in relation to the macro-slope, and are equivalent to horizontal and vertical symmetry for a level macro-slope. The parallel symmetry index exceeds 0·5 if the longer semi-wavelength lies on the upslope side. The perpendicular symmetry index is 0·5 where the crests and troughs have the same shape. It lies above 0·5 for isolated 'hills', e.g. termite mounds, and below 0·5 for a slope with isolated depressions, e.g. rill channels (Fig. 74*B*).

If recurrent features are not apparent the slope is surveyed in detail and wave forms identified by Fourier analysis (Stone and Dugundji, 1965). The values of roughness index for different wavelengths provide a means of comparing surface form features of different orders of magnitude; for the macro-relief, the roughness index is given by valley depth/valley width.

The most widely distributed micro-relief features are *terracettes*, also known as 'sheep tracks'. They are sub-parallel, approximately horizontal steps, with nearly level treads 10-50 cm wide and risers that are steeper than the macro-slope. Regular terracette rows, as distinct from isolated steps, usually occur on steep slopes, over 32°

(Young, 1961; Kerney, 1964; Clark, 1965). They occur mainly under grassland. The turf is sometimes scarred at the base or on the face of the riser, and may overhang at the top.

There is a distinct and less common type of terracette, for which no recognized name exists, in which the risers are nearly vertical and bare of vegetation. They are less regularly aligned along the slope than normal terracettes (Sharpe, 1938, Plate VII; Meynier, 1951). They occur on steep slopes where the regolith is deep, for example on clay or loess, usually where there is undercutting at the base. This type of terracette is produced by subsidence, with some rotation, along vertical to steeply-angled fracture planes.

Normal terracettes are superficial features, and may occur where the regolith is as thin as 30 cm. Ødum (1922) observed horizontal cracks in the turf cover on steep slopes, and suggested that curved fracture planes, concave to the ground surface, developed from these, and that terracettes originated by miniature rotational slumping along these planes. Such fractures, if present, should be detectable by excavation, but have never been observed. That terracettes are primarily associated with trampling by cattle or sheep remains the most probable explanation. Rahm (1962) noted that grassed areas that were totally inaccessible to animals lacked terracettes, and they are absent from woodlands. Kerney (1964) obtained evidence, from valley deposits, that terracette formation in Kent, England, post-dated woodland clearance in *c.* 1000 B.C. Fenced forestry areas of 50 or more years' age might provide evidence on this point. A controlled experiment would be possible, destroying terracettes, re-seeding the ground, and then making part only of the area accessible to livestock. The mechanism of formation is that stock tread preferentially on an initial slightly flattened area, or the upper side of a grass tuft; this compacts the soil below, producing a flat tread and straining the turf above. The process is then self-reinforcing. The development of horizontal rows is probably because stock prefer to walk in this way, although fracture mechanisms may also be involved. Whether the lower limiting angle for terracettes is related to the strength of the turf or the regolith is not known. (Ødum, 1922; Sharpe, 1938; Meynier, 1951; Demangeot, 1951; Thomas, 1959; Rahm, 1962.)

Solifluction terraces or *lobes* occur in polar climates on slopes of 10°-30°; they have broader treads than terracettes. If the volumetric downslope movement by solifluction is greater where the regolith is thicker, an initial random thickening would override the thinner regolith below, so accentuating the initial thickness difference; self-reinforcement of this mechanism would produce a lobe. Mechanisms involving vegetation protection or destruction may also be involved (Watt and Jones, 1948; Metcalfe, 1950). *Micro-scarps* and *seepage steps* have been noted above (p. 72). *Termite mounds* produce substantial micro-relief in the tropics. *Earth pyramids* occur where a soft deposit containing boulders, e.g. till, is rapidly dissected by rainsplash and surface wash. The boulders act as a protective capping, and become isolated on pillars. Earth pyramids have an historical interest as the subject of one of the earliest descriptions, by Lyell in 1830, of the action of surface processes. A recent study has been made by Becker (1963). There are many micro-relief features of human origin, including lynchets, ridge-and-furrow, and ephemeral features such as are produced by ploughing.

XVI PEDIMENTS AND INSELBERGS

THERE can be no more common misconception in geomorphology than the belief that tropical landscapes consist exclusively of pediments and inselbergs. This impression, conveyed by much published work, arises partly from the historical accident that early work was carried out in the arid American West where such features are common, and partly because, as identifiable and often striking components of the landscape, they have attracted greater attention than other landforms. The attraction, for geomorphologists, of the isolated Ayer's Rock in central Australia is an extreme example. But with the exception of arid regions, most of the land surface in the tropics is formed, as in temperate lands, of valleys—steeply, moderately or gently sloping. As such, the general discussions concerning slope evolution are as applicable to such valleys as to landforms of the temperate and polar zones. There is no branch of systematic slope theory applicable exclusively to the tropics.

Pediments and inselbergs, however, are common in all tropical climates, from arid to rainforest, and in the Mediterranean environment. Moreover, on any objective, form-based definition they can also be identified in the temperate and polar zones. There is no substantial form difference between the tor and the kopje, whilst the term monadnock is applied to hills that in the tropics would certainly be called inselbergs. Pediments occur in Millstone Grit strata in northern England. The following is essentially a systematic discussion of the form and origin of these particular slope forms and not an account of slopes under tropical climates (see p. 230 ff.).

A *pediment* is a gentle slope, formed in bedrock, that occurs below a substantially steeper slope, and is separated from the latter by a relatively rapid change in angle (Fig. 75*A*). Pediments are usually, but not by definition, concave in profile. A slope superficially resembling a pediment but formed by deposition is a *bahada* (for an account of related terminology see Tator, 1953). There is no established term for the steep slope above a pediment, which will here be called the *hillslope*. The hillslope may or may not include a free face and a debris slope. The transitional zone between hillslope and pediment is the *piedmont zone*. The angle made by the lower part of the hillslope and the upper part of the pediment is the *piedmont angle*. Essential features in the definition of a pediment are first, that it occurs in conjunction with a hillslope; secondly, that there shall be a substantial contrast in angle between hillslope and pediment; and thirdly, that ground at intermediate angles occupies a relatively short profile distance. A profile across a hillslope and pediment thus possesses a bimodal angle frequency distribution. Suggested quantitative limits are that the mean angle of the pediment must be less than 10°, and the mean angle of the hillslope must be at least

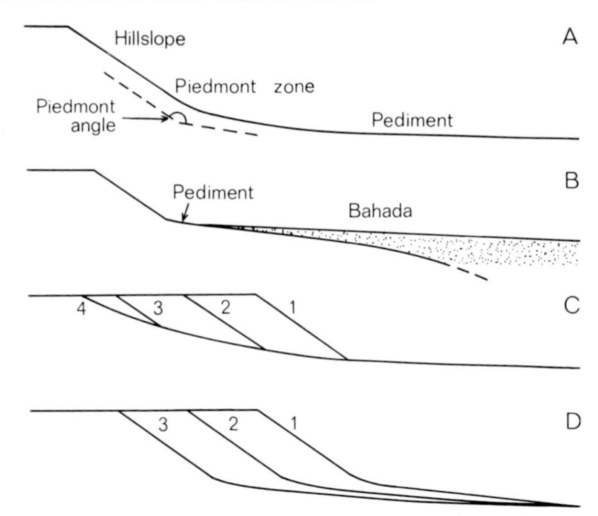

FIG. 75. A = terminology of pediments; B = origin of the pediment as a buried form; C = hillslope and pediment evolution by slope replacement; D = hillslope and pediment evolution by parallel retreat with pediment regrading.

twice that of the pediment. There is no minimum angle for a pediment, but it must possess a consistent direction of slope away from the hillslope (excluding micro-relief). Most pediments are in fact less than 6°, most hillslopes more than 20°, and the piedmont zone usually less than 50 m long. On the above definition, a gentle slope of identical form to a pediment but not surmounted by a hillslope is not a pediment. A slope resembling a pediment but steeper than 10° may be referred to as a footslope.

An *inselberg* is a steep-sided isolated hill rising relatively abruptly above gently-sloping ground. Suggested quantitative limits are that the maximum angle of the sides of the inselberg must average at least 20°, and that around at least 75% of its circumference there shall be, on ground sloping away from the inselberg (i.e. prior to reaching a drainage line or depression), a minimum slope of less than half the average maximum angle of the sides of the inselberg. For example, if the minimum slope exceeds 15° on 25% of the circumference, the mean maximum angle of the inselberg must be at least 30°.

On the above definitions, a pediment necessarily occurs in conjunction with a hillslope and piedmont zone; the hillslope above a pediment may, but need not necessarily, be an inselberg, and an inselberg may, but need not, have a pediment below it. In practice, pediments frequently occur below linear hillslopes, unassociated with inselbergs, whereas most inselbergs are surrounded by a pediment.

The origin of the hillslope-piedmont zone-pediment landform will first be considered, the arguments being applicable whether or not the hillslope is an inselberg. This is followed by a discussion of inselbergs.

HILLSLOPES, PEDIMENTS AND THE PIEDMONT ZONE

Where a short pediment leads downslope to a bahada, a possible origin of the pediment is by progressive burial beneath accumulating and advancing debris (Fig. 75*B*). The manner of formation is similar to that of a buried rock core beneath a scree (p. 105), except that fine debris accumulating at a low angle takes the place of the scree. This ingenious mechanism, suggested in 1915 by Lawson, is only applicable to pediments on which a bahada is present, as in some interior basins of arid regions. Even for such cases it is unproven, and to investigate it would require borehole data on regolith depth.

Where the pediment is a rock-cut feature, it is possible for the hillslope and pediment to be evolving either by slope replacement or by parallel retreat. In both cases the hillslope itself must retreat approximately parallel; whether it becomes slightly gentler or steeper is not central to the problem. If the pediment does not undergo further ground loss subsequent to its formation, slope replacement must occur (Fig. 75*C*), and the height of the hillslope will be progressively reduced as it retreats further from the drainage line. *Pediment regrading* refers to continued slow ground loss from the pediment, accompanied by a slight decrease with time in its angle at any given point. If pediment regrading occurs, then parallel retreat of the landform as a whole is possible (Fig. 75*D*). Intermediate cases can occur, where there is slight pediment regrading, but insufficient to prevent some reduction in height of the hillslope. Inductive comparison of observed forms suggests that both slope replacement and parallel retreat are to be found. The key problems of origin, however, are equally applicable to either mode of slope retreat. These problems are first, how the hillslope angle remains steep; secondly, how the form of the pediment is determined; and thirdly, how the relatively abrupt curvature of the piedmont zone is maintained.

The most generally applicable explanation is that the retreat of the hillslope is subject to control by weathering, and that of the pediment to control by removal.* If the hillslope consists entirely of bare rock, then all parts of it are equally exposed to weathering, the debris formed by weathering is removed quasi-instantaneously, and the whole slope retreats by an equal amount. Such cases are rare. If the hillslope consists of a free face over a debris slope, the same argument applies to the free face. The angle of the scree or debris slope is controlled mainly by accumulation, but its rate of retreat is restricted by the protection given by the free face above. Conversely, regolith accumulation on the free face is prevented by the constant tendency for it to be undercut. A third possibility is that the hillslope is boulder-controlled (p. 133), and that its angle is related to the angle of repose of the coarse fragments covering it. In this case, the retreat of the bedrock is subject to control by removal, although this is not brought about by transport of the boulders as such, but through their reduction to fine material and subsequent transport. Assuming that the processes of transport are more than able to remove all the fine material supplied by weathering of the boulders, then the retreat of such a hillslope is ultimately subject to control by weathering, even though its angle is differently determined. In all of the above cases, the essential condition is that fine debris shall not accumulate on the steepest part, or the whole, of the hillslope. It is thus implied that even at the point where the debris supply is greatest, in the piedmont zone, the processes of transport are able to remove downslope all the material supplied by weathering, so maintaining the condition of control by weathering on the hillslope

* The theory outlined here has been stated in part by Fair (1948b) and Savigear (1960).

above. It is the existence of this excess of potential debris-removal capacity over actual debris supply in the piedmont zone that determines that a hillslope-pediment landform, and not a more gradually curved slope, is formed.

The pediment is covered by a regolith, and its retreat is subject to control by removal. The angle at each point on its surface is close to the minimum angle at which material supplied from upslope will be removed by transport downslope. If at any point supply temporarily exceeds removal, accumulation will occur, increasing the angle immediately downslope and so increasing the rate of removal and restoring its equality with supply. If removal locally exceeds supply, some of the regolith *in situ* will be removed, causing ground loss, a gentler downslope angle and a decreased rate of removal. In the theoretical condition of an exact equality between supply and removal at all points the pediment becomes a transportation slope.

The angle of each point on the pediment is conditioned by the fact that, considered over a relatively short period, it is very nearly a transportation slope. Over the longer term, however, as denudation progresses on the hill mass or slope above, the rate at which material is supplied to the top of the pediment decreases. Gentler angles therefore become sufficient for it to be transported, and there is slow net ground loss from the pediment. This is the mechanism of *pediment regrading*.

The smoothness of the pediment surface, formed by sheetwash acting on the regolith, is transmitted to the bedrock beneath by the mechanism of *mantle-controlled planation* (Mabbutt, 1966). The bedrock/regolith junction lies at varying depths as a result of lithologically-induced differential penetration by weathering. When, in the course of pediment regrading, parts of the bedrock are brought close to, or reach, the ground surface, they become subject to more rapid weathering. Thus the bedrock is kept planate overall despite its irregularity in detail.

This does not account for cases where the uppermost fringe of the pediment consists of bare rock, at a low angle. Such rock fringes may be explained either as relict or as short term features. They may be relict forms, resulting from the exhumation of a previous regolith-covered pediment, following climatic change or accelerated erosion. Alternatively, if time-independent, they may result from short-period variations in the rate of supply of debris from the hillslope, becoming buried when weathering conditions are exceptionally severe, and exposed during the intervening, longer, periods.

The change in slope angle in the piedmont zone is associated with a change in the particle size of the regolith. Coarse material accumulates on the hillslope, giving to it a steep angle. The rate of supply of fine debris from the hillslope to the piedmont zone, however, is limited; and the angle at the top of the pediment is such that the coarser particles in such debris can just be removed. As the contrast in particle size between the regolith material on the hillslope and on the pediment increases, the piedmont zone becomes shorter and the piedmont angle greater. Under semi-arid and arid conditions the piedmont zone is particularly narrow on granites, which weather first to large boulders and then to sand; for the same climatic conditions, rocks which weather first into smaller stones, for example basalts and schists, give a broader and more gently curved piedmont zone (Rahn, 1966; Mabbutt, 1966). The profile curvature of the piedmont zone is greatest in deserts, and becomes less as the rainfall increases; for example, on sandstones in semi-arid Karroo, the typical curvature is 20-30°/100 m, whereas in sub-humid Natal it is 7-15°/100 m (Fair, 1948b).

Other hypotheses for the origin and maintenance of the piedmont zone are:

(*i*) It is associated with a change in lithology, and caused by a difference in rock resistance.

(*ii*) It is maintained by intermittent lateral corrasion by streams, with inconstant courses, issuing from the hillslope.

(*iii*) It is associated with a change in the manner of water flow, and consequently in the transporting or eroding capacity, between the hillslope and pediment from (*a*) sheetwash to stream-flow (Bryan, 1935); (*b*) rillwash to sheetwash (King, 1949); (*c*) supercritical flow to subcritical flow, the transition being termed the hydraulic jump (Schultz, 1955); (*d*) turbulent flow to laminar flow (King, 1957).

(*iv*) It is associated with augmented chemical weathering at the foot of the hillslope.

(*v*) It is an exhumed feature, originating by weathering.

The first of these hypotheses is only applicable where a lithological change closely coincides with the piedmont zone. Most field studies indicate that whilst lateral planation may locally steepen the base of the hillslope, this process is not applicable as a general explanation (see Cotton, 1947, Chapter IV for discussion and references). None of the third group of explanations have been substantiated by field observations of processes. The fourth and fifth hypotheses are discussed below in relation to inselbergs.

In 1930, Davis pointed to analogies between the pediments of the arid cycle and the concavities of humid landscapes, and this topic was reconsidered by Holmes (1955). The fundamental contrast lies in the narrowness of the piedmont zone in arid regions, temperate and tropical, compared with the gradual concavities of most humid climate landforms. This contrast arises because under arid conditions all fine material is rapidly removed from steep slopes, so maintaining the condition of the control of slope retreat by weathering. Under humid conditions part of this material accumulates, on any angle below *c.* 35°, making the retreat of the slope subject at least in part to control by removal (cf. Savigear, 1960). The existence in deserts of a marked change in particle size between the regolith on steep and gentle slopes is a consequence of the difference between control by weathering and by removal, and not its cause. This regolith contrast is absent from humid landforms because, apart from steep cliffs, slopes subject to control by weathering do not occur. Where in humid regions structural conditions give rise to a free face, the retreat of the latter is subject to control by weathering, and a pediment, or a long concavity somewhat steeper than 10°, may occur. The basic cause of the contrast between arid and humid landscapes is the retention of a regolith on slopes of moderate steepness under humid conditions; and this feature is in itself caused mainly by climatically-induced differences in vegetation.

THE FORM OF PEDIMENTS

Nearly all pediments are concave in profile. The mean angle is usually in the range 0·5°-5·0°, and less frequently up to 10° (Tator, 1952; Balchin and Pye, 1955). The curvature decreases downslope. From the base upwards, the rate of increase in angle remains very low for the greater part of the length of the pediment, increasing in an approximately exponential manner close to its upper limit. Pediment angles are least in deserts, and become steeper with increase in rainfall. In central Australia, Mabbutt (1966) found pediments on schists to be steeper and less regular than those on granites. But in a survey of 54 pediments in arid western United States, Mammerickx (1964) found no significant relation between mean angle and either lithology or length of the pediment. Some pediments have a smooth ground surface, whilst others are dissected by shallow channels.

The most commonly described transport process on pediments is surface wash, flowing as sheet-wash, fine rills, or shallow and braided channels. In some cases the profile appears to be related to the gradient of sub-parallel streams (Gilluly, 1937). Evidence has also been given for the downslope transport of clay by subsurface wash (Ruxton, 1958; Mabbutt, 1966). Whatever the exact process, the volume of flow increases downslope on the pediment, enabling the same amount of debris to be transported across a progressively gentler angle. Reduction in size of the debris in course of transport will augment this effect. Thus the concave form of a pediment is consistent with the view that its evolution is subject to control by removal, and that over a short period it is virtually a transportation slope (p. 96).

Where a pediment occurs below an inselberg, continued hillslope retreat may cause the reduction in size and subsequent removal of the inselberg. For so long as even the smallest inselberg remains, the pediment remains concave. But it is unusual to find multi-concave slopes lacking a crest inselberg; hence it appears that as soon as the inselberg is eliminated, a broad, gentle convexity is formed relatively rapidly (Davis, 1933; Sharp, 1957; Mammerickx, 1964).

INSELBERGS

The following types of inselbergs may be distinguished:

(*i*) Butte. Formed of a caprock, usually outcropping as a free face, over a debris slope; found with horizontal strata, or where laterite forms a caprock (Fig. 76); also present, often as outliers beyond a scarp, in temperate regions.

(*ii*) Conical hill. Rectilinear sides, with irregular micro-relief, and a relatively sharp angle at the crest; found in arid regions (Fig. 83, p. 230).

(*iii*) Convex-concave hill, entirely regolith-covered; transitional with hills not falling within the definition of an inselberg.

(*iv*) Rock crest over regolith-covered slope, the latter often a steep debris-slope; probably the most common type in savanna regions (Fig. 77).

(*v*) Rock dome ('sugar loaf'). Frequently with nearly vertical sides, although low domes with gentle angles, called ruwares by Thomas (1965), do occur; typical of rain-forest regions, but found also in savannas (Fig. 78).

(*vi*) Kopje or tor. Formed, at least in part, of a heap of large boulders, although a bedrock core may be present; found in temperate and tropical regions.

Fig. 76. Butte formed by an outlier to a laterite-capped plateau, with a rectilinear debris slope. Formed in sandstone under a semi-arid climate, Niger, W. Africa.

Fig. 77. Inselberg, with rock dome, debris slope, and relatively steep pediment. Central Malawi, savanna climate.

Fig. 78. Left foreground—rock dome, partly colonized by vegetation, with pediment below. Background—Mount Hora, a rock dome crest over a debris slope. Northern Malawi, savanna climate.

The above descriptions refer to examples with approximately equi-dimensional plan form. Very often all types except the first possess a ridge-like form, due to structural orientation. Types (*v*) and (*vi*) are particularly characteristic of granites, although not confined to them. Savigear (1960) distinguished between smoothly-curved and faceted types. Detailed form descriptions of individual examples are given by Twidale (1964) and Thomas (1965, 1967).

The following mechanisms may contribute to the origin, or modification of form, of inselbergs:

(*i*) Differential erosion, structurally induced.
(*ii*) Hillslope retreat.
(*iii*) The slower weathering of exposed rock compared with rock beneath a regolith cover.
(*iv*) An augmented rate of weathering in the piedmont zone.
(*v*) Exfoliation caused by pressure release.
(*vi*) Differential weathering beneath the regolith, followed by exhumation.

A basic question is whether the rock of which the inselberg is composed is different from that underlying the surrounding pediment, and if so, whether the structural junction coincides with the piedmont zone. The paucity of information on this point will surprise no one who has worked in the tropics. It arises from the difficulty of obtaining unweathered specimens that are typical of the bedrock beneath a deeply-weathered pediment; if a rock outcrops at the surface, it is probably atypical. Many inselbergs are undoubtedly lithologically distinct; for example, many of those rising above the plain of central Malawi are formed from syenite intrusions into the surrounding Basement Complex. Gneisses and schists usually have ridge-like inselbergs, and on such rocks the difference may be one of mineral composition, frequently not indicated by conventional geological mapping. On granites it is commonly supposed that the joints are less closely spaced within the inselberg than in the surrounding rock, but this is difficult to test objectively. To establish whether structural differences exist, and if so where their boundaries lie in relation to the landforms, does not in itself solve the problem of origin of an inselberg, but there is little purpose in proceeding with other studies until this aspect has been investigated.

If it can be demonstrated that a structural boundary coincides with the margin of an inselberg, then subaerial differential erosion is the most likely mode of origin. The surrounding rocks suffer more ground loss, leaving the resistant rock projecting; its form may be subsequently modified by differential weathering mechanisms.

Origin by hillslope retreat is probable in most cases of buttes and conical inselbergs. It is clearly the case in regions where successive stages of plateau, plateau remnant, mesa, butte and conical inselberg are represented in the present landscape. A similar origin may apply if the inselberg is formed of a more resistant rock, hillslope retreat being slowed down on reaching this rock. Unless retreat was virtually halted, however, the structural boundary would be expected to lie some way out from the present base of the inselberg, giving a difference in deduced form from that of the differential erosion hypothesis.

The two mechanisms of differential weathering at, or close to, the ground surface do not account for the origin of an inselberg, but could modify its form. The hypothesis that, under tropical conditions, the weathering of bare rock is slower than that of rock beneath a regolith cover requires experimental testing. It is based on the prima facie argument that rock beneath the regolith will be kept moist for much longer than bare rock. If true, it would imply that once a rock became exposed it would tend to grow, rather than diminish, in height. Thus low, gently-sloping rock hills could be embryo inselbergs, rather than senile or weakly-developed ones. The hypothesis that the piedmont zone coincides with a belt of augmented weathering is supported by field evidence from savanna and semi-arid environments (Clayton, 1956; Ruxton, 1958; Twidale, 1962, 1967, 1967b). The cause is the higher moisture content in this zone, caused by runoff from the hillslope or inselberg. Such a mechanism could produce a progressive steepening of the lower side slopes of an inselberg; examples of steepening beyond the vertical are described by Twidale (1962).

The rounded form of some rock dome inselbergs in massive rocks, especially granites, is probably associated with exfoliation consequent upon pressure release. This mechanism is discussed by Ollier (1969).

There remains the controversial view that some inselbergs of humid tropical regions, and possibly also tors of temperate latitudes, were shaped by differential weathering beneath a thick regolith cover, and subsequently exhumed. This is part of the general hypothesis of *double planation surfaces* (Büdel, 1957; Ollier, 1959, 1960). It is held that during a period of prolonged planation the basal surface of weathering penetrates several tens or hundreds of metres below the surface of subaerial planation, to a depth which is very uneven, due to variations in lithology and, especially, joint spacing. Following uplift, the deeply-weathered plain is dissected; the incoherent regolith is partially stripped away, leaving upstanding rock dome or kopje inselbergs. A similar mechanism for the tors of temperate lands has been proposed by Linton (1955, 1964), Wilhelmy (1958) and Cunningham (1969), and opposed by Palmer (1956), Palmer and Radley (1961) and King (1958). This hypothesis, the implications of which are far-reaching, will not be discussed here (cf. Ollier, 1969, pp. 198-212). The fine, and unresolved, disputation between King (1948, 1966) and Thomas (1965, 1966, 1966b), as to whether rock domes were shaped beneath the ground or formed sub-aerially, may be consulted.

XVII ENVIRONMENT: STRUCTURE

ENVIRONMENTAL FACTORS

THE distinction between treating slopes as time-dependent and time-independent phenomena has already been noted (p. 21). In the time-dependent, or historical, approach, present form is explained by reference to its geomorphological history, for example changes in base level, climate, and the rate of basal river erosion. The time-independent approach seeks to account for the features of slope form in terms of the present-day environmental factors: lithology and structure, climate, aspect, ground-water conditions, soil and vegetation.

In principle it would be possible to carry out a large-scale multi-variate study, based on a random sample of slopes from all parts of the world. The form of each slope, including regolith properties, would be surveyed, and independent observations made of the environmental factors. The various properties of slope form would be taken as dependent variables, and the other environmental variables as independent. Multiple regression would yield a measure of statistical 'explanation' of each of the slope form properties. A study of this type, concerned not only with slopes but with morphometric properties in general, has been carried out by Melton (1957). Besides the large amount of data needed, there are three limitations to this approach. First, to account for the values assumed by a dependent variable in terms of statistical correlation with other variables is not in itself an explanation of the causes of these values. Secondly, there is a substantial element of feedback between the slope and 'non-slope' variables, particularly with respect to soil properties. Thirdly, there is in practice no doubt that historical, time-dependent influences are of substantial importance on most slopes, and whilst it is not impossible to include these influences as variables, there are problems in doing so owing to the difficulty of making independent observations of phenomena no longer in existence. The alternative to a massive multi-variate treatment is to find groups of slopes on which all environmental conditions except the factor under study are similar. This is clearly difficult to achieve in practice.

In some cases the relation between two environmental factors is one of cause and effect operating in one direction only; thus solid geology influences slope form, but there is no converse effect. In other cases, in particular slope/soil relations, there is a mutual interaction. The two directions of cause and effect usually operate on different time-scales. For example, slope angle influences soil moisture conditions, and therefore denudational processes, over a time-scale of a few days or weeks; but for the slope steepness itself to be appreciably altered as a consequence of this difference in moisture and processes takes a period of the order of 10 000-100 000 years.

In this and the following chapter the factors of structure, soil, vegetation and climate are discussed in relation to surface processes, slope form and evolution. The treatment in each section is mainly a brief summary of results, with references to relevant sources. In many instances the generalizations given are based on relatively sparse data, and are liable to modification following further studies. To avoid repetition the qualification 'other factors being equal' has been omitted, and may be taken to apply throughout.

STRUCTURE *SENSU STRICTO*

As a generalization it may be said that structure, *sensu stricto,* has a greater influence on slope form than lithology; specifically, the effects arising from the super-position of strata of differing lithology are greater than those caused by each individual rock type. A simple slope on sandstone has many features in common with one on shale, but both differ markedly from a compound slope formed of sandstone overlying shale.

Simple slopes, composed of one rock type, tend to evolve in a different manner from compound slopes. On simple slopes, a free face appears only if stream erosion is very rapid. As soon as the intensity of basal erosion slackens the free face is eliminated and the profile becomes convex-concave or, more frequently, convex-rectilinear-concave. Evolution is then by slope decline. The less resistant the rock, the less probable it becomes that a free face, possessing long-term stability, will be produced.

Compound or complex slopes formed by a resistant caprock overlying less resistant strata are more likely to possess a free face and to evolve by parallel retreat. The initial appearance of a free face requires that stream erosion should be sufficiently rapid, in relation to slope retreat, to produce steep valley sides. Below the free face a debris slope is formed. For the free face to be maintained during slope retreat there must be unimpeded basal removal of material from the debris slope, so that retreat is then parallel. If basal removal from the debris slope is wholly or partly impeded the debris slope grows in height and eliminates the free face. The ensuing convex-recti-linear-concave slope may, however, continue to evolve by parallel retreat. A stage is necessarily reached when two opposed retreating slopes meet, reducing the caprock to a narrow ridge; at this stage the profiles of the landscape are predominantly concave. Further retreat leads to the elimination of the caprock, leaving a short convexity of the underlying weaker rock. The mode of evolution now changes to slope decline; the divide is lowered, the convexity extends in length, and the maximum angles of the slopes become progressively gentler.

This distinction between the typical form and manner of evolution of simple and compound slopes was established by Fair (1947, 1948), using the method of inductive comparison of profile forms. Fair's work was based on landforms in Natal, under sub-tropical climates with a marked summer concentration of rainfall. In interior Natal, with a rainfall of 600-800 mm, slopes are formed from horizontally bedded sandstones and shales, with 60-100 m dolerite sills capping 'nearly every hill'. Wherever the caprock remains, the form of the upper parts of the profile is similar, regardless of the length of the pediment. This similarity applies whether or not a free face is exposed. The length of the pediment is assumed to be proportional to the amount of retreat since the slope was first formed by dissection. As profile form is apparently unaffected by the length of pediment, parallel retreat is inferred. In coastal Natal conditions are more humid, with a rainfall of 900-1100 mm; Table Mountain Sandstone overlies tillite and shales. Under these conditions a free face is present only where there is rapid, active basal erosion, particularly at gully heads, and as soon as the rate of erosion decreases the free face is eliminated. The slope does not then decline, but assumes a constant profile form and retreats parallel, such that the thicker the sandstone bed, the longer the convexity. Dwyka Tillite and Ecca Shales rarely show a free face, and evolve by slope decline. Granites in the Valley of a Thousand Hills are an exception to the generalization that simple slopes decline; their slopes are kept predominantly concave, and steep in their upper parts, by a system of closely-spaced gullies.

Everard (1963, 1964), obtained similar results for landforms under a semi-arid climate in Cyprus. Slopes on marls and shales have convex-concave profiles and evolve by decline ('downwearing'). The presence of a caprock produces predominantly concave profiles and parallel retreat ('backwearing'). The caprock in this instance is commonly calcrete, or secondary limestone. Other materials of secondary origin may also act as caprocks (p. 198). Frye (1959) compared three areas with horizontally-bedded Mesozoic rocks of similar lithology; under a humid temperate climate in Ohio, slopes were convex-concave, but under drier climates in Kansas and Texas, free faces and pediments were common. The occurrence of a free face on compound slopes is not confined to arid and semi-arid conditions. It occurs under a humid temperate oceanic climate in the English Pennines, on Millstone Grit (Carboniferous) strata, where hard and coarse-grained sandstone beds lie horizontally above shales and softer sandstones. Even under rainforest conditions, for example in Malaya, a free face frequently persists at the outcrop of a bed of quartzite.

Where the strata are almost horizontal, the summit area of mesas is not always formed by the caprock. Part of the weaker rocks above may be preserved, with the caprock outcropping a short distance down the surrounding scarp (King, 1968).

The effect of structure on slope is demonstrated in the contrast between scarp and dip slopes. Macar and Lambert (1960) investigated the relation between dip and slope angle on limestones in Belgium. On dips of 30°-65°, slopes inverse to the dip were found to be steeper than those conformable with it, although the values overlapped. Both inverse and conformable slopes tended to steepen with increasing dip up to 65°; this agrees with the finding of Lambert (1961) for the Condroz area. Where the dips rose above 65° there was no further increase in slope angles, either on inverse or conformable slopes, nor was there a significant difference between these two classes of slope. Flohr (1962) found that in very steeply-dipping limestones the morphological 'scarp' faces were formed by the bedding planes, with gentler opposed slopes where successive strata outcropped.

Jahn (1964) attributed slope and ridge trenches to the subsidence of regolith into widened tensional cracks. Folding, faulting, imbricated structures, overthrusts and other tectonic features may affect slope form in regions of complex structure (Pippan, 1964).

Savigear (1960) suggested, with reference to humid and arid tropical climates in West Africa, that slope form is affected by the massiveness of rocks, through its influence upon the relative rates of regolith production and removal. On non-massive rocks, with closely-spaced interstices, rock is quickly reduced to debris, and the slope becomes blanketed by a thick regolith; form is then convex-concave, and evolution is by slope decline. On massive rocks the reduction of rock to waste blocks is slower, and the regolith likely to be thin; slopes tend to consist of angularly separated facets, and evolution is by slope replacement. Parallel retreat occurs only where a caprock, often of ferricrete, is present.

LITHOLOGY

On simple slopes, lithology influences profile form and angle. In central Belgium, Fourneau (1960; cf. Macar and Fourneau, 1960) found that convexities form the greater part of the profile on sandstones, about half on limestones, and less than half on shales. A rectilinear maximum segment occurred on most shale slopes, on less than half of those on sandstones, but was usually absent from limestone profiles; segments tended to occur more frequently on less coherent rocks. A basal concavity was present on most limestone and shale slopes but on less than half the sandstone slopes. The convexities on limestones were more smoothly curved than those on sandstones. Lambert (1961) however, found the converse to be true for slopes in the Condroz area of Belgium, and he attributed the irregularity of the limestones to the effect of solution. Slopes on chalk are very smoothly curved (Clark, 1965; Lewin, 1969). Granites, where the surface is not broken by tors, may also have very smoothly curved profiles. Irregular profiles occur where the strata consist of frequent alternations of sandstones and shales, as in the Coal Measures.

With respect to the influence of lithology on angle, Fourneau found the following mean values of maximum slope angles: sandstones 21°, limestones 20°, shales 9°, chalk $5\frac{1}{2}$°, clays 5°, sands 5°. Thus coherent rocks have steeper slopes than non-coherent. More massive rocks have steeper angles than thin-bedded strata of similar lithology (Pippan, 1963). Lithological effects are equally clear in regions of generally gentle slopes; in Brabant the median slope was found to be 4·75° on sands and 1·67° on clays (de Béthune and Mammerickx, 1960). For a given region, characteristic angles may occur in pairs, the steeper on the more resistant rock (p. 168). Beds of similar lithology give different angles according to their relation to other rocks. Where a given rock type overlies a less resistant bed, angles on it will be steeper than where it overlies a more resistant bed. A rock type which on simple slopes gives gentle angles may be characterized by steep slopes if it outcrops in mid-slope between two more resistant rocks, particularly if the overlying stratum is more resistant than the underlying (Gregory and Brown, 1966). Comparing valley-side slope angle with various parameters in a multi-variate study, Melton (1957) found the highest correlation to be with the infiltration capacity of the regolith ($r = +0{\cdot}75$), the infiltration capacity being itself strongly influenced by bedrock lithology.

The term *resistant* has been employed above in accordance with normal usage. A rock is described as resistant if it is associated with hills, scarps or other landforms projecting above the surrounding relief; the term is applicable only in the relative sense of one rock being more resistant than another. The implication is that the more resistant a rock, the slower is the retreat of slopes formed from it. The term, defined in this way, thus refers to the observed expression of the rock in landforms, and not to its lithologically intrinsic properties. It is, for example, possible for the relative resistance of two rock types to differ according to the climate under which they are found.

The properties that determine the resistance of a rock depend on whether the slope is subject to control by weathering or by removal. If the former, it is those properties of the rock that determine the rate at which it is weathered that are significant, but if the latter, the resistance is governed by the rate at which the regolith is removed. Rock properties that may affect resistance are hardness, permeability, mineral composition, and frequency of planes of weakness or fracture. Mineral composition is of primary importance under control by weathering. The effects of regolith properties on processes of downslope transport have been discussed above (p. 197).

It is not at present possible to place all major rock types in order of decreasing resistance. For rocks of approximately equal age there is usually a decreasing order limestone > sandstone > shale or clay. A comparison with marine cliffs suggests that the rates of retreat of unconsolidated or weakly-consolidated rocks, such as clays, marls and younger shales, probably differ by order of 10^2-10^3 from those of crystalline rocks and lower Palaeozoic sedimentaries. An example in the dating and comparison of retreat of two rock types is described by Bout *et al.* (1960). In experiments on frost-shattering, Potts (1970) found the order of shattering susceptibility to be igneous rock < sandstone and mudstone < shale; and Wiman (1963) the order quartzite < granite, gneiss and schist < slate; in each case the relative susceptibility of the middle group of rocks differed according to the intensity of freezing. Where mineral composition is similar, Palaeozoic rocks are usually more resistant than Mesozoic or Tertiary; indeed, a hard lower Palaeozoic shale may be more resistant than a younger sandstone. Even unconsolidated sands may be quite resistant, first, because being composed predominantly of quartz they are little affected by weathering, and secondly, because both the rock and the derived regolith have a high permeability. The subject of rock and regolith properties in relation to rates of weathering and denudation is so basic, both to slopes and to geomorphology in general, that systematic investigations of it should be a research priority.

GRANITE

Two rock types, granite and limestone, are the subjects of specialized branches of geomorphology. They may occur as large masses, offering conditions for the investigation of slopes in relatively homogeneous lithology. On both rock types, some material is lost from the slope by removal in solution, but they differ in that the insoluble portion of the rock, requiring removal by downslope transport, is small on limestones but substantial on granites. Variations of slope form associated with climatic differences are as great as or greater, on both granites and limestones, than the similarities of form arising from lithology.

The following slope forms have been described on granites:

(*i*) Bare rock domes, either smoothly rounded or faceted, found mainly in savanna and rainforest climates.

(*ii*) The steep and irregular bare rock slopes of tors and kopjes, tending towards rectangular forms conditioned by jointing; these occur in all climates from polar to tropical.

(*iii*) Predominantly concave slopes with a free face near the crest, and with gullies on the steeper part of the concavity; reported from sub-tropical and rainforest climates.

(*iv*) The succession downslope of free face, debris slope (boulder covered) and pediment, with abrupt breaks of slope between these units, reported under semi-arid conditions.

(*v*) Slopes with irregular, stepped micro-relief, concave or approximately rectilinear in macro-form, occurring in arid and semi-arid climates, in temperate mountain regions and in the polar zone.

(*vi*) Smoothly convex-concave profiles with a continuous regolith cover; the most extensive form on granite in humid climates, both temperate and tropical (Fig. 79).

(Fair, 1947, 1948; Ruxton and Berry, 1957, 1961; Ruxton, 1958; Wilhelmy, 1958; Demek, 1964; Dumanowski, 1964; Wahrhaftig, 1965.)

FIG. 79. Granite landscape under a high-altitude savanna climate, 2500 m, the Nyika Plateau, northern Malawi.

Following chemical weathering, part of the feldspar material in granites is removed in solution. Joint spacing is important in determining relative rock resistance and therefore differential relief. There is evidence of selective removal of clays, from between the coarser residual quartz particles, by subsurface flow. Under tropical conditions, mass movements and gullying are important processes. A succession of weathering zones has been recognized in the regolith derived from granite: Zone IV, weathered rock (>90% solid rock); Zone III, core-stones with fine material (50-90% rock); Zone II, debris with core-stones (<50% rock); and Zone I, debris without core-stones (Ruxton and Berry, 1957). In Hong Kong, on concave profiles the zones outcrop at the surface successively downslope, whilst on convex-concave profiles Zone II occurs at the surface over all but the lowest part of the slope (Ruxton and Berry, 1957; see Figs. 64-5 in Ollier, 1969). It is probable that the rate of slope retreat is inversely proportional to the degree of reduction of the surface layer of the regolith. This suggests that evolution is by parallel retreat on the steep concave slopes, and slope decline on the convex-concave profiles.

Granite is noteworthy for the occurrence of both bare rock forms and slopes on which the regolith is unusually thick. It is one of the few rock types on which a regolith over 10 m thick occurs in temperate climates. If such slopes are related to present conditions, the implication is that the rock slopes are subject to control by weathering, but the portions with a thick regolith to control by removal. This difference in control would be self-reinforced if, as has been held, chemical weathering is more rapid beneath the regolith than on exposed rock. It is alternatively possible that such regolith features are in part relict from hot and humid and/or periglacial conditions (see discussion in Ollier, 1969, p. 120 ff.).

LIMESTONE

The properties of limestones that give rise to their different morphology are their susceptibility to chemical weathering, low proportion of insoluble material, and, in hard limestones only, the possession of sufficient rock strength to maintain long-term stability on near-vertical cliffs. It is where a free face is present that the most distinctive forms occur; if a continuous regolith cover exists, the slopes are more like those on siliceous rocks.

In Britain, most slopes on both hard and soft limestones are regolith-covered. Although sometimes appearing convex-concave, profiling usually shows that a rectilinear maximum segment is present; the convexity is longer than the concavity. On soft limestones, particularly chalk, most slopes are of this form (Clark, 1965; Lewin, 1970). On hard limestones a near-vertical free face may occur, either near the crest or in mid-slope, below which occurs a 30°-35° debris slope, either of bare scree or vegetated. It has not been established whether the cliff and debris slope are steady-state forms, or whether they are relict from glacial or periglacial conditions and the cliff is currently being eliminated by upward growth of the debris slope (cf. Sweeting, 1966). In Yugoslavia, Terzaghi (1958) found a prevalence of 30°-34° slopes, this corresponding to the upper limiting angle for a regolith cover. The regolith was of uniform thickness. Terzaghi suggested that this condition was maintained by regolith control of the rate of solution, any local thinning of the regolith causing more rapid solution and therefore release of additional insoluble residue. He further suggested that the rate of solution, and therefore of slope retreat, was less on bare rock than beneath the regolith; hence an initial rock outcrop would 'grow' by differential lowering of the surrounding ground (cf. p. 212).

Under semi-arid conditions, bare rock slopes are abundant. There is an abrupt break of slope between steep cliffs and rock-cut pediments (Jennings and Sweeting, 1963).

Some of the most striking slopes in the world occur on limestones under tropical rainforest conditions. These are smooth high rock walls, of 60°-90°, usually surrounding isolated hills. They are not rock domes, but are surmounted by a regolith-covered convexity. Caves and low overhanging slopes with solutional micro-forms are common at the foot of the rock walls. This suggests that their steepness is produced by solutional undercutting, combined with the rock strength sufficient to maintain high cliffs. The concentration of chemical weathering at the base is caused by the position of the water table. Two other forms are common: profiles with a convex crest (sometimes with stepped micro-relief) over a 30°-40° segment (Fig. 58, p. 156), and, where the maximum angle is less than 30°, convex-concave or convex-rectilinear-concave profiles. (Lasserre, 1956; Sweeting, 1958; Paton, 1964; Tjia, 1969.)

XVIII ENVIRONMENT: CLIMATE AND VEGETATION

THE basic hypothesis of climatic geomorphology is that each type of climate is characterized by different landforms. These differences apply to the processes, forms and evolution of both rivers and slopes. The characteristics of slope form and evolution are brought about mainly by differences in surface processes; but they are also influenced by the relative rates of linear fluvial erosion and areal slope denudation. Climate has some direct effects upon processes, for example through the influence of intensity of rainfall upon rainsplash and surface runoff. Climatic influence upon surface processes mainly operates indirectly, through vegetation. Under limited circumstances, vegetation may be treated independently of climate, as, for example, in comparing denudation of forested and grassed catchments or experimental plots. But it is rare for a similar vegetation cover to occur naturally under different climates, and hence the climatic factor cannot be considered in isolation from vegetation.

The concept that landforms evolve differently according to the climate originated at least as early as the publication of Davis's arid cycle in 1905. Following Davis, three climatically-related groups of landforms were recognized: 'normal' (humid temperate), arid and glacial. In the 1930s the main areas of contrasted landforms to be studied were in the semi-arid and arid American West. There was little opportunity for study of the humid tropics; a notable exception is German work in South America (e.g. Freise, 1932, 1935, 1938; Sapper, 1935), but this received little attention outside continental Europe. The general problem of pediments and inselbergs was formulated, mainly with reference to arid landscapes. A pioneer attempt at the comparison of landforms under different climates is Davis's discussion of rock floors in arid and humid climates (1930). By 1947, in *Climatic accidents in landscape-making*, Cotton was able to distinguish landform cycles in arid, semi-arid and savanna climates, and some of the distinctive features of rainforest relief had also been recognized (Cotton, 1941, pp. 155-8).

The 'European' school of climatic geomorphology was developed in Germany and France, commencing in the late 1940s. Contributory to its origin was German work on processes and landforms in polar and montane environments (Troll, 1943, 1944, 1947, 1948; Büdel, 1948b), and French work in the humid tropics (de Martonne, 1940; de Martonne and Birot, 1944; Birot, 1949). The first synthesis was produced in 1948 by Büdel, who appears to have coined the term climatic morphology although the concept had clearly been anticipated by Troll. Büdel distinguished submarine, subglacial and subaerial form-groups, the latter divided into seven climatic-morphological zones: the frost-debris zone, the tundra zone, the extra-tropical *Ortsbodenzone* (zone of *in situ* soils), the Mediterranean transition zone, the dry-debris zone, the sheetwash zone, and the tropical *Ortsbodenzone*. On the continent the concept was taken up for systematic textbook treatment from the 1950s onwards. In the English-speaking world Peltier's tentative formulation of morphogenetic regions (1950) failed to give rise to further systematization, and there was no translation of the continental system until 1968. There were, however, numerous, detailed English and American studies of landforms under different climates, notably American work in polar environments and British work in Africa.

General accounts of climatic geomorphology are available in standard texts (e.g. Derruau, 1956; Birot, 1960; Louis, 1961; Cailleux and Tricart, 1965; Tricart and Cailleux, 1965); a review is given by Stoddart (1969b). The summaries for individual climates which follow are confined to studies of slopes. A refined climatic division would be incommensurate with the limited present state of knowledge on landforms. The following generalized climatic classes are employed, giving equivalents in the Köppen classification in brackets: humid temperate (*Cfb*, *D*), polar (*E*), montane (*H*), arid (*BW*), semi-arid (*BS*), Mediterranean(*Cs*), savanna (*Aw*, *Cw*) and rainforest (*Af*, *Am*). The only other climatic zone of major extent, excluding glacierized areas, is the humid sub-tropical (*Cfa*), but this does not appear to be an identifiable climo-morphological zone. Different parts of it have affinities with the humid temperate, savanna and rainforest zones, but the climatic parameters separating these different parts have not been established. A consideration of the effects of vegetation on surface processes is a necessary preliminary to the discussion of climate and slopes.

VEGETATION AND SURFACE PROCESSES

The main influence of vegetation upon surface processes is the protection from rainsplash and surface wash that it gives to the soil. Other effects are the retention of the regolith on slopes through the action of roots, the improvement (indirectly) of soil structure through the supply of organic matter, and the contribution of organic acids to rock weathering, the importance of which has not yet been quantitatively evaluated. Chorley (1964) compared penetrometer resistance in soils, finding that a greater proportion of the variation was accounted for by root weight than by particle size.

The total protective effect of a plant community against rainsplash and wash is made up of the protection given by the tree and shrub layers, the field layer (grasses and herbs), the ground layer (mosses), plant remains (leaf litter, dead grass) and the organic horizons of the soil (partly decomposed and fully humified organic matter). The leaves of the tree and shrub layer shield the ground from direct raindrop impact, and the other components either slow down surface flow or prevent it coming into contact with the mineral soil. The relation between percentage ground cover and soil loss by surface wash is non-linear, showing an abrupt increase when the cover is reduced below a critical value (Costin and Gilmour, 1970).

Forest communities give the greatest protection. When plant growth is unrestricted by either drought or cold, competition for light ensures that the various strata of the plant community together give a fairly complete ground cover. This applies to evergreen coniferous forest (pine, spruce) and to tropical and sub-tropical rainforest. Under tall equatorial rainforest, however, rainsplash may still occur, since the field layer is sparse, and water drops from leaves over 15 m above the ground attain close to their terminal velocity. Temperate deciduous woodland, both broadleaved and coniferous (larch), provides a close leaf cover in summer only, but in winter this is supplemented by the accumulation of dead leaves, forming a litter several centimetres thick and almost entirely covering the soil. Under rainforest the litter cover is incomplete, and is itself subject to wash, some bare soil therefore being exposed to water dripping from the canopy. In temperate woodlands the mineral soil particles are further protected by a continuous horizon of soil organic matter. This also acts as a sponge, having the capacity to absorb moisture exceeding its own dry weight. Consequently under temperate woodlands, deciduous and coniferous, neither rainsplash nor surface wash can occur except where the vegetation and surface soil has been temporarily disturbed, for example by fire, treefall and landslides. After heavy rain, subsurface seepage can be observed at the boundary between mineral and organic soil, but its velocity is very much lower than that of surface wash. Some forms of sub-tropical and tropical semi-deciduous forest, and tropical montane forest, also give almost complete protection. Under tropical evergreen rainforest, however, decomposed organic matter becomes rapidly incorporated into the soil, and mineral particles are exposed at the ground surface. In conjunction with the sparser field layer and litter, the consequence is that some rainsplash erosion and surface wash can occur.

Grassland communities of humid temperate oceanic climates also give complete protection. The grasses are of the turf-forming kind, giving a continuous and dense mat of roots, below which there is an organic soil horizon. However, such grasslands rarely formed the natural climax vegetation prior to woodland clearance.

In the prairie, savanna and steppe communities most grasses have a tufted habit, growing as compact clumps and leaving bare soil between. Both in African savanna and its botanically richer South American equivalent, *cerrado*, the ground coverage by the tree and shrub layers is fairly open in the wet season, and largely absent at the end of the dry season when, at the first onset of the rains, the heaviest storms occur. Under savanna in Northern Australia, M. A. J. Williams (1969) found that for comparable rainfall, soil movement by surface wash varied seasonally between periods of sparse and dense plant cover by a factor of 20. Except at high altitudes there is no wholly organic soil horizon. In these communities, both direct raindrop impact and surface runoff between the grass tufts can be seen during heavy rain (Fig. 80). The drier the climate, the higher is the proportion of bare soil; thus savanna woodland and tall-grass prairie give greater protection than short-grass prairie and steppe.

Protection becomes still less where plant growth is severely restricted by drought, as in thorn scrub and other semi-arid communities. Trees, where present, and shrubs are widely spaced owing to competition for moisture, and grasses are short and have a tufted habit. There is correspondingly little plant litter, and no organic soil horizon. Much of the ground is therefore exposed to direct raindrop impact and is potentially subject to surface wash. Mediterranean vegetation types (where not altered by man) are transitional between the close cover of temperate woodlands and the open nature of semi-arid communities. Bare ground is also exposed where growth is limited by cold and by snow cover, as in tundra communities and montane environments above the tree-line; surface wash can therefore occur relatively unhindered following snow-melt. Plant protection is reduced to low proportions in extreme polar conditions, and to nil in deserts.

FIG. 80. Sandy soil exposed to rainsplash in *cerrado*, the Mato Grosso, Brazil.

The following list ranks vegetation communities in order of decreasing protection against rainsplash and surface wash:

Temperate deciduous woodland
Temperate coniferous forest
Humid temperate grassland
Sub-tropical rainforest
Tropical semi-deciduous rainforest
Tropical evergreen rainforest
Mediterranean communities
Savanna woodland
Tree and shrub savanna
Steppe
Thorn scrub and other semi-arid communities
Tundra and montane communities
Extreme polar communities
Desert

This ranking is based mainly on *a priori* considerations. It would be possible, by comparative quantitative studies, to evaluate the passive effect of different plant communities, establishing indices of protection. Methods developed in plant ecology for measuring light penetration and ground coverage could usefully be applied. By combining such indices with the active climatic factor, namely the frequency of high-intensity rainfall, the combined effects of climate and vegetation on the rates of rainsplash and surface wash could be evaluated.

Clearance of the natural vegetation increases the rate of surface denudation by one or several orders of magnitude. This applies not only to the extreme case of accelerated soil erosion, but also to land under normal agricultural use (Starkel, 1962; Douglas, 1967; Meade, 1969). Man-induced changes of vegetation on uncultivated land can also have considerable effects. In the Southern Uplands of Scotland, Tivy (1957) attributed the formation of terracettes, bare scree and clitter slopes and gullies to accelerated erosion following forest clearance in Neolithic times and sheep-farming from the eighteenth century onwards. Heather-burning can convert moorland into a condition resembling the tundra. In Cyprus, Everard (1963, 1964) attributed a change in slope evolution from downwearing to backwearing partly to vegetation clearance in *c.* 4000 B.C. It is therefore essential that studies of surface processes, unless specifically concerned with man-induced effects, should be conducted in nature reserves or other areas that approximate as closely as possible to the presumed natural vegetation.

CLIMATE

THE HUMID TEMPERATE ZONE

The humid temperate environment is described geomorphologically as 'normal' because of the historical accident that it was the basis of the Davisian cycle of erosion. The epithet cannot be justified on grounds of predominance in world areal extent, in which it is matched by the desert zone, nor on any absolute criteria of normality. King's argument that sub-tropical environments with seasonal rainfall should be regarded as intrinsically normal or typical is equally invalid. There is nothing 'abnormal' about other climatic regions. The only justification for calling the humid temperate region normal is that most geomorphological studies have been made there; it therefore serves as a convenient standard of reference, with which features of other environments may be contrasted.

Features of the humid temperate climate of geomorphological significance are the evenly-distributed rainfall, occurring mainly as prolonged falls of low intensity, an excess of rainfall over potential evapotranspiration for most months of the year, and winter coldness sufficient to freeze only a shallow top layer of the soil. These features are most fully developed in the oceanic varieties of the climate, becoming modified under the summer rainfall maximum and greater winter coldness of continental interiors. The soil is permanently moist in depth, and weathering is mainly chemical. The depth of freezing is insufficient for solifluction, and under natural conditions rainsplash and surface wash are largely inhibited by the woodland vegetation. Hence solution loss and soil creep are probably the main processes of denudation. There is some experimental evidence that transport by creep exceeds that by wash (p. 87); the humid temperate may be the only climatic type for which this is so, possibly excepting rainforest for which there is no data. The question of relict periglacial features in the temperate region is discussed below (p. 241).

Humid temperate slopes are typically convex-rectilinear-concave, with convexities frequently forming half or more of the total profile. Many of the convex and concave slope elements are smoothly curved, although faceted forms also occur. A high proportion of all slopes under 35° possess a continuous regolith cover. The fact that this cover is usually only 20-200 cm thick and yet rarely absent is noteworthy, as it suggests, if it is a time-independent feature, that slope retreat is subject to control neither wholly by weathering nor by removal. It appears likely that if the regolith thickens, the rate of reduction of the underlying rock to soil decreases, so counteracting the increased thickness; if the regolith is thinned by net removal, the rate of rock weathering is increased. This mechanism of negative feedback, linking rock weathering, net regolith removal and regolith thickness, is unproven, but process-response models based upon it give surface form and regolith features which match those that are observed in the field.

Inductive studies suggest that once long-term stability has been attained, the evolution of slopes in homogeneous rocks is by slope decline (Fig. 81). Many compound

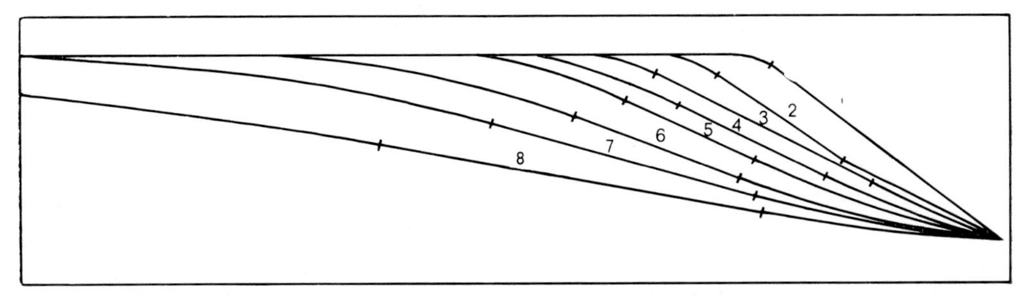

FIG. 81. The course of slope evolution on non-calcareous Palaeozoic rocks under a humid temperate climate in Britain. The divisions between the convexity, maximum segment and concavity are marked. After Young (1963).

slopes, including structural scarps, are also convex-concave in form, but evolve by parallel retreat. A free face may be present as a relict feature from glacial erosion, and one also appears where basal erosion is exceptionally rapid, as for example on marine cliffs. With these exceptions, a free face only occurs on some compound slopes, where the upper stratum is permeable and resistant to weathering, and the lower stratum relatively impermeable. Palaeozoic limestones and sandstones overlying shales may give rise to a persistent free face, but softer limestones and sandstones do not. Where a structurally-conditioned free face is present a vegetated debris slope occurs beneath it, passing without an abrupt break of slope into a concave or rectilinear pediment or a slope similar to a pediment but steeper. Such slopes occur only under exceptional structural conditions. For the humid temperate landscape as a whole, the frequency distribution of angles is unimodal unless cyclic complications occur. Gentle and moderate slopes, of 2°-10°, are the most extensive, but a feature which contrasts with most other climatic regions is the relatively high frequency of slopes in the moderate and moderately steep, 5°-18°, range.

The typical relief found in Britain and western Europe, with broad valleys and extensive undissected slopes, contrasts with a relief type which is common in New Zealand, also under a humid temperate climate, in which the main valley sides are dissected by frequent insequent streams. The latter is called *feral relief* (Cotton, 1958b). The typical slope form in areas of feral relief is a short convexity of high curvature succeeded by a long and fairly steep rectilinear segment, with little or no concavity (Fig. 1, p. 1). Similar landforms have been described from Chile (Mortensen, 1959) and Virginia (Hack and Goodlett, 1960). Two theories have been advanced to explain the rarity of feral relief in Europe and its frequency in New Zealand. The first is that the difference is climatic in origin; in New Zealand there is a winter rainfall maximum, giving higher rainfall at the time of lowest evaporation, hence greater runoff and a tendency towards slope dissection. The second is that the tendency towards slope dissection is less under periglacial conditions, and Europe is more influenced by relict periglacial forms (Cotton, 1958, 1958b, 1962, 1963; Mortensen, 1959). Hack and Goodlett (1960) give details of slope forms in profile and plan in an area of feral relief on sandstones in the Appalachians.

POLAR AND MONTANE CLIMATES

Within the polar zone, conveniently taken as that lying beyond the polar limit of forests, various internal climatic differences of morphological significance have been suggested. The first is that between the frost-debris zone and the tundra zone (Büdel, 1948b). In the *frost-debris zone* vegetation is very sparse except for lichens, and the regolith consists of coarse, frost-shattered material; stone stripes are common on 2°-15° slopes. In the *tundra zone* there is more fine material in the regolith, but solifluction is checked by a root mat; 2°-15° slopes are frequently covered by a micro-relief of solifluction terraces. A second climatic division is between the *Icelandic*, or oceanic, regime, with temperatures frequently crossing 0°C but not falling to very low levels, and the *Siberian*, or continental, regime, with very low winter temperatures. Frost-shattering is greater under the Icelandic regime (Wiman, 1963; Potts, 1970). Thirdly, there is a division between dry and humid polar climates. Dry polar climates are those with less than 20 cm precipitation; they coincide only in part with areas of the Siberian temperature regime. Moisture is essential to all the main processes of polar environments: frost-shattering, solifluction, surface wash, slush-avalanching and solution. Hence slope retreat is slower in dry polar regions. Two unresolved questions are concerned with the relative importance of solifluction and surface wash, and the absolute rates of slope retreat. For a given type of polar climate, the relative importance of wash probably increases as the slope becomes steeper. The rate of slope retreat is commonly assumed (when discussed in a periglacial context) to be greater in polar than in temperate climates, but this has not been substantiated. It has been suggested that in dry polar climates morphological change is extremely slow (Malaurie, 1960; Souchez, 1967).

Two contrasted types of slope occur in polar regions: irregular faceted forms and smoothly-curved forms. Some slopes consist of approximately rectilinear segments with angular intersections, the segments having irregular micro-relief. Pissart (1966) comments that the abruptness of the change in angle at the foot of steep slopes resembles that found in arid lands (Fig. 82). The frequency of cliffs, and consequently of screes, is in part relict from glaciation. The detailed form of steep slopes is markedly influenced by structure, with cliff-scree associations occurring in mid-slope at sandstone or other outcrops. Limestones are non-resistant, being susceptible to solution at low temperatures. Gullying is common on slopes of moderate steepness. Rectilinear slopes have been noted at 12°-14° (Souchez, 1967) and 6°-8° (Pissart, 1966).

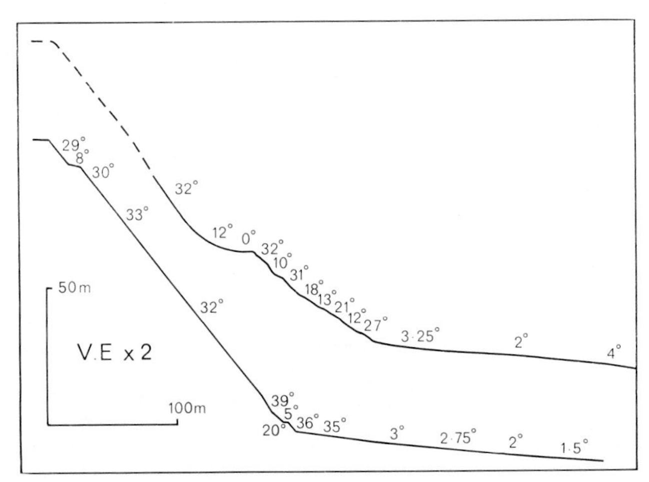

FIG. 82. Slopes in a polar environment. Prince Patrick Island, Canadian Archipelago. The break of slope is at the junction between Devonian and Cretaceous formations. Based on Pissart (1966).

In contrast with these angular forms are smoothly-curved slopes. Convexities occur above some cliffs, and Wood's four hillslope elements have been identified (Twidale, 1956). Flat-floored valleys with entirely convex slopes have been described (Poser, 1936; cf. Fig. 87, 7). Steep slopes, relict from glaciation, evolve by slope replacement; growing screes replace cliffs, and are in turn replaced by gentler slopes mantled by congeliturbate. Richter denudation-slopes occur. For a given polar climate there is an upper limiting angle of the retention of a fine-textured regolith. As soon as slopes steeper than this angle have been eliminated by replacement from below, slope retreat becomes mainly subject to control by removal. Convexities form by the mechanism of downslope decrease in net removal. Subsequent evolution is by slope decline. (Troll, 1944, 1947, 1948; Peltier, 1950; Rudberg, 1963; Tricart, 1970.)

The montane environment is not climo-morphologically the same as the polar. It is characterized by diurnal crossings of 0°C for much of the year, having a high frost-change frequency (Troll, 1943). For a given region there is an altitudinal zone of maximum frost-change frequency, e.g. *c.* 1400 m in the Alps, and 3950-4500 m on

Mount Kenya. In parts of the equatorial Andes the frequency exceeds 300 days per year. In such conditions, patterned ground features occur on a micro-scale.

THE ARID ZONE

In deserts, the main transporting process is sheetwash following rare storms; there are eye-witness descriptions of the occurrence of surface flow carrying fine particles (McGee, 1897; Grove, 1960). Wind is of only minor importance. There is much evidence that present forms are in part relict from more humid conditions. Slope retreat is probably slower in deserts than under any other climate (Mortensen, 1956; Coque, 1960; Peel, 1960, 1966).

Most desert slopes are either steep or very gentle. The steep slopes include bare rock cliffs of up to 90°, screes, and boulder-covered slopes of 20°-35°. These contrast with slopes of 2° and less, including pediments (Fig. 83). Angles of 3°-20° are infrequent, giving a markedly bimodal angle frequency distribution. Concavities (other

Fig. 83. The contrast between steep and very gentle slopes in desert relief, Egypt. Structurally-produced ledges in horizontal sandstones have been largely eliminated by rectilinear slopes at the scree angle.

than pediments) and convexities are uncommon, although not entirely absent. Mountain areas are dissected by closely-spaced gullies with steep slopes and narrow crests. Where mountains adjoin plains, the typical slope form is a steep hillslope above a very gentle pediment. The piedmont zone is narrow, often less than 5 m; it is sometimes possible to sit on the pediment and lean against the hillslope. The steep slopes lack a continuous regolith, other than coarse fragments. Because of the absence of a regolith, structural differences are expressed as irregular micro-relief; stepped slopes, with repeated cliff and scree units, are common. Early travellers reported inselbergs rising from flat ground, without a pediment, which is the impression given by distant views. A pediment, sometimes both gentle and narrow, is always found however, unless the slope is undercut by wadi erosion or buried by sand accumulation. (Peel, 1939, 1960, 1966; Mabbutt, 1955; Grove, 1960; Dumanowski, 1960; Ollier and Tuddenham, 1962; Daveau, 1964; Butzer, 1965; Williams and Hall, 1965; Pachur, 1970.)

Owing to the absence of a regolith, the retreat of steep slopes is wholly subject to control by weathering. Usually they are reduced to *c.* 35° by the mechanism of boulder control, maintaining this angle during continued retreat; massive rocks may retain almost vertical cliffs. Lowering of the pediment is subject to control by removal. Because of the slowness of weathering, the amount of fine debris reaching the piedmont zone is small, and consequently a very gentle slope is sufficient for it to be removed by wash. Where screes are present, it is possible that in some cases a time-independent condition between the heights of cliff and scree is attained (p. 124); in other cases the cliffs are relict features, undergoing replacement by growing screes. With this exception, slope evolution in deserts is by parallel retreat. The essential feature is the absence of vegetation, and consequent failure to retain a fine regolith on steep slopes, so that the retreat of these slopes is controlled by weathering, and boulder-control sets a lower limit to their angle. A consequence is the abrupt transition from hillslope to pediment, which in its turn causes the contrast in regolith between coarse fragments and a fine material.

SEMI-ARID CLIMATES

Over the climatic transition from deserts to humid climates there is a progressive increase in the frequency of high-intensity rainfall, the active factor in rainsplash and surface wash, but the increasing density of vegetation gives a corresponding decrease in the proportion of bare soil exposed. Curves could be constructed to show the influence of each of these factors on the rate of wash, although to scale the curves correctly would require considerably more observational data than is at present available. If correctly scaled, however, the intersection of the two curves would show the climatic conditions at which the rate of transport by rainsplash and surface wash reaches a maximum. Present evidence suggests that this intersection occurs in the semi-arid zone, where the vegetation cover is relatively sparse, but heavy falls of rain occur in every year. Some support for this conclusion is provided by the data for rates of erosion within basins in the southwest United States. Langbein and Schumm (1958) concluded that both the rate of accumulation of sediment in reservoirs and the rate of discharge of suspended sediment were at a maximum at an effective precipitation (as measured by runoff) of about 250 mm/year.

The semi-arid zone comprises tropical and sub-tropical areas with approximately 250-600 mm rainfall. Rainsplash and surface wash are the main transporting processes, although solution loss, subsurface wash and soil creep also occur. The rate of creep may, indeed, be faster than in the temperate zone, but the volume of material affected is insignificant compared with that moved by wash (Leopold *et al.*, 1956). A distinction from deserts is the retention of a regolith cover on most slopes. This cover, together with the high temperatures and appreciable soil moisture for part of the year, permits substantial chemical weathering. It is possible that the total rate of slope denudation is

FIG. 84. Hillslopes in a semi-arid climate, Himalayan foothills, north of Rawalpindi, West Pakistan. Denudation is rapid, with thin soils and much loose, coarse debris on the ground surface. Slopes predominantly rectilinear at *c.* 30°, with clear structural influence.

higher in the semi-arid zone than in any other climate, although present evidence is inconclusive.

The most frequent slope form under semi-arid conditions is again the hillslope and pediment. Pediments are steeper than in deserts, typically 1° over the lower half of the profile, 2° over most of the upper half, steepening to 5° close to the piedmont zone. Two tendencies appear in this zone: first, the development of a broader concave piedmont zone in place of an abrupt angular junction, and secondly, the appearance of convex-concave slopes in place of the hillslope-pediment landform under certain structural conditions. There is no simple relation between rainfall and the occurrence of these critical changes. A piedmont zone (taken as the 20°-10° transition) 400 m long may occur on sandstones under a rainfall as low as 350 mm (Fig. 85), yet on granites under a 650 mm rainfall in Sudan, the transition from a 24° hillslope to a 5° pediment occurs as 12 m at 13° followed by a 13°-5° concavity occupying only 2 m (Ruxton, 1958). Massive rocks usually have a hillslope and pediment, whilst weak, closely-bedded or jointed rocks tend towards a convex-concave form (Savigear, 1960). In a *Mediterranean* climate with a 250-400 m rainfall, simple slopes in argillaceous rocks are convex-concave, but a caprock invariably gives a free face above a dominantly concave

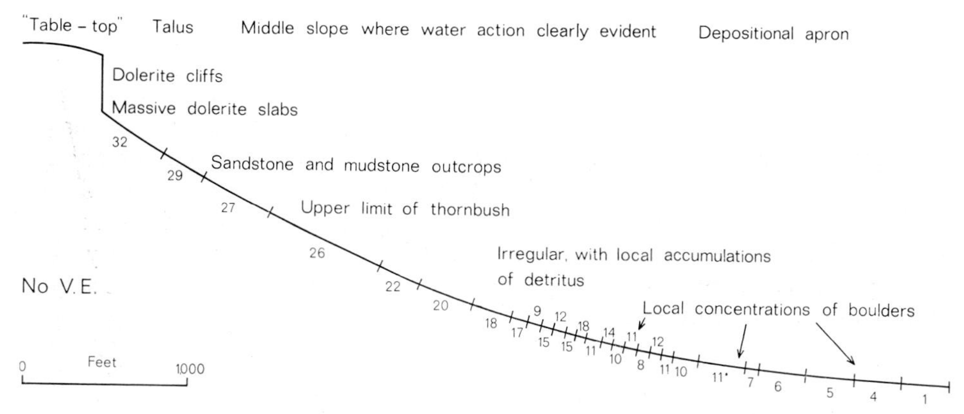

FIG. 85. Slope in dolerite overlying Karroo sandstones and mudstones under a semi-arid climate, Cape Province, South Africa. Based on Robinson (1966).

slope (Everard, 1963, 1964). Pediments are quite common in the Mediterranean zone, and inselbergs occur locally. Louis (1961) states that in the *steppe zone*, slopes at intermediate angles, neither steep nor gentle, are common. Thus in the range of environments between deserts and humid regions (savanna and temperate), most compound slopes have a free face and retreat parallel; some simple slopes, especially in massive rocks, retain the hillslope-pediment form, also evolving by parallel retreat, but weaker homogeneous rocks may have convex-concave profiles and evolve by slope decline.

THE SAVANNA ZONE

Surface wash is also important in the savanna zone, where bare soil is exposed between the tufted grasses. It is especially active during the intense storms that characterize the onset of the rainly season, when the vegetation cover is at a minimum. Rainsplash is also important (Christofoletti, 1968; de Ploey and Savat, 1968). On permeable soils the volume of material transported by rainsplash may exceed that carried by wash. For example, soils derived from sandstones in part of the Mato Grosso, Brazil, have an infiltration capacity exceeding 500 mm/hr, making surface runoff (theoretically) impossible, but after storms the ground is pitted with raindrop hollows, and traps show that sand grains are thrown 20 cm and more (Fig. 24, p. 68). Creep also occurs, and termites move large volumes of soil (M. A. J. Williams, 1968). The regolith is permanently moist in depth, giving rapid chemical weathering and consequently solution loss. On crystalline rocks the regolith is thick, typically 10-50 m, the depth of weathering being very irregular (Thomas, 1966); all but the uppermost part, about 1-2 m, of this remains *in situ*, as indicated by quartz veins. Nearly all slopes of less than 35° have a regolith cover.

Many accounts of slope form in the savannas are misleading. Most of the land surface comprises valley slopes of convex-rectilinear-concave form. This is particularly true of the extensive plains which characterize the shield areas occupying much of this zone, whether in Africa (Louis, 1964), the Deccan of India (Büdel, 1965) or South America. The convexity and concavity are smoothly curved, and there is usually a maximum segment of substantial length. Slopes of this type occur with maximum segments from 30° to as low as 2°. On steep slopes the concavity may be absent, the maximum segment is long, but there is still a convex crest. Some examples of the relative proportions of the convexity, maximum segment and concavity on moderate and very gentle slopes are illustrated in Fig. 86. Convex slopes usually occupy a greater area than concave. Low rock outcrops, no higher than the vegetation, are sometimes found on and near interfluve crests.

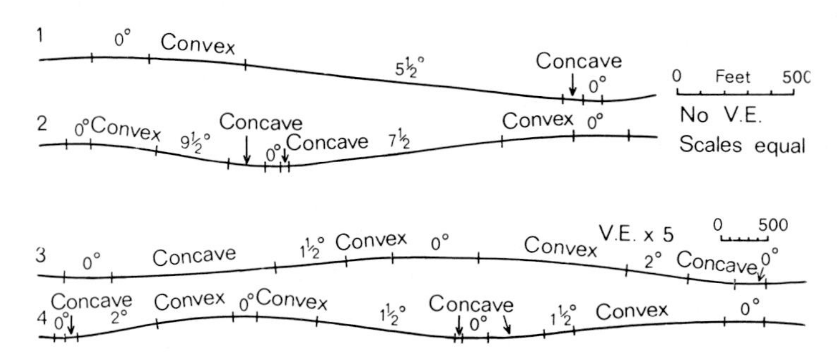

Fig. 86. Slope profiles across erosion surfaces in Malawi. Profiles 1 and 2, the Nyika Plateau, granite, 2500 m, montane savanna climate; cf. Fig. 2, p. 2. Profiles 3 and 4, the Lilongwe Plain, Basement Complex rocks, 1100 m, savanna climate; cf. Fig. 3, p. 3.

A different slope assemblage is found where a hill mass adjoins a plain or where inselbergs are present. The hillslope-pediment association then occurs, and slopes are predominantly concave. Compared with the semi-arid zone, there is much less tendency for a free face to be formed, and it is absent even from some compound slopes. Where a free face is present, as for example beneath slopes capped by dolerite sills or thick massive laterite, the debris slope beneath it consists of vegetation-covered soil with partially embedded boulders, sometimes with outcrops of rock *in situ*. The debris slope is 20°-35°, and concave rather than rectilinear. In some cases there is a relatively narrow piedmont zone, or a hillslope without a free face, between the debris slope and the pediment. Elsewhere they are separated by a long concavity, within which one or more segments at intermediate angles may occur. Pediments of 2°-6° are typical. (Pugh, 1956; Clayton, 1956; Sparrow, 1966; Mabbutt and Scott, 1966.)

All types of inselberg occur in the savanna zone, but the most common form is that of a domed rock crest over a steep regolith-covered slope. Structurally-controlled orientation is common. Some clearly originate as outliers beyond a retreating mountain front. The origin of others, particularly the high, steep mountains isolated amid extensive plains, is not established (p. 211). There is a marked contrast in slope form between adjacent interfluves with and without a crest inselberg. The former have concave to rectilinear pediments, whilst slopes on the latter are predominantly convex.

Similarly, compound slopes capped by horizontal sandstones are predominantly concave, in contrast to the convex-concave form on crystalline rocks (Fig. 87, 3).

A common landform of savanna plains is the *dambo*, a Nyanja vernacular word literally meaning 'meadow' but applied geomorphologically to valley floors that are continuously concave in cross-profile (Ackermann, 1936; Louis, 1964). Most dambos lack a stream channel, but if one is present there is no level flood-plain, and the channel does not substantially interrupt the concavity (Figs. 87, 1-3, and 88). The regolith is

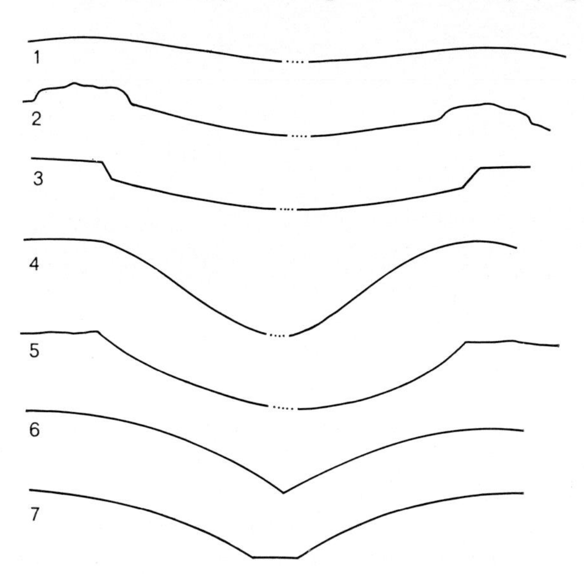

FIG. 87. Schematic valley forms, after Louis (1964): 1 = *Flachmuldental* (trough valley); 2 = *Flachmuldental* with inselbergs, in crystalline rocks; 3 = *Flachmuldental* with inselbergs, in horizontally-bedded sandstones; 4 = *Kehltal* (throat-like valley) in crystalline rocks; 5 = *Kehltal* in horizontally-bedded sandstones; 6 = *Kerbtal* (incised valley); 7 = *Sohlenkerbtal* (flat-floored incised valley). Types 1-5 are characteristic of savanna landforms, and types 6 and 7 of the rainforest zone.

clayey in the valley centre, but lens-shaped sandy belts occupy the upper parts of the valley-side concavities, giving place to a normal soil on the maximum segment and convexity above. There are characteristic lobate expansions at the valley heads, giving amphitheatre-shaped areas, convex in profile and plan, with a sandy regolith. It is possible that dambos are relict forms, previously occupied by actively-eroding streams but now, following some climatic change, being infilled by surface wash. More probably, however, they are formed under the present environmental conditions. After heavy rains the entire valley floor is flooded, allowing the evacuation of sediment downvalley by a process intermediate between surface wash and channel flow.

Little evidence is available on the manner of slope evolution in the savannas. Hillslope-pediment forms must evolve either by replacement or parallel retreat. The forms assumed by the more extensively developed convex-concave valley slopes are consistent with slope decline.

THE RAINFOREST ZONE

Chemical weathering reaches its maximum intensity in the rainforest zone, and even quartz grains are slowly attacked. Consequently solution loss is an important process of

FIG. 88. Gently concave valley head, resembling an African *dambo* in form but with gallery forest in the valley centre. Mato Grosso, Brazil, savanna climate.

removal, possibly accounting for more material than is carried by all processes of downslope transport. Both creep and wash occur, but their relative magnitude is not established. The frequency of landslide scars, particularly from debris avalanches, has often been noted; in nearly all cases, however, such descriptions refer to slopes of 35°-50°. The importance of landsliding is therefore in part a result of the high frequency of steep slopes in this zone (Wentworth, 1943; White, 1949; Berry and Ruxton, 1960; Ruxton, 1967; Simonett, 1967; Meis and da Silva, 1968). A regolith-stripping cycle has been suggested (p. 78). An extended account of processes is given by Rougerie (1960).

The regolith is usually, but not always, thick. On granites in Hong Kong it is mainly 15-30 m, reaching a maximum of 100 m (Berry and Ruxton, 1960), but on 35°-40° slopes in Papua the soils are thin and contain primary minerals other than quartz, indicating that an equality between the rates of weathering and removal is attained with only a shallow regolith (Ruxton, 1967). On sandstones and shales in Malaya, fresh rock may occur within 5 m of the surface even on gentle slopes. Regolith-covered slopes of up to 70° have been reported, the soil being held in place by tree roots, but such cases are exceptional and slopes do not normally exceed 50°.

There are numerous insequent streams, giving a high and unusually uniform drainage density. Over large areas, no point on the ground is more than 500 m from a drainage line. Where stream erosion is rapid, this results in *ridge-and-ravine topography*, a relief type particularly characteristic of this zone. The entire surface is made up of valley slopes with fairly narrow but smoothly curved convex crests above long 30°-40° segments (Fig. 89; cf. Fig. 95, p. 251). Where the rate of stream erosion is less the valley floor is occupied by a flood plain, flat as a whole but with channel and levee micro-relief. Above this floor, the convexity may form the entire slope (Figs. 87, 7,

Fig. 89. Rainforest relief, central Pahang, Malaya. Narrow convex crests and steep valley sides.

and 90). The crests are gently but continuously convex; as the angle steepens downslope so the curvature also increases, until at the base a 20°-50° slope abruptly joins the valley floor, with a concavity of less than a metre. In a *sub-tropical rainforest* climate in Natal, Sparrow (1966) has described very long convexities, reaching up to 70° without a free face, below which are short concavities of high curvature. More remarkably, such dominance by the convexity may occur in gently-sloping relief. On sandstones in part of the Rio Suiá-Missu basin in the Mato Grosso, Brazil, at the margin of the rainforest zone, the mean angle is approximately 1°; 95% of all slopes are less than 2°, and yet the convexity occupies 95-99% of the profile (cf. Fig. 65, p. 174). Concave elements are short or absent, except beneath structurally-controlled mountain masses and inselbergs. Angle frequency distributions are approximately log-normal, both in steeply and gently-dissected areas (Young, 1970b).

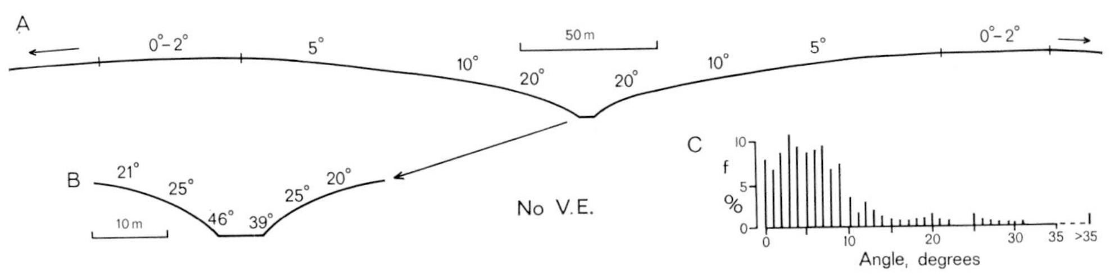

Fig. 90. Slope profiles under a rainforest environment, andesite, central Pahang State, Malaya. *C* = angle frequency distribution.

Except on inselbergs, rock outcrops are infrequent. The topographic expression of structural differences in slope micro-relief is less than in any other climatic zone. Thick quartzite beds may give rise to a free face, but sandstones over shales normally produce only a convex-concave slope. The rock dome type of inselberg is characteristic. They are usually found to be composed of different rock to their surrounds, whilst the contrary has never been demonstrated. It has been suggested that weathering is slower on bare rock than beneath the permanently moist regolith; if true, this could cause rock inselbergs, once exposed, to increase in relative height.

In ridge-and-ravine topography, the similarity of angle between different valleys indicates that evolution is by uniform ground loss, giving time-independence of form (p. 21). The manner of retreat of the wholly-convex slopes is not established. The convex form is maintained by lateral basal erosion; the stream channel only takes normal discharge, and after storms the entire valley floor becomes flooded to a depth of as much as a metre with rapidly-flowing water. Process-response models show that solution loss gives parallel retreat, whilst downslope transport gives slope decline (p. 114), but which of these modes of removal is quantitatively greater is not known. On the basis of observations in southern Malaya, Swan (1970b, 1970c) suggested a difference in profile form and manner of evolution according to whether the bedrock contains free quartz. On rocks in which quartz is lacking, for example basic volcanics, the regolith contains very little sand; most removal of regolith is in suspension or solution, and slopes remain largely or entirely convex. On granites and argillaceous sedimentaries, which yield an unweatherable quartz fraction, convex-concave profiles occur and retreat is by slope decline. The occurrence of convex slopes on sandstones, however, conflicts with this theory.

Climatic conditions similar to those of the present are thought to have persisted in rainforest areas for a sufficiently long period to make the existence of relict forms unlikely, possibly the only climo-morphological zone of which this can be said. There are still large areas with relatively unaltered vegetation. Slope form and the surface processes that have produced it can therefore be studied in conjunction. Neither the relative importance of solution loss, creep and wash, nor that of rapid mass-movements in relation to the continuous processes, is established. Similar slope forms occur repetitively over large areas, permitting statistically-based generalizations on form to be obtained. The manner of slope evolution has not been established. Of all the climatic zones, the rainforest environment offers the greatest opportunities for the study of slopes.

XIX INHERITED AND RELICT FEATURES; VALLEY ASYMMETRY

MANY features of existing landforms originated under environmental conditions substantially different from the present. There are two main respects in which the external factors acting upon a slope can change: the conditions at the base, and the active surface processes. Changes in basal erosion and removal are associated in part with relative vertical movements of the land and sea, as for example in rejuvenation. Different surface processes are brought about by climatic change, acting both indirectly, through its effects on vegetation, and directly. Features of form originating at a time of differing basal conditions will be termed *inherited features*; in conventional terms these are features inherited from earlier epicycles of erosion. Forms dating from periods of different climate are *relict features*.

The requirement that the conditions producing an inherited or relict feature should have differed *substantially* from the present is qualitative, but in practice not difficult to interpret; the qualification is necessary since neither basal erosion nor climate is ever completely constant, and it could therefore otherwise be argued that all slopes are both inherited and relict. The latter, extreme, view illustrates, however, that the concepts are only relative. All landforms carry evidence of the past, and all have been modified, to a greater or lesser extent, under present-day conditions. Recognition of this permits a reconciliation between the time-dependent and time-independent approaches. 'The influence of a past process having acted upon a geomorphic system is proportional to the intensity and duration of its action but inversely proportional to the time elapsed since its action. That is to say, the amount of information within landforms about their historical changes decreases the more remote the past. Some landform features reflect only recent changes of environment, whereas others yield a certain amount of information about the more distant past.' (Howard, 1965.)

INHERITED FEATURES

Polycyclic slopes are the commonest type of inherited feature. It has long been recognized that valley-in-valley forms result from an increase in the rate of river erosion caused by a fall in base-level. The slope profiles of such valleys fall into two or more portions, each containing a convexity, with or without part of the former maximum segment and, less frequently, the concavity. Assuming evolution by slope decline, the curvatures of the convex elements and the angles of the segments below them will both increase downslope with successive epicycles, for if the duration of the later epicycle exceeds that of the earlier, evidence of the earlier is obliterated. This interpretation has been applied both to valleys (Fig. 91) and to seaward slopes (Fig. 48, p. 126). A further consequence is that the angle frequency for such a region is polymodal, with characteristic angles corresponding to the modal angle range developed under each epicycle (p. 168).

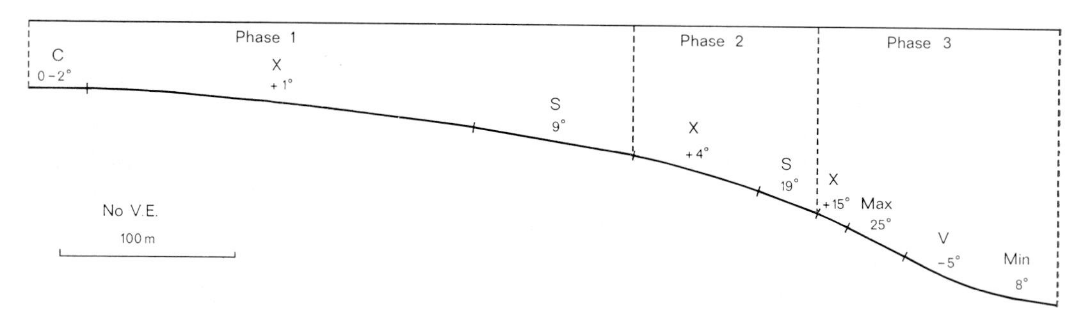

FIG. 91. Idealized slope units on a valley-side slope of polycyclic development. The portions of the slope formed primarily under three successive epicycles are shown as Phases 1-3. The units are generalized from observed profiles in the Heddon Basin, south-west England. Curvatures are given in degrees per 100 m. For meaning of lettering, see Fig. 54. Based on Young (1963).

It is rarely possible to investigate slope evolution without reference to the denudation chronology of the area studied. As a generalization, the contribution of historical geomorphology to the explanation of existing slope forms is as great as that made by climatic geomorphology; both contribute less than does explanation in terms of structure. Conversely, slopes can contribute to regional studies of denudation chronology. The latter have long been heavily dependent on evidence from erosion surface remnants, yet these frequently occupy only a small proportion of the land surface area. Slopes are a more extensive potential source of evidence, obtainable by profile analysis. To interpret profiles in terms of epicycles requires, it is true, assumptions on the manner of slope evolution, but insofar as present form is to be used as evidence of past evolution, such inter-dependence is inescapable. There is mutual benefit to be obtained from a closer integration of historical geomorphology and systematic work on slope evolution.

RELICT FEATURES

During the past two million years, climatic changes sufficient to have a substantial effect on surface processes have occurred in most parts of the world, with the probable exception of the rainforest zone (Büdel, 1953). The main changes have been:

(*i*) Warmer climates in the temperate zone.
(*ii*) Glacierization of cool temperate and montane regions.
(*iii*) Periglacial conditions in the temperate zone.
(*iv*) More humid climates in arid and semi-arid regions.
(*v*) Drier climates in parts of the humid tropics.

Warmer climates existed in the temperate zone during the late Tertiary and the Penultimate Interglacial. The suggestion that present landforms preserve features relict from such conditions was first applied to the Hercynian massifs of Europe (Von Freyburg, 1923, 1932; Jessen, 1938; Louis, 1935, 1957; Neef, 1955). Louis (1957) postulates that areal denudation predominates over linear erosion in savanna climates, whilst the reverse is true in the temperate zone. Consequently, where relatively narrow valleys dissecting extensive gently-sloping surfaces occur, climatic change as well as uplift is a possible cause. A deeply-weathered regolith, containing core-stones and having the reddish colour associated with tropical weathering, occurs in places on igneous and metamorphic rocks. In Scotland, a layer of weathered rock up to 12 m thick and overlain by the earliest till has been described (Fitzpatrick, 1963). Deep weathering in tropical climates, followed by regolith stripping under periglacial conditions, is one hypothesis of tor formation.

Slopes relict from glacierization are common in cool temperate and montane regions. Study of their origin lies in the field of glacial geomorphology, which is not considered here.

By far the most frequently described relict climatic landforms are periglacial phenomena, formed during glacial periods under cold climatic conditions existing beyond the limits of glacierization. The historical growth of the recognition of periglacial features is traced by Bryan (1949), and a bibliography of earlier work given by Smith (1949). Among classic works, responsible in part for the massive post-war growth of periglacial geomorphology in continental Europe, are the studies of the Mittelgebirge by Büdel (1937) and of the Massif Central by Beaujeau-Garnier (1951). Summaries of periglacial features in Britain are given by Fitzpatrick (1956) and Galloway (1961). The last seven chapters of Embleton and King (1968) provide a useful summary.

From numerous studies it is clear that relict periglacial features can be identified over large parts of Europe and North America. They are particularly widespread in Poland, where it is customary to interpret the entire landscape in periglacial terms (e.g. Dylik, 1956). This widespread occurrence raises the disturbing question of whether it is meaningful to relate landforms of temperate regions to contemporary processes, or whether the present slopes are largely relict, an extreme view that has been voiced by L. C. King (p. 36). This view is not proven, however, by the fact that some periglacial features have been identified in most regions. Periglacial studies are necessarily selective, and if the search for such features yielded a negative result it is unlikely that such a finding would be reported. In relation to slopes, the main questions that arise are the nature and extent of slope deposits, the manner of profile change and, in particular, the magnitude of slope retreat accomplished under periglacial conditions.

Periglacial slope deposits are abundant in the Palaeozoic shale uplands of Wales and Southern Scotland (Young, 1958b, 1963; Tivy, 1962; Watson, 1965; Ragg and Bibby, 1966). Layers of sharply angular stones, bedded approximately parallel to the slope, are found on concavities (Fig. 92). Two separate types of deposit have been identified, coarse and fine fragments, the modal lengths of the longest axes being of the order of 5-10 and 0·5-1 cm respectively. They are identifiable as congelifractate by the

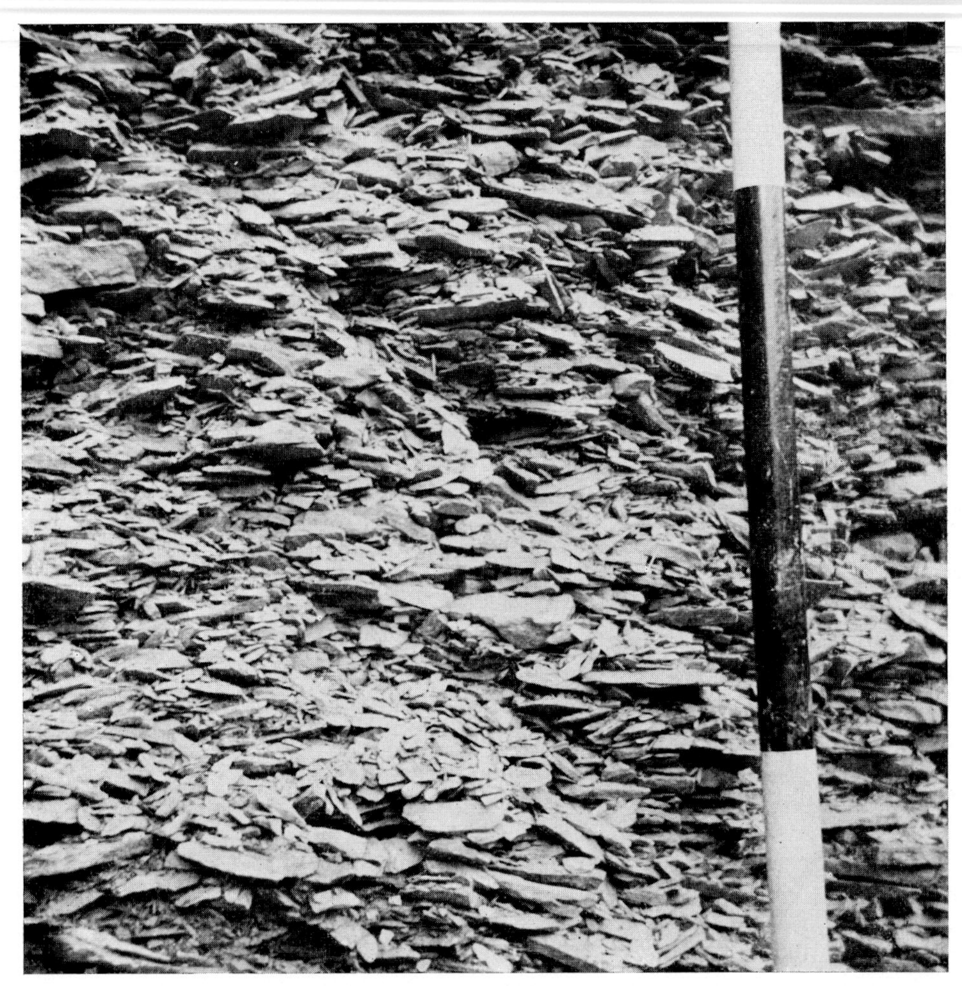

FIG. 92. Angular congelifractate, derived from Palaeozoic shales, on a 28° slope. Central Wales.

extreme angularity, the presence of involuted structures, and a distinctive particle-size distribution, more sharply peaked in one size-grade than any layer of the contemporary regolith. In central Wales the coarse deposit occurs on slopes of 28°-34°, whilst the fine is associated with concave elements with angles of 13°-26° and curvatures of 30-100°/100 m. The probable origin is that in a cold climate, all existing soil was removed, exposing the rock to frost-shattering. The shattered material was carried downslope, coming to rest at angles related to its composition and the intensity of the processes then operative. The consequence is that the form of the concavities in these regions can only be interpreted in periglacial terms.

By no means all concavities in cool temperate regions carry periglacial deposits.

Many are cut in bedrock, with a regolith of insufficient thickness to have a substantial effect upon their form (Peel and Palmer, 1956; Young, 1963).

Stratified slope deposits have been described at many localities in east and central Europe, comprising sequences of regolith layers with differing composition, including palaeosols. Complex chronologies have been constructed on the basis of such deposits, relating successive climatic phases of the Pleistocene to surface processes and their effects on the slope (e.g. Dylik, 1960; Pécsi, 1967). The basic pattern is an alternation of periglacial phases with warmer periods. On the upper parts of slopes there is intensified denudation, with regolith stripping, in cold periods, and relative stability, with soil formation, in warm intervals. The deposits on concavities, laid down during cold periods, consist first of the soil from the upper slopes, followed by the frost-shattered material. Excavation of stratified deposits may reveal fossil slopes (Dylik, 1969).

The change in slope form under periglacial conditions is generally held to be a rounding and smoothing of the profile, with net erosion on the upper, convex, part and deposition on the concavity. Starkel (1964) compared assumed rates of downslope transport with resulting profile modifications (Fig. 93). He assumed a warm semi-arid

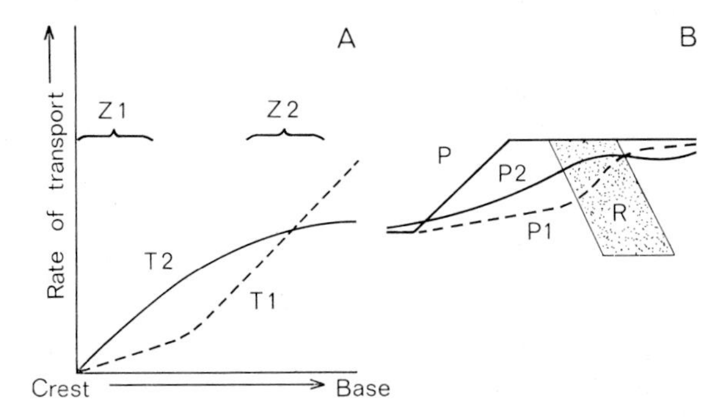

FIG. 93. Manner of action of surface processes and the resulting slope evolution in the Flysch Carpathians, according to Starkel (1964, modified). *A* = variation in rates of surface transport with distance downslope; *T1* = in warm semi-arid climate of the Upper Pliocene; *T2* = in periglacial climates of the Quaternary; *Z1* = zone of maximum ground loss in cold climate; *Z2* = zone of maximum degradation in warm semi-arid climate; *B* = slope evolution; *P* = initial slope; *P1* = mature profile typical of the Upper Pliocene; *P2* = mature profile typical of Quaternary morphogenesis; *R* = resistant rock.

climate in the Upper Pliocene, during which pedimented slopes with angular convexities originated. Ground loss is assumed to result from slope wash, the rate of which increases slowly near the crest but linearly with distance downslope thereafter. Under solifluction in cold periods, the increase in intensity of transport is greatest near the crest. Net ground loss is proportional to the slope of curves in Fig. 93*A*; it is consequently greatest on the lower slope in the dry climate but near the crest under cold conditions. The result is a transformation from a previous hillslope-and-pediment profile to a convex-concave form. Although containing many unverified assumptions, this study is an attempt to relate form change to the manner of action of processes under periglacial conditions.

R. B. G. Williams (1968) made estimates of the total thicknesses of material removed by denudation under periglacial conditions in Britain. He concluded that amounts were less in the west than in the east. 'In Devon and Cornwall up to 3-4·5 metres may have been removed from granite and gabbro, and up to 7·5-9 metres from slates, shales and schists. In south-central England 9-10·5 metres were apparently lost on the Chalk through solifluction. This figure does not allow for solution.' These estimates refer to the combined effects of all Pleistocene cold periods. Watson and Watson (1970) give thicknesses of coastal periglacial deposits in the Cotentin Peninsula, France, together with the lengths of the seaward slopes above them. From their figures, the maximum net removal is equivalent to a layer 4 m thick, whilst on other sites mean thicknesses are 0·2-2 m. The total time elapsed since the onset of the earliest Pleistocene glaciation has been estimated as about 600 000 years, of which half may have been glacial periods. The amount of denudation during some 300 000 years of periglacial conditions depends heavily on whether it is assumed that ground loss in cold climates is substantially more rapid than in other climo-morphologic regions. There is no evidence that this is so. Hence if the median values for rate of denudation given on p. 89 are adopted, ground loss during all periglacial periods would be 14 m on normal relief and 150 m on steep slopes. This is greater than the values obtained from geological evidence. The tentative indications at present are that the total slope retreat under periglacial conditions has been of the order of 1-10 m.

Features relict from the last glaciation have been little modified in post-glacial time. This has sometimes been interpreted (e.g. Rapp, 1967) as indicating a greatly decreased intensity of denudation under present compared with periglacial conditions. It is, however, sufficiently accounted for by the shortness of the period: at an assumed rate of 50 mm/1000 years, the expected ground loss in post-glacial time is less than one metre. A further complication is that in some periglacial areas measurements suggest that the post-glacial landforms present cannot have formed in the time available since deglaciation, suggesting that at some stage in the past (perhaps immediately after deglaciation) rates were appreciably higher than today (Rapp, 1960b).

There is much evidence that arid and semi-arid zones have experienced more humid conditions in the recent past (e.g. Peel, 1966); whether these pluvial periods correspond to the northern-hemisphere glaciations will not be discussed. In view of the slowness of contemporary slope retreat in deserts (p. 230), many desert slopes may be in large part relict. Ahnert (1960) describes bare rock slopes in sandstones and shales under a semi-arid climate in Colorado (Fig. 94). Two contrasted form elements occur: vertical cliffs, and gently-sloping, sometimes rounded, rock slopes. The cliffs sharply truncate the gentler slopes, hence the formation of the two elements is not simultaneous but successive. Ahnert considers that these scarps were formed under a series of alternating humid and dry climates; the cliffs originated from groundwater seepage during

for land that is ploughed and harvested annually. Moderately steep slopes, up to 18°, can be cultivated with difficulty and at greater cost, but such land is usually under permanent grass. Slopes of over 18° are usually under grass, afforestation or natural woodland. This no longer applies where labour is plentiful and land pressure becomes great. Thus in the mountains of northern Spain, slopes of up to 25° are cultivated without terracing, whilst in Java and Nepal, terraced level fields, sometimes no wider than 2 m, are constructed on slopes of up to 35°.

In most climates, with the exception of the cool temperate oceanic, the danger of accelerated soil erosion and the cost of its prevention are a major restriction on land use. Where erosion hazard is severe, only very gentle slopes, of less than 2°, can be cultivated without the need for conservation measures. With increase in angle there is a more than linear increase in liability to erosion (p. 65), and consequently in the cost of bunds and other works. Increase in slope length also adds to the erosion hazard, through its effect on the volume of runoff. In the tropics, annual crops are commonly grown on slopes of up to 10°, but the cost of conservation works on moderate, 5°-10° slopes is considerably more than on gentle slopes of less that 5°. High-value perennial crops can bear the cost of terracing when grown under efficient management (Fig. 95), and tea in particular is frequently grown on steep slopes. But even with perennials, the cultivation of steep slopes without terracing can lead to severe sheet erosion, as has occurred, for example, on tea grown on smallholdings in Ceylon. In some tropical

FIG. 95. Oil palm plantation on slopes of up to 18°, with terracing and a dense cover crop. The level skyline of the forest canopy in the background obscures the dissected relief. Central Pahang, Malaya.

countries it is illegal to cultivate slopes of more than a certain angle, for example 18½° in Malaya.

A consequence is that land classification schemes make substantial use of slope angle as a limiting factor. This applies to the land-capability classification of the U.S. Department of Agriculture (Klingebiel and Montgomery, 1961), the many adaptations of this scheme used in tropical countries, and the classification employed by the British Soil Survey (Bibby and Mackney, 1969). A comparison between maps of slope angle, soils and land classification frequently shows land class to be determined to a greater extent by slope than by soil properties, particularly in the highest category of classification. Examples have been described showing the substantial effects of slope angle upon type of farming (Hidore, 1963) and land values (Paget, 1961).

Methods used to describe slope in the course of soil surveys are reviewed by Curtis *et al.* (1965). The necessary observations are angle, profile curvature, plan curvature, and position with respect to the slope as a whole. The properties may be estimated at sites of auger observations, and measured instrumentally only at soil pits. These parameters record the slope characteristics at a point. For the regional description of slopes, the qualitative methods used in many soil surveys are inadequate, in view of the significance of slope for land use potential. Since most areas contain slopes covering a wide range of angle, a method based on angle frequency distribution is necessary. The following is a possible procedure.

The area to be surveyed is divided into landform regions, by air photograph interpretation. Within each region, randomly-selected profile lines are obtained, using the method given on p. 145 or a simplified adaptation. These profiles are surveyed; angle measurements in one direction only, to the nearest whole degree, are adequate. For all profiles within a region, the combined lengths of ground at each angle value are obtained, and converted into percentages; slopes of 0°-2° are summed separately according to whether they occur on crest or valley-floor sites. An angle classification is selected (p. 173), and the percentage frequencies in each angle class obtained for the region. Slope angle regions are defined by an angle such that 90% of the ground surface has gentler slopes; thus in a 0°-18° region, up to 10% of the land may exceed 18°. For each slope angle region obtained in this way, the percentage frequencies of land within each angle range are tabulated, as in Table 15. Maps of slope angle regions may be used in conjunction with soil maps to obtain land classification.

The appropriate designations for slope angle regions are normally of the form 0°-5°, 0°-10°, 0°-18°, etc. It is only correct to describe a region as, e.g., 10°-18° if less than 10% of the land is gentler than 10°, which is rarely the case. By plotting a graph of percentage frequency (ordinate, log scale) against angle (abscissa), the proportion of ground in each region above and below any given angle may be read off.

SOIL

The earliest investigation into the relation between slope and soil properties was by Norton and Smith (1930), studying soils derived from loess on Illinois. They found an inverse correlation between angle and depth of the textural *B* horizon, as well as

Table 15. Estimated percentage frequency of slopes according to slope angle region

Slope angle region		Percentage of slopes							Percentage of surveyed area occupied by slope angle region
Name	Over 90% of ground in range	Level to very gentle 0°–2° (Valley floor)	Level to very gentle 0°–2° (Crest)	Gentle 2°–5°	Moderate 5°–10°	Moderately steep 10°–18°	Steep 18°–30°	Very steep >30°	
$S1$	0°–2°	100	0	0	0	0	0	0	10
$S2$	0°–5°	20	20	50	10	0	0	0	2
$S3$	0°–10°	5	10	40	35	7	3	0	45
$S4$	0°–18°	4	5	10	35	35	10	1	26
$S5$	0°–30°	2	3	5	10	30	45	5	13
$S6$	0°–45°	1	2	5	7	20	50	15	4
Percentage of surveyed area in angle class		14	7	24	20	17	11	7	100

relations between slope and soil colour, texture, structure and consistence. Correlations between parameters of slope and soil have since been demonstrated by Troeh (1964), Acton (1965), Ruhe and Walker (1968), Walker *et al.* (1968) and Furley (1968).

The two principal effects of slope are upon site drainage and soil thickness. The influence on drainage may be divided into the effects of position and of the shape of the ground surface. Sites near the slope base have poor drainage owing to the position of the water table; soils on concavities frequently possess a gleyed horizon, which occurs progressively nearer to the ground surface lower on the concavity. Higher on the slope, above the permanent water table, both profile concavity and plan concavity tend to give surface water gleying.

Steep slopes tend to have thin and stony soils. If the soil thickness has attained a time-independent condition (the equilibrium balance of denudation, p. 101), the rate of net soil removal will be equal to the rate at which the underlying rock is weathered into soil (cf. Nikiforoff, 1949). Due to removal of regolith, a given portion of the rock material passes progressively upwards through the soil, being in addition transported downslope in the later stages. If the rate of denudation is high, a rock particle reaches the surface soil horizon in a relatively short period, and hence in a less highly weathered state. This effect is of great importance in the humid tropics. On level to gently-sloping land, removal of material is slow; weathering penetrates deeply, and the upper horizons attain a state in which no weatherable minerals remain. Where slopes are steeper, weathered material is constantly being removed and the rock minerals renewed from beneath. Hence at corresponding depths, horizons in soils on steeply-sloping land are less highly weathered than on gentle slopes (Fig. 96). This is the mechanism which, on erosional relief, determines the stage of weathering of the soil To refer to this stage of weathering as the 'age' of the soil is incorrect. The age of a soil, i.e. time as a factor, may only be considered as an independent variable in circumstances where a time-independent state has not been attained, for example on recently-fixed sand dunes or glacial moraines.

These relatively direct influences, on drainage and stage of weathering, lead to secondary effects upon other soil parameters. Drainage status affects colour and organic matter, and stage of weathering affects texture and cation exchange capacity. Furley (1968) compared topsoil carbon, nitrogen and pH with slope angle on soils in Oxfordshire. Good correlations ($r \simeq \pm 0{\cdot}9$) were found for sites on convexities, but on concavities there was relatively little relation to angle. On the convexities there was a decrease in carbon and nitrogen with increase in angle, possibly as a result of improved drainage or greater topsoil erosion. The soil reaction became less acid with increase in angle on siliceous rocks, but on calcareous rocks the steeper slopes had higher pH values. The latter result could be explained in terms of the mechanism described above, the topsoil having undergone less decalcification where ground loss is slower, on the steeper slopes.

Ruhe (1960) and Mulcahy (1961) discussed the hypothetical effects on soils resulting from different types of slope evolution. Mulcahy suggested that with parallel retreat of a scarp above a pediment, the time during which the pediment has been exposed to weathering, and consequently the stage of weathering of the soil, will increase pro-

Fig. 96. Slope, soil and vegetation relations at the margin of Malosa Plateau, southern Malawi; syenite, 1700 m. On the undulating plateau surface, soils are intensely weathered; the montane grassland is possibly fire-induced. On the steep slopes, under savanna woodland, soils are shallower and richer in weatherable minerals, owing to the higher rate of natural denudation.

gressively away from the scarp base. This, however, does not distinguish between the time-scales involved in slope retreat and in soil formation. An alternative hypothesis is that soils on a pediment are related to near-contemporaneous geomorphological conditions, comprising surface processes and the dynamics of the regolith upon the pediment. That is, with respect to soil evolution, the relief may be taken as constant without substantial error. The critical questions in deciding between these alternative approaches are the relative rates of slope retreat, downslope regolith transport, and pedogenic processes. In a study of soils above and below laterite breakaways in Nigeria, Moss (1968) demonstrated the relation of soil formation to contemporary processes of scarp retreat and regolith transport.

Slope form units are routinely used in soil survey to extrapolate auger observations. Bridges and Doornkamp (1963) independently produced a morphological map and a soil map of an area of alternating sandstones and shales in northern England, and subsequently compared the two maps. They found that major breaks of slope nearly always indicate soil changes, but many minor changes of slope are insufficient to affect soil type, except where they coincide with changes in parent material. On gently-sloping glacial drift topography, for example in East Anglia, frequent and substantial changes in soil properties occur without any indication in the surface form of the ground. A number of studies have shown that landform units interpreted from air photographs have only a limited predictive value for soil properties (e.g. Webster and Beckett, 1968; Webster and Wong, 1969; Jarvis, 1969). This result is not unexpected, since the hypothesis investigated is that soil distributions on a detailed scale can be predicted from landforms surveyed by reconnaissance methods. The predictive ability of landforms can be increased by applying quantitative methods of slope survey and analysis, at an intensity comparable to that at which soil information is desired. Such intensive use of slope as a predictor will not always be justified in terms of survey efficiency. The circumstances under which it is most likely to be beneficial are where there are repeating patterns of slope, complex in detail, over extensive areas. Thus on morainic topography in Canada, Acton (1965) was able to relate soil types to precise slope angles and curvature; on this basis he constructed a table showing the percentage frequencies of five soil types in each of six landform regions, the latter mapped from air photographs.

The effects of slope upon soil are expressed most clearly in the soil catena. In explaining the differentiation of soil properties within a catena, two kinds of effect may be distinguished. The first is that caused by differences in the site conditions immediately proximate to the soil; examples are the steepness of slope and depth to the water table. The second type of effect arises from the position of the site within the catena as a whole, and refers to the consequences arising from the lateral movement of solid particles, water and material in solution from the upper to lower parts of the slope. It may be possible to find means of isolating these two kinds of effect, and comparing their relative importance under different environmental circumstances. The downslope transfer of material by lateral movement in solution affects soil fertility. In the humid tropics in particular, valley-floor gleys may have a substantially higher nutrient status than the upper members of the catena.

A further possibility is the establishment of a system of catenary types. The commonest example in temperate latitudes is the sequence, in order downslope, of brown earth/gleyed brown earth/gley. In upland Britain there is a repeating sequence consisting, where all its members are present, of hill peat/thin iron-pan podzol/iron podzol/shallow acid brown soil, becoming deeper downslope/gleyed acid brown soil/gley/basin peat (Taylor, 1960). A frequent catena on gently-undulating relief under savanna climates is that first identified in the classic descriptions by Milne (1947). A reddish soil on the crest becomes progressively yellower on the convexity; this leads to a lens of mottled and more sandy material on the concavity, which gives place to a heavy black clay in the valley centre. On different parent materials the properties of each member of the catena differ, but there is the same pattern of changes with slope. The gradual downslope changes contrast with a catenary type common in rainforest relief, in which the upper, freely-drained, soil type persists unchanged over the whole of the convexity, giving place abruptly to a gley at the valley-floor margin (Young, 1968).

Some aspects of the relations between slopes and soils are of direct practical application, whilst others contribute to the understanding of soil formation. The soil catena is an expression of the identity between geomorphological and pedogenetic processes. Soil creep, surface and subsurface wash, and solution loss are in part processes of soil formation. Conversely, properties normally regarded as pedological, such as the type of clay minerals and degree of structural aggregation, influence the processes of surface transport. Slope retreat involves the dynamic relations of the rock and regolith with the forces acting upon it; the same forces are responsible for soil formation, for which the mobile regolith provides the parent material. There is opportunity for a continuing exchange of information between soil science and the geomorphology of slopes.

BIBLIOGRAPHY

The Reports of the Commission on Slope Evolution of the International Geographical Union, although published independently, have been treated here as a serial publication, abbreviated as *Slopes Comm. Rep.* The full references are as follows:

1st Report: *Premier rapport de la Commission pour l'Étude des Versants, Union Géographique Internationale.* Amsterdam, 1956.
2nd Report: Contributions internationales à la morphologie des versants. *Z. Geomorph. Suppl.*, **1**, 1960.
3rd Report: *Neue Beiträge zur internationalen Hangforschung*, Vandenhœck and Ruprecht, Göttingen, 1963.
4th Report: Fortschritte der internationale Hangforschung. *Z. Geomorph. Suppl.*, **5**, 1964.
5th Report: *L'évolution des versants.* Univ. Liège. 1967.
6th Report: New contributions to slope evolution. *Z. Geomorph. Suppl.*, **9**, 1970.

ACKERMANN, E. 1936. Dambos in Nordrhodesien. *Wiss. Veröff. Museums Länderkunde Leipzig*, **4**, 147-57.

ACKERMANN, E. 1959. Der Abtragungsmechanisms bei Massenverlagerungen an der Wellenkalk-Schichtstufe. *Z. Geomorph.*, **3**, 283-304.

ACTON, D. F. 1965. The relationship of pattern and gradient of slopes to soil type. *Canadian J. soil Sci.*, **45**, 96-101.

AHNERT, F. 1954. Zur Frage der rückschreitenden Denudation und das dynamischen Gleichgewichts bei morphologischen Vorgängen. *Erdkunde*, **8**, 61-4.

AHNERT, F. 1960. The influence of Pleistocene climates upon the morphology of cuesta scarps on the Colorado Plateau. *Ann. Ass. Am. Geogr.*, **50**, 139-56.

AHNERT, F. 1966. Zur Rolle der elektronische Rechenmaschine und des mathematischen Modells in der Geomorphologie. *Geogr. Z.*, **54**, 118-33.

AHNERT, F. 1967. The role of the equilibrium concept in the interpretation of landforms of fluvial erosion and deposition. *Slopes Comm. Rep.*, **5**, 23-41.

AHNERT, F. 1970. An approach towards a descriptive classification of slopes. *Slopes Comm. Rep.*, **6**, 71-84.

AHNERT, F. 1970b. A comparison of theoretical slope models with slopes in the field. *Slopes Comm. Rep.*, **6**, 88-101.

AHNERT, F. 1970c. Functional relationships between denudation, relief, and uplift in large mid-latitude drainage basins. *Am. J. Sci.*, **268**, 243-63.

AITCHISON, C. D., and GRANT, K. 1967. The P.U.C.E. programme of terrain description, evaluation and interpretation for engineering purposes. *Proc. 4th Regl. Conf. for Africa on Soil Mech. and Foundation Engng, Cape Town*, **1**, 1-8.

ALEXANDRE, J. 1967. L'action des animaux fouisseurs et des feux de brousse sur l'efficacité érosive du ruissellement dans une région de savane boisée. *Slopes Comm. Rep.*. **5**, 43-9.

ALLEN, J. R. L. 1969. The maximum slope angle attainable by surfaces underlain by bulked equal spheroids with variable dimensionable ordering. *Bull. geol. Soc. Am.*, **80**, 1923-30.

ALLUM, J. 1966. *Photogeology and regional mapping.* Pergamon, Oxford.

AMERICAN GEOPHYSICAL UNION. 1941. Symposium on dynamics of land erosion. *Trans. Am. geophys. Un.*, **22**, 236-320.

ANDERSSON, J. G. 1906. Solifluction, a component of subaerial denudation. *J. Geol.*, **14**, 91-112.

ANDRE, J. E., and ANDERSON, H. W. 1961. Variation of soil erodibility with geology, geographic zone, elevation and vegetation type in northern California wildlands. *J. geophys. Res.*, **66**, 3351-8.

ANTEVS, E. 1952. Arroyo-cutting and filling. *J. Geol.*, **60**, 375-85.

ARBER, M. A. 1949. Cliff profiles of Devon and Cornwall. *Geogrl J.*, **114**, 191-7.

BAILEY, R. G., and RICE, R. M. 1969. Soil slippage: an indicator of slope instability on chaparral watersheds of southern California. *Prof. Geogr.*, **21**, 172-7.

BAILEY, R. W. 1935. Epicycles of erosion in the valleys of the Colorado Plateau Province. *J. Geol.*, **43**, 337-55.

BAILEY, R. W. 1941. Land erosion—normal and accelerated—in the semi-arid West. *Trans. Am. geophys. Un.*, **22**, 240-50.

BAIN, G. W. 1931. Spontaneous rock expansion. *J. Geol.*, **39**, 715-35.

BAKKER, J. P. 1956. Niederländische Schriften über Hangentwicklung. *Slopes Comm. Rep.*, **1**, 56-65.

BALCHIN, W. G. V., and PYE, N. 1955. Piedmont profiles in the arid cycle. *Proc. geol. Ass.*, **66**, 167-81.

BALK, R. 1939. Disintegration of glaciated cliffs. *J. Geomorph.*, **2**, 305-34.

BALL, D. F. 1966. Late-glacial scree in Wales. *Biul. Peryglac.*, **15**, 151-63.

BASS, M. A. 1956. The characteristics of some major areas of landsliding in the southern Pennines. *Slopes Comm. Rep.*, **1**, 130.

BAULIG, H. 1940. Le profil d'équilibre des versants. *Ann. Geogr.*, **49**, 81-97. Reprinted with additions in *Essais de Géomorphologie*, Paris, 1950.

BAVER, L. D. 1938. Ewald Wollny—a pioneer in soil and water conservation research. *Proc. soil Sci. Soc. Am.*, **3**, 330-3.

BEATY, C. B. 1956. Landslides and slope exposure. *J. Geol.*, **64**, 70-4.

BEATY, C. B. 1959. Slope retreat by gullying. *Bull. geol. Soc. Am.*, **70**, 1479-82.

BEATY, C. B. 1962. Asymmetry of stream patterns and topography in the Bitterroot Range, Montana. *J. Geol.*, **70**, 347-54.

BEAUJEU-GARNIER, J. 1953. Modèle périglaciare dans le Massif Central français. *Rev. Géomorph. Dyn.*, **4**, 251-81.

BECKER, H. 1963. Über die Entstehung von Erdpyramiden. *Slopes Comm. Rep.*, **3**, 185-94.

BECKETT, P. H. T. 1968. Soil formation and slope development. 1. A new look at Walther Penck's Aufbereitung concept. *Z. Geomorph.*, **12**, 1-24.

BECKETT, P. H. T., and WEBSTER, R. 1965. A classification system for terrain (M.E.X.E., No. 872). Field trials of a terrain classification system: organisation and methods (M.E.X.E., No. 873). Field trials of a terrain classification system: statistical procedure (M.E.X.E., No. 874). Minor statistical studies on terrain evaluation (M.E.X.E., No. 877). *Milit. Engng. Exptl Establishment*, Christchurch.

BEHRE, C. H. 1933. Talus behaviour above timber in the Rocky Mountains. *J. Geol.*, **41**, 622-35.

BERRY, L., and RUXTON, B. P. 1959. Notes on weathering zones and soils on granitic rocks in two tropical regions. *J. soil Sci.*, **10**, 54-63.

BERRY, L., and RUXTON, B. P. 1960. The evolution of Hong Kong harbour basin. *Z. Geomorph.*, **4**, 97-115.

BESKOW, G. 1930. Erdfliessen und Strukturboden der Hochgebirge im Lichte der Frosthebung. *Geol. För. Stockh. Förh.*, 52.

BIBBY, J. S., and MACKNEY, D. 1969. Land use capability classification. *Soil Survey, Tech. Monogr.* 1, Rothamsted.

BIGARELLA, J. J., and MOUSINHO, M. R. 1966. Slope development in south eastern and southern Brazil. *Z. Geomorph.*, **10**, 150-9.

BIGARELLA, J. J., MOUSINHO, M. R., and DA SILVA, J. X. 1965. Consideracões a respeito da evolução das vertentes. *Bol. Paranaese Geogr.*, **16**, 85-116.

BIGARELLA, J. J., MOUSINHO, M. R., and DA SILVA, J. X. 1969. Processes and environments of the Brazilian Quaternary. In Péwé (1969), *q.v.*, 417-87.

BIROT, P. 1949. *Essai sur quelques problèmes de morphologie générale*. Lisbon.

BIROT, P. 1960. *Le cycle d'érosion sous les différents climats*. Rio de Janeiro. English translation: *The cycle of erosion in different climates*. Batsford, London, 1968.

BIROT, P. 1963. Evolution de versants à corniche dans la série miocène au sud de Teruel (Espagne). *Slopes Comm. Rep.*, **3**, 67-70.

BISHOP, W. W. 1962. Gully erosion in Queen Elizabeth National Park. *Uganda J.*, **26**, 161-5.

BJERRUM, L., and JÖRSTAD, F. 1963. Correspondence: "Stability of steep slopes on hard unweathered rock" by Karl Terzaghi. *Géotechnique*, **13**, 171-3.

BLACHE, J. 1942. Des versants aux rivières. *Rev. Geogr. alp.*, **30**, 1-50.

BLACKWELDER, E. 1912. The Gros Ventre slide, an active earth-flow. *Bull. geol. Soc. Am.*, **23**, 487-92.

BLACKWELDER, E. 1928. Mudflow as a geologic agent in semi-arid mountains. *Bull. geol. Soc. Am.*, **39**, 465-83.

BLACKWELDER, E. 1942. The process of mountain sculpture by rolling debris. *J. Geomorph.*, **5**, 325-8.

BLENK, M. 1963. Eine kartographische Methode der Hanganalyse, erläutert an zwei Beispielen: N.W.-Harz und Salinstal, Kalifornien. *Slopes Comm. Rep.*, **3**, 29-44.

BLONG, R. J. 1970. The development of discontinuous gullies in a pumice catchment. *Am. J. Sci.*, **268**, 369-83.

BONNIARD, F. 1929. Représentation graphique de la pente moyenne d'un bassin-versant. *Rev. Géogr. phys. Géol. dynam.*, **2**, 247-52.

BOUT, P., DERRUAU, M., and FEL, A. 1960. Utilisation des cônes et des coulées volcaniques du Massif Central français pour évaluer le recul des versants cristallins. *Slopes Comm. Rep.*, **2**, 133-9.

BRADLEY, W. H. 1940. Pediments and pedestals in miniature. *J. Geomorph.*, **2**, 244-55.

BRAMMER, H. 1956. Visit to Haute Volta, 30th January-3rd March 1955. *Tech. Rep. No. 9, Gold Coast Dept. Soil Land Use Survey*, Kumasi.

BRAMMER, H. 1956b. A note on former pediment remnants in Haute Volta. *Geogr. J.*, **122**, 526-7.

BRIDGEMAN, P. W. 1936. Shearing phenomena at high pressures of possible importance in geology. *J. Geol.*, **44**, 653-69.

BRIDGEMAN, P. W. 1938. Reflections on rupture. *J. appl. Phys.*, **9**, 517-28.

BRIDGES, E. M., and DOORNKAMP, J. C. 1963. Morphological mapping and the study of soil patterns. *Geography*, **48**, 175-81.

BRINK, A. B. A., and WILLIAMS, A. A. B. 1964. Soil engineering mapping for roads in South Africa. *C.S.I.R. Res. Rep.*, 227, Pretoria.

BRINK, A. B., *et al.* 1966. Report of the working group on land classification and data storage (M.E.X.E., No. 940). *Military Engng. Expertl. Establishment*, Christchurch.

BRYAN, K. 1925. The Papago Country, Arizona. *U.S.G.S. Water Supply Paper*, 499.

BRYAN, K. 1925b. Date of channel cutting (arroyo cutting) in the arid south-west. *Science*, **62**, 338-44.

BRYAN, K. 1935. Processes of formation of pediments at Granite Gap, New Mexico. *Z. Geomorph.*, **9**, 125-35.

BRYAN, K. 1940. The retreat of slopes. *Ann. Ass. Am. Geogr.*, **30**, 254-68.

BRYAN, K. 1940b. Gully gravure—a method of slope retreat. *J. Geomorph.*, **3**, 89-107.

BRYAN, K. 1948. Cryopedology—the study of frozen ground and intensive frost-action with suggestions of nomenclature. *Am. J. Sci.*, **244**, 622-42.

BRYAN, K. 1949. The geologic implications of cryopedology. *J. Geol.*, **57**, 101-4.

BRYAN, K., and WILSON, G. W. 1931. The W. Penck method of analysis in Southern California. *Z. Geomorph.*, **6**, 287-91.

BRYAN, R. B. 1969. The relative erodibility of soils developed in the Peak District of Derbyshire. *Geogr. Annlr*, **51A**, 145-59.

BUCKHAM, A. F., and COCKFIELD, W. E. 1950. Gullies formed by sinking of the ground. *Am. J. Sci.*, **248**, 137-41.

BÜDEL, J. 1937. Eiszeitliche und rezente Verwitterung und Abtragung im ehemals nicht vereisten Teil Mitteleuropas. *Pett. Geogr. Mitt.*, Erg., **229**.

BÜDEL, J. 1948. Das System der klimatischen Morphologie. *Verh. Deutsch. Geographentages*, **27**, 65-100.

BÜDEL, J. 1948b. Die klima-morphologischen Zonen der Polarländer. *Erdkunde*, **2**, 22-53.

BÜDEL, J. 1953. Die "periglazial"-morphologischen Wirkungen des Eiszeitklimas auf der ganzen Erde. *Erdkunde*, **7**, 249-66.

BÜDEL, J. 1957. The "doppelten Einebnungsflächen" in den feuchten Tropen. *Z. Geomorph.*, **1**, 201-28.

BÜDEL, J. 1965. *Die Relieftypen der Flächenspulzone Süd-Indiens am Ostabfall Dekans gegen Madras.* Dümmlers, Bonn.

BUNTING, B. T. 1960. Bedrock corrosion and drainage initiation by seepage moisture on a gritstone escarpment in Derbyshire. *Nature, Lond.*, **185**, 447.

BUNTING, B. T. 1961. The role of seepage moisture in soil formation, slope development and stream initiation. *Am. J. Sci.*, **259**, 503-18.

BUNTING, B. T. 1964. Slope development and soil formation on some British sandstones. *Geogrl J.*, **130**, 506-12.

BUTLER, B. E. 1959. Periodic phenomena in landscapes as a basis for soil studies. *C.S.I.R.O., Soil Publ.*, **14**, Melbourne.

BUTLER, B. E. 1967. Soil periodicity in relation to landform development in south-eastern

Australia. In Jennings and Mabbutt (1967), *q.v.*, 231-55.

BUTZER, K. W. 1965. Desert landforms at the Kurkur Oasis, Egypt. *Ann. Ass. Am. Geogr.*, **55**, 578-91.

CAILLEUX, A., and TRICART, J. 1950. Un type de solifluction: les coulées boueuses. *Rev. Géomorph. dyn.*, **1**, 4-47.

CAILLEUX, A., and TRICART, J. 1965. *Introduction à la géomorphologie climatique.* S.E.D.E.S., Paris.

CAINE, N. 1963. The origin of sorted stripes in the Lake District, northern England. *Geogr. Annlr*, **45**, 172-9.

CAINE, N. 1967. The texture of talus in Tasmania. *J. sed. Petrol.*, **37**, 796-803.

CAINE, N. 1968. The fabric of periglacial blockfield material on Mt. Barrow, Tasmania. *Geogr. Annlr*, **50**A, 193-206.

CAINE, N. 1968b. The log-normal distribution and rates of soil movement: an example. *Rev. Géomorph. dyn.*, **18**, 1-7.

CAINE, N. 1969. A model for alpine talus slope development by slush avalanching. *J. Geol.*, **77**, 92-100.

CALEF, W., and NEWCOMB, R. 1953. An average slope map of Illinois. *Ann. Ass. Am. Geogr.*, **43**, 305-16.

CARSON, M. A. 1967. The magnitude of variability in samples of certain geomorphic characteristics drawn from valley-side slopes. *J. Geol.*, **75**, 93-100.

CARSON, M. A. 1969. Models of hillslope development under mass failure. *Geogrl Anal.*, **1**, 76-100.

CARSON, M. A., and PETLEY, D. 1970. The existence of threshold hillslopes in the denudation of the landscape. *Trans. Inst. Br. Geogr.*, **49**, 71-96.

CHALLINOR, J. 1931. Some coastal features of north Cardiganshire. *Geol. Mag.*, **68**, 111-21.

CHALLINOR, J. K. 1948. A note on convex erosion-slopes, with special reference to north Cardiganshire. *Geography*, **33**, 27-31.

CHAMBERLIN, T. C., and SALISBURY, R. D. 1904. *Geology, Vol. 1, Geologic processes and their results.* Holt, New York.

CHAPMAN, C. A. 1952. A new quantitative method of topographic analysis. *Am. J. Sci.*, **250**, 428-53.

CHORLEY, R. J. 1959. The geomorphic significance of some Oxford soils. *Am. J. Sci.*, **257**, 503-15.

CHORLEY, R. J. 1962. Geomorphology and general systems theory. *U.S.G.S. Prof. Paper*, (500-*B*).

CHORLEY, R. J. 1964. Geomorphological evaluation of factors controlling resistance of surface soils in sandstone. *J. geophys. Res.*, **69**, 1507-16.

CHORLEY, R. J. 1964b. The nodal position and anomalous character of slope studies in geomorphological research. *Geogrl J.*, **130**, 503-6.

CHORLEY, R. J. 1965. A re-evaluation of the geomorphic system of W. M. Davis. In Chorley, R. J. and Haggett, P. (ed.), *Frontiers in geographical teaching*, 21-38. Methuen, London.

CHORLEY, R. J. (ed.) 1969. *Water, earth and man.* Methuen, London.

CHORLEY, R. J., *et al.* 1964. *The history of the study of landforms.* Vol. 1, *Geomorphology before Davis.* Methuen, London.

CHRISTOFOLETTI, A. 1968. O fenômeno morfogenêtico no Município de Campinas. *Notícia Geomorf.*, **8**, 1-97.

CLARK, M. J. 1965. The form of Chalk slopes. *Southampton Res. Ser. in Geogr.*, **2**, 3-34.

CLAUZON, G., and VAUDOUR, J. 1969. Observations sur les effets de la pluie en Provence. *Z. Geomorph.*, **13**, 390-405.

CLAYTON, R. W. 1956. Linear depressions (Bergfussneiderungen) in savannah landscapes. *Geogrl Stud.*, **3**, 102-126.

COMMON, R. 1954. A report of the Lochaber, Appin, and Benderloch floods, May 1953. *Scott. Geogr. Mag.*, **70**, 6-20.

COORAY, P. G. 1958. Earthslopes and related phenomena in the Kandy District. *Ceylon Geogr.*, **12**, 75-90.

COPPINGER, R. W. 1881. On soilcap motion. *Qtrly J. Geol. Soc.*, **37**, 348-50.

COQUE, R. 1960. L'évolution des versants en Tunisie présaharienne. *Slopes Comm. Rep.*, **2**, 173-77.

CORBEL, J. 1959. Vitesse de l'érosion. *Z. Geomorph.*, **3**, 1-28.

COSTIN, A. B., and GILMOUR, D. A. 1970. Portable rainfall simulator and plot unit for use in field studies of infiltration, runoff and erosion. *J. appl. Ecol.*, **7**, 193-200.

COTTON, C. A. 1926. *Geomorphology of New Zealand. Part I—Systematic.* Wellington.

COTTON, C. A. 1941. *Landscape. As developed by the processes of normal erosion.* Cambridge Univ. Press.

COTTON, C. A. 1947. *Climatic accidents in landscape-making.* Hafner, New York.

COTTON, C. A. 1951. Atlantic gulfs, estuaries and cliffs. *Geol. Mag.*, **88**, 113-28.

COTTON, C. A. 1952. The erosional grading of convex and concave slopes. *Geogrl J.*, **118**, 197-204.

OTTON, C. A. 1958. Fine-textured erosional relief in New Zealand. *Z. Geomorph.*, **2**, 187-210.

COTTON, C. A. 1958b. Alternating Pleistocene morphogenetic systems. *Geol. Mag.*, **95**, 125-36.

COTTON, C. A. 1962. The origin of New-Zealand feral (fine-textured) relief. *N.Z. J. Geol. Geophys.*, **5**, 269-70.

COTTON, C. A. 1963. Development of fine-textured landscape relief in temperate pluvial climates. *N.Z. J. Geol. Geophys.*, **6**, 528-33.

COTTON, C. A. 1967. Plunging cliffs and Pleistocene coastal cliffing in the southern hemisphere. In Sporcy, J. (ed.), *Mélanges de géographie physique, humaine, économique, appliquée, offerts à M. Omar Tulippe, Vol.* 1. Duclot, Gembloux.

CRANDELL, D. W., and VARNES, D. J. 1961. Movement of the Slumgullion earthflow near Lake City. *U.S.G.S. Prof. Paper*, (424-*B*), 136-9.

CROZIER, M. J. 1969. Earthflows and related environmental factors of eastern Otago. *J. Hydrol.* (N.Z.), **7**, 4-12.

CULLING, W. E. H. 1963. Soil creep and the development of hillside slopes. *J. Geol.*, **71**, 127-61.

CULLING, W. E. H. 1964. Theory of erosion on soil-covered slopes. *J. Geol.*, **73**, 230-54.

CUNNINGHAM, F. F. 1969. The Crow Tors, Laramie Mountains, Wyoming, U.S.A. *Z. Geomorph.*, **13**, 56-74.

CURTIS, L. F., DOORNKAMP, J. C., and GREGORY, K. J. 1965. The description of relief in field studies of soils. *J. soil Sci.*, **16**, 16-30.

DALRYMPLE, J. B., BLONG, R. J., and CONACHER, A. J. 1968. A hypothetical nine-unit landsurface model. *Z. Geomorph.*, **12**, 60-76.

DARWIN, C. 1881. *The formation of vegetable mould, through the action of worms.* Murray, London.

DAVEAU, S. 1964. Façonnement des versants de l'Adrar Mauritanien. *Slopes Comm. Rep.*, **4**, 118-30.

DAVIS, W. M. 1892. The convex profile of badland divides. *Science*, **20**, 245.

DAVIS, W. M. 1896. Bearing of physiography on uniformitarianism. *Bull. Geol. Soc. Am.*, **7**, 8-9.

DAVIS, W. M. 1898. The grading of mountain slopes. *Science*, **7**, 1449.

DAVIS, W. M. 1899. The geographical cycle. *Geogrl J.*, **14**, 481-504.

DAVIS, W. M. 1899b. The peneplain. *Am. Geologist*, **23**, 207-39.

DAVIS, W. M. 1902. Base-level, grade, and peneplain. *J. Geol.*, **10**, 77-111.

DAVIS, W. M. 1905. The geographical cycle in an arid climate. *J. Geol.*, **13**, 381-407.

DAVIS, W. M. 1909. *Geographical essays.* Ginn, New York.

DAVIS, W. M. 1912. *Die erklärende Beschriebung den Landformen.* Teubner, Leipzig.

DAVIS, W. M. 1930. Rock floors in arid and humid climates. *J. Geol.*, **38**, 1-27, and 136-58.

DAVIS, W. M. 1932. Piedmont benchlands and Primärrumpfe. *Bull. geol. Soc. Am.*, **43**, 399-440.

DAVIS, W. M. 1933. Granitic domes of the Mohave Desert, California. *Trans. San Diego Soc. nat. Hist.*, **7**, 211-58.

DAVIS, W. M. 1938. Sheetfloods and streamfloods. *Bull. geol. Soc. Am.*, **49**, 1337-416.

DAVIS, W. M., and SNYDER, W. H. 1898. *Physical geography.* Ginn, Boston.

DAVISON, C. 1888. Note on the movement of scree-material. *Qtrly J. geol. Soc.*, **44**, 232-8 and 825-6.

DAVISON, C. 1889. On the creeping of the soil cap through the action of frost. *Geol. Mag.*, **36**, 255-61.

DEBANO, L. F., and KRAMMES, J. S. 1966. Water repellent soils and their relation to wildfire temperatures. *Bull. int. Ass. scient. Hydrol.*, **11**, 14-19.

DE BÉTHUNE, P. 1967. Sur la développement de la convexité sommitale des versants. *Slopes Comm. Rep.*, **5**, 89-100.

DE BÉTHUNE, P., and MAMMERICKX, J. 1960. Etudes clinometriques du laboratoire géomorphologique de l'Université de Louvain (Belgique). *Slopes Comm. Rep.*, **2**, 93-102.

DE LA NOË, G., and DE MARGERIE, E. 1888. Les formes du terrain. *Serv. géogr. de l'Armée*, Paris.

DEMANGEOT, J. 1951. Observations sur les "sols en gradins" de l'Appenin central. *Rev. Géomorph. dyn.*, **2**, 110-19.

DE MARTONNE, E. 1940. Problèmes morphologiques du Bresil tropical Atlantic. *Ann. Géogr.*, **49**, 1-27 and 106-29.

De Martonne, E., and Birot, P. 1944. Sur l'evolution des versants en climat tropicale humide. *C. r hebd. Séanc. Acad. Sci., Paris*, **218**, 529-32.

Demek, J. 1964. Slope development in granite areas of Bohemian Massif (Czechoslovakia). *Slopes Comm. Rep.*, **4**, 82-106.

Denevan, W. M. 1967. Livestock numbers in nineteenth-century New Mexico and the problem of gullying in the Southwest. *Ann. Ass. Am. Geogr.*, **57**, 691-703.

De Ploey, J., and Savat, J. 1968. Contribution à l'étude de l'érosion par le splash. *Z. Geomorph.*, **12**, 174-93.

Derruau, M. 1956. *Précis de geomorphologie.* Masson, Paris.

De Swart, A. M. J. 1964. Lateritisation and landscape development in parts of equatorial Africa. *Z. Geomorph.*, **8**, 313-33.

Dimblebey, G. W. 1962. *The development of British heathlands and their soils.* Clarendon Press, Oxford.

Dixey, F. 1962. Applied geomorphology. *S. Afr. geogr. J.*, **44**, 3-24.

Douglas, I. 1967. Man, vegetation and sediment yields of rivers. *Nature, Lond.*, **215**, 925-8.

Dowling, J. W. F. 1968. Land evaluation for engineering purposes in northern Nigeria. In Stewart (1968), *q.v.*, 147-59.

Dumanowski, B. 1960. Notes on the evolution of slopes in an arid climate. *Slopes Comm. Rep.*, **2**, 178-89.

Dumanowski, B. 1964. Problem of the development of slopes in granitoids. *Slopes Comm. Rep.*, **4**, 30-40.

Dutton, C. E. 1880-1. The physical geology of the Grand Canyon district. *U.S.G.S. 2nd Ann. Rep.*, 47-166.

Dutton, C. E. 1882. Tertiary history of the Grand Canyon region. *U.S.G.S. Monographs*, **2**.

Dylik, J. 1956. Coup d'œil sur la Pologne périglaciare. *Biul. Peryglac.*, **4**, 195-238.

Dylik, J. 1960. Rhythmically stratified slope waste deposits. *Biul. Peryglac.*, **8**, 31-41.

Dylik, J. 1967. Solifluxion, congelifluxion and related slope processes. *Geogr. Annlr*, **49***A*, 167-77.

Dylik, J. 1969. Slope development affected by frost fissures and thermal erosion. In Péwé (1969), *q.v.*, 365-86.

Eckblaw, W. E. 1918. The importance of nivation as an erosive factor, and of soil flow as a transporting agency in northern Greenland. *Proc. nat. Acad. Sci.*, **4**, 288-93.

Ellison, W. D. 1944. Studies of raindrop erosion. *J. agric. Eng.*, **25**, 131-6, and 181-2.

Ellison, W. D. 1945. Some effects of raindrops and surface flow on soil-erosion and infiltration. *Trans. Am. geophys. Un.*, **26**, 415-29.

Ellison, W. D. 1947. Soil erosion studies. Parts I-VII. *Agric. Eng.*, **28**, 145-6, 197-201, 245-8, 297-300, 349-51, 402-5, and 442-4.

Ellison, W. D. 1948. Soil detachment by water in erosion process. *Trans. Am. geophys. Un.*, **29**, 499-502.

Ellison, W. D. 1950. Soil erosion by rainstorms. *Science*, **111**, 245-9.

Ellison, W. D. 1952. Raindrop energy and soil erosion. *Emp. J. exp. Agric.*, **20**, 81-97.

Embleton, C., and King, C. A. M., 1968. *Glacial and periglacial geomorphology.* Edward Arnold, London.

Evans, A. C. 1948. Studies on the relationship between earthworms and soil fertility. II. Some effects of earthworms on soil structure. *Ann. appl. Biol.*, **35**, 1-13.

Everard, C. E. 1963. Contrasts in the form and evolution of hill-side slopes in central Cyprus. *Trans. Inst. Br. Geogr.*, **32**, 31-47.

Everard, C. E. 1964. Climate change and man as factors in the evolution of slopes. *Geogrl J.*, 498-502.

Everett, K. R. 1963. Slope movement, Neotoma Valley, Southern Ohio. *Rep. Inst. Polar Stud., Ohio State Univ.*, **6**.

Everett, K. R. 1966. Slope movement and related phenomena. In Wilimovsky, N. J. and Wolfe, J. N. (ed.) *Environment of the Cape Thompson region.* U.S. Atomic Energy Commission, 172-220.

Everett, K. R. 1966b. Instruments for measuring mass-wasting. *Proc. permafrost Int. Conf.*, 1963, Wash. D.C., 136-9.

Everett, K. R. 1967. Mass-wasting in the Taseriaq area, West Greenland. *Medd. om Grønland*, **165**.

Fair, T. J. 1947. Slope form and development in the interior of Natal. *Trans. geol. Soc. S. Afr.*, **50**, 105-20.

Fair, T. J. 1948. Slope form and development in the coastal hinterland of Natal. *Trans. geol. Soc. S. Afr.*, **51**, 37-53.

Fair, T. J. 1948b. Hillslopes and pediments of the semi-arid Karoo. *S. Afr. Geogr. J.*, **30**, 71-9.

FAO. n.d. *Guidelines for soil description.* Rome.

FARMER, I. W. 1968. *Engineering properties of rocks.* Spon, London.

FENNEMAN, N. M. 1908. Some features of erosion by unconcentrated wash. *J. Geol.*, **16**, 746-54.

FINNEY, H. R. 1962. The influence of microclimate on the morphology of certain soils of the Allegheny Plateau of Ohio. *Proc. soil Sci. Soc. Am.*, **26**, 287-92.

FISHER, O. 1866. On the disintegration of a chalk cliff. *Geol. Mag.*, **3**, 354-6.

FITZPATRICK, E. A. 1956. Grande Bretagne. Progress report on the observations of periglacial phenomena in the British Isles. *Biul. Peryglac.*, 4, 99-115.

FITZPATRICK, E. A. 1963. Deeply weathered rock in Scotland, its occurrence, age, and contribution to the soils. *J. soil Sci.*, **14**, 33-43.

FLOHR, E. F. 1962. Schichtenbau und Hanentwicklung in Kalkstein des Gross-Weissach-Tals. *Z. Geomorph.*, **6**, 279-95.

FOURNEAU, R. 1960. Contribution a l'étude des versants dans le sud de la Moyenne Belgique et dans le nord de l'Entre Sambre et Meuse. Influence de la nature du substratum. *Ann. Soc. géol. Belg.*, **84**, 123-51.

FREISE, F. W. 1932. Beobachtungen über Erosion an Urwaldgebirgsflüssen des brasilianischen Staates Rio de Janeiro. *Z. Geomorph.*, **7**, 1-9.

FREISE, F. W. 1935. Erscheinungen de Erdfleissens in Tropenurwalde. Beobachtungen aus brasilianischen Küstenwäldern. *Z. Geomorph.*, **9**, 88-98.

FREISE, F. W. 1938. Inselberge und Inselberg-Landschaften im Granit und Gneisgebiete Brasiliens. *Z. Geomorph.*, **10**, 137-68.

FRYE, J. C. 1959. Climate and Lester King's 'Uniformitarian nature of hillslopes'. *Geol. J.*, **67**, 111-13.

FURLEY, P. A. 1968. Soil formation and slope development. 2. The relationship between soil formation and gradient angle in the Oxford area. *Z. Geomorph.*, **12**, 25-42.

GABERT, P. 1964. Premiers résultats des mesures d'érosion sur des parcelles expérimentales dans la région d'Aix-en-Provence (Bouches du Rhône—France). *Slopes Comm. Rep.*, **4**, 213-14.

GALLOWAY, R. W. 1961. Periglacial phenomena in Scotland. *Geogr. Annlr*, **43**, 348-53.

GARDNER, J. 1969. Observations of surficial talus movement. *Z. Geomorph.*, **13**, 317-23.

GEIKIE, A. 1868. On modern denudation. *Trans. geol. Soc. Glasgow*, **3**, 153-90.

GELLERT, J. 1967. Further works on the unification of signs and signatures of geomorphological detail maps. *Z. Geomorph.*, **11**, 506-9.

GERBER, E. 1934. Zur Morphologie wachsender Wände. *Z. Geomorph.*, **8**, 213-23.

GERLACH, T. 1963. Les terrasses de culture comme indice des modifications des versants cultivés. *Slopes Comm. Rep.*, **3**, 239-49.

GERLACH, T. 1967. Évolutions actuelles des versants dans les Carpathes, d'après l'example d'observations fixes. *Slopes Comm. Rep.*, **5**, 129-38.

GIBBS, H. S. 1945. Tunnel gulley erosion on the Wither Hills, Marlborough. *N.Z. J. Sci. Tech.*, *A*, **27**, 135-46.

GIFFORD, J. 1953. Landslides on Exmoor caused by the storm of 15th August 1952. *Geography*, **38**, 9-17.

GILBERT, G. K. 1877. *Report on the geology of the Henry Mountains.* Washington. Page citations refer to the second edition, 1880.

GILBERT, G. K. 1909. The convexity of hilltops. *J. Geol.*, **17**, 344-50.

GILEWSKA, S. 1967. Different methods of showing relief on detailed geomorphological maps. *Z. Geomorph.*, **11**, 481-90.

GILLULY, J. 1937. Physiography of the Ajo region, Arizona. *Bull. geol. Soc. Am.*, **48**, 323-48.

GLOCK, W. S. 1932. Available relief as a factor of control in the profile of a land form. *J. Geol.*, **40**, 74-83.

GÖTZINGER, G. 1907. Beiträge zur Entstehung der Bergrückenformen. *Geogr. Abhl.*, **9**.

GOULD, S. J. 1965. Is uniformitarianism necessary? *Am. J. Sci.*, **262**, 223-8.

GREGORY, K. J., and BROWN, E. H. 1966. Data processing and the study of land form. *Z. Geomorph.*, **10**, 237-63.

GREYSUKH, V. L. 1967. The possibility of studying landforms by means of digital computers. *Soviet Geogr.*, **8**, 137-49.

GROVE, A. T. 1953. Account of a mudflow on Bredon Hill, Worcestershire, April 1951. *Proc. geol. Ass.*, **64**, 10-13.

GROVE, A. T. 1960. Geomorphology of the Tibetsi region with special reference to western Tibetsi. *Geogr. J.*, **126**, 18-31.

HACK, J. T. 1960. Interpretation of erosional topography in humid temperate regions. *Am. J. Sci.*, **258-A**, 80-97.

HACK, J. T., and GOODLETT, J. C. 1960. Geomorphology and forest ecology of a mountain region in the central Appalachians. *U.S.G.S. Prof. Paper*, (347).

HADLEY, R. F., and ROLFE, B. N. 1955. Development and significance of seepage steps in slope erosion. *Trans. Am. geophys. Un.*, **36**, 792-804.

HAINES, W. B. 1923. The volume-changes associated with variations of water-content in soil. *J. agric. Sci.*, **13**, 296-310.

HAUSER, A., and ZÖTL, J. 1955. Die morphologische Bedeutung der unterirdischen Erosion durch Gesteinausspülung. *Pett. Geogr. Mitt.*, **99**, 18-21.

HAY, T. 1942. Physiographical notes from Lakeland. *Geogr. J.*, **100**, 165-73.

HEIM, A. 1882. Ueber Bergstürze. *Neujahrsblatt Naturf. Gesell. Zürich*, 84.

HEIM, A. 1932. Bergsturz und Menschenleben. Vierteljahrsschrift. *Nat. Forsch. Gesell. Zurich*, 77.

HENKEL, L. 1926. Einwände gegen wichtige Punkte in W. Penck's Erosionstheorie. *Pett. Geogr. Mitt.*, **72**, 263-4.

HEY, R. W. 1963. Pleistocene screes in Cyrenaica (Libya). *Eiszeitalter und Gegenwart*, **14**, 77-84.

HICKS, L. E. 1893. Some elements of land sculpture. *Bull. geol. Soc. Am.*, **4**, 133-46.

HIDORE, J. J. 1963. The relationship between cash-grain farming and landforms. *Econ. Geogr.*, **39**, 84-9.

HIRANO, M. 1968. A mathematical model of slope development. *J. Geosci. Osaka*, **11**, 13-52.

HOGBOM, B. 1914. Über die geologische Bedeutung des Frosts. *Bull. geol. Inst. Upsala*, **12**, 255-390.

HÖLLERMANN, P. 1963. Beispiele für anthropogen verstärkte Hangabtragungs-und-formungsvorgänge in inneralpinen Tälern. *Slopes Comm. Rep.*, **3**, 251-73.

HOLLINGWORTH, S. E. 1934. Some solifluction phenomena in the northern part of the Lake District. *Proc. geol. Ass.*, **45**, 167-88.

HOLLINGWORTH, S. E., TAYLOR, J. H., and KELLAWAY, G. A. 1944. Large-scale superficial structures in the Northampton ironstone field. *Qtrly J. geol.*, **100**, 1-34.

HOLMES, C. D. 1955. Geomorphic development in humid and arid regions: a synthesis. *Am. J. Sci.*, **253**, 377-90.

HORMANN, K. 1969. Geomorphologische Kartenanalyse mit Hilfe elektronischer Rechenanlagen. *Z. Geomorph.*, **13**, 75-98.

HORTON, R. E. 1945. Erosional development of streams and their drainage basins: hydrophysical approach to quantitative morphology. *Bull. geol. Soc. Am.*, **56**, 275-370.

HOWARD, A. D. 1965. Geomorphological systems—equilibrium and dynamics. *Am. J. Sci.*, **263**, 302-12.

HUBBERT, M. K. 1937. Theory of scale models as applied to the study of geologic structures. *Bull. geol. Soc. Am.*, **48**, 1460-1515.

HUNTER, J. M. 1961. Morphology of a bauxite summit in Ghana. *Geogr. J.*, **127**, 469-76.

HUTCHINSON, J. N. 1961. A landslide on a thin layer of quick clay at Furre, central Norway. *Géotechnique*, **11**, 69-94.

HUTCHINSON, J. N. 1967. The free degradation of London Clay cliffs. *Proc. Geotech. Conf. Oslo*, 1967, **1**, 113-18.

HUTCHINSON, J. N. 1968. Mass movement. In Fairbridge, R. W. *The encyclopedia of geomorphology*, 688-96. Reinhold, New York.

HUTTON, J. 1788. Theory of the earth. *Trans. Roy. Soc. Edin.*, **1**, 209-304.

HUTTON, J. 1795. *Theory of the earth, with proofs and illustrations.* 2 vols., Edinburgh.

INTERNATIONAL GEOGRAPHICAL UNION. 1968. The unified key to the detailed geomorphological map of the world, in 1 : 25,000 to 1 : 50,000 scale. *Folia Geogr., Ser. Geogr. Phys.*, **2**.

IRELAND, H. A., SHARPE, C. F. S., and EARGLE, D. H. 1939. Principles of gully erosion in the Piedmont of South Carolina. *U.S. Dept. Agric. Tech. Bull.*, 633.

JACKSON, G., and SHELDON, J. 1949. The vegetation of Magnesian Limestone cliffs at Markland Grips near Sheffield. *J. Ecol.*, **37**, 38-50.

JAHN, A. 1954. Denudacyjny bilans stoku. *Czas. Geogr.*, **25**, 38-64.

JAHN, A. 1960. Some remarks on evolution of slopes on Spitzbergen. *Slopes Comm. Rep.*, **2**, 49-58.

JAHN, A. 1963. Importance of soil erosion for the evolution of slopes in Poland. *Slopes Comm. Rep.*, **3**, 229-38.

JAHN, A. 1964. Slopes as morphological features resulting from gravitation. *Slopes Comm. Rep.*, **4**, 59-72.

JAHN, A. 1967. Some features of mass movement on Spitzbergen slopes. *Geogr. Annlr*, **49*A***, 213-25.

JAHN, A. 1968. Denudational balance of slopes. *Geogr. Polnica*, **13**, 9-29.

JAHN, A. 1968b. Morphological slope evolution by linear and surface degradation. *Geogr. Polnica*, **14**, 9-21.

JARVIS, M. G. 1969. Terrain and soil in north Berkshire. *Geogr. J.*, **135**, 398-403.

JEFFREYS, H. 1918. Problems of denudation. *Phil. Mag.*, **36**, 179-90.

JEFFREYS, H. 1932. Scree slopes. *Geol. Mag.*, **69**, 383-4.

JENNINGS, J. N., and MABBUTT, J. A. 1967. *Landform studies from Australia and New Guinea*. Cambridge Univ. Press.

JENNINGS, J. N., and SWEETING, M. M. 1963. The limestone ranges of the Fitzroy Basin, Western Australia. *Bonn. Geogr. Abh.*, 32.

JESSEN, O. 1938. Tertiärklima und Mittelgebirgsmorphologie. *Z. Ges. Erdk. Berlin*, 36-49.

JEWELL, P. A. (ed.). 1963. The experimental earthwork on Overton Down, Wiltshire 1960. *Brit. Ass. Adv. Sci. Pubn.*

JEWELL, P. A., and DIMBLEBY, G. W. 1966. The experimental earthwork on Overton Down, Wiltshire, England: the first four years. *Proc. prehist. Soc.*, **32**, 313-42.

JOHNSON, D. W. 1933. Available relief and texture of topography. A discussion. *J. Geol.*, **41**, 293-305.

JOHNSON, D. W. 1939-40. Studies in scientific method. III. The inductive method. IV. The deductive method of presentation. *J. Geomorph.*, **2**, 366-72, and **3**, 59-64.

JOHNSON, R. H. 1965. A study of the Charlesworth landslides near Glossop, north Derbyshire. *Trans. Inst. Br. Geogr.*, **37**, 111-26.

JUNGERIUS, P. D. 1965. Some aspects of the geomorphological significance of soil texture in eastern Nigeria. *Z. Geomorph.*, **9**, 332-45.

JUTSON, J. T. 1919. Sheet-flows, or sheet-floods in Western Australia. *Am. J. Sci.*, **48**, 435-9.

KEEPING, W. 1878. Notes on the geology of the neighbourhood of Aberystwyth. *Geol. Mag.*, **5**, 532-47.

KENT, P. E. 1966. The transport mechanism in catastrophic rock falls. *J. Geol.*, **74**, 79-83.

KERNEY, M. P., *et al.* 1964. The Late-glacial and Post-glacial history of the Chalk escarpment near Brook, Kent. *Phil. Trans. Roy. Soc. Lond., B*, **248**, 135-204.

KERR, W. C. 1881. On the action of frost in the arrangement of superficial earthy material. *Am. J. Sci.*, **21**, 345-58.

KIESLINGER, A. 1960. Residual stress and relaxation in rocks. *Int. geol. Congr. Copenhagen*, Session 21, 270-6.

KING, C. A. M. 1956. Scree profiles in Iceland. *Slopes Comm. Rep.*, **1**, 124-5.

KING, L. C. 1948. A theory of bornhardts. *Geogr. J.*, **112**, 83-7.

KING, L. C. 1949. The pediment landform: some current problems. *Geol. Mag.*, **86**, 245-50.

KING, L. C. 1951. *South African scenery*. 2nd edn. Oliver and Boyd, Edinburgh.

KING, L. C. 1953. Canons of landscape evolution. *Bull. geol. Soc. Am.*, **64**, 721-51.

KING, L. C. 1956. Drakensberg scarp of South Africa: a clarification. *Bull. geol. Soc. Am.*, **67**, 121-2.

KING, L. C. 1957. The uniformitarian nature of hillslopes. *Trans. Edin. geol. Soc.*, **17**, 81-102.

KING, L. C. 1958. The problem of tors. *Geogr. J.*, **124**, 289-92.

KING, L. C. 1962. *Morphology of the earth*. Oliver and Boyd, Edinburgh.

KING, L. C. 1966. The origin of bornhardts. *Z. Geomorph.*, **10**, 97-8.

KING, L. C. 1968. Scarps and tablelands. *Z. Geomorph.*, **12**, 114-15.

KIRKBY, M. J. 1964. In Slope profiles: a symposium. *Geogr. J.*, **130**, 86.

KIRKBY, M. J. 1967. Measurement and theory of soil creep. *J. Geol.*, **75**, 359-78.

KIRKBY, M. J. 1969. Infiltration, throughflow, and overland flow. In: Chorley (1969), *q.v.*, 215-28.

KIRKBY, M. J. 1969b. Erosion by water on hillslopes. In Chorley (1969), *q.v.*, 229-38.

KIRKBY, M. J., and CHORLEY, R. J. 1967. Throughflow, overland flow and erosion. *Bull. Int. Ass. scient. Hydrol.*, **12**, 5-21.

KITTLER, G.-A. 1955. Merkmale, Verbreitung und Ausmass der schleichenden Bodenerosion. *Pett. Geogr. Mitt.*, **99**, 269-73.

KLATKOWA, H. 1967. L'origine et les étapes d'évolution des vallées sèches et des vallons en berceau—examples des environs de Lodz. *Slopes Comm. Rep.*, **5**, 167-74.

KLIMAZEWSKI, M. (ed.). 1963. *Problems of geomorphological mapping*. Inst. Geogr. Polish Acad. Sci., Warsaw.

KLINGEBIEL, A. A., and MONTGOMERY, P. H. 1961. *Land Capability classification.* Soil Conserv. U.S. Dept. Agric., Handbook 210.

KOONS, D. 1955. Cliff retreat in south-western United States. *Am. J. Sci.*, **253**, 44-58.

LAKE, P. 1928. On hill slopes. *Geol. Mag.*, **65**, 108-16.

LAMBERT, J. L. M. 1961. Contribution a l'étude des pentes du Condroz. *Ann. Soc. géol. Belg.*, **84**, 241-50.

LANGBEIN, W. B., and SCHUMM, S. A. 1958. Yield of sediment in relation to mean annual precipitation. *Trans. Am. geophys. Un.*, **39**, 1076-84.

LANGE, A. L. 1963. Planes of repose in caves. *Cave Notes*, **5**, 41-8.

LASSERRE, G. 1956. Evolution des versants calcaires de Grande Terre et Marie Galante (Guadeloupe). *Slopes Comm. Rep.*, **1**, 134-6.

LAWSON, A. C. 1915. Epigene profiles of the desert. *Bull. Calif. Univ. Dept. geol. Sci.*, **9**, 23-48.

LAWSON, A. C. 1932. Rain-wash erosion in humid regions. *Bull. geol. Soc. Am.*, **43**, 703-24.

LEACH, W. 1930. A preliminary account of the vegetation of some non-calcareous British screes (Gerolle). *J. Ecol.*, **18**, 321-32.

LEBLANC, F. 1842. Observations sur le maximum d'inclinasion des talus dans les montagnes. *Bull. Soc. géol. Fr.*, **14**, 85-98.

LEHMANN, O. 1918. Die Talbildung durch Schuttgerinne. In *A. Penck Festband*, 48-65. Englehorns, Stuttgart.

LEHMANN, O. 1922. Beiträge zur gesetzmässigen Erfassung des Formenablaufs bei ständig bewegter Erdrinde und fliessendem Wasser. *Mitt. Geogr. Ges. Wien*, **65**, 55-78.

LEHMANN, O. 1931. Über die Bewegungsenergie des Regenwassers. *Z. Geomorph.*, **6**, 223-54.

LEHMANN, O. 1933. Morphologische Theorie der Verwitterung von Steinschlagwänden. *Viertel-jahrsschrift Naturforsch. Ges. Zurich*, **78**, 83-126.

LEHMANN, O. 1934. Ueber die morphologischen Folgen der Wandwitterung. *Z. Geomorph.*, **8**, 93-9.

LEOPOLD, L. B., EMMETT, W. W., and MYRICK, R. M. 1966. Channel and hillslope processes in a semiarid area, New Mexico. *U.S.G.S. Prof. Paper*, (352-G), 193-253.

LEOPOLD, L. B., WOLMAN, M. G., and MILLER, J. P. 1964. *Fluvial processes in geomorphology.* Freeman, San Francisco.

LEWIN, J. 1969. The Yorkshire Wolds. A study in geomorphology. *Univ. Hull occas. Pubn Geogr.*, 11.

LEWIS, G. M. 1959. Some recent American contributions in the field of landform geography. *Trans. Inst. Br. Geogr.*, **26**, 25-36.

LEWIS, G. M. 1962. Changing emphases in the description of the natural environment of the American Great Plains area. *Trans. Inst. Br. Georg.*, **30**, 75-90.

LEWIS, W. V., and MILLER, M. M. 1955. Kaolin model glaciers. *J. Glac.*, **2**, 533-8.

LINTON, D. L. 1955. The problem of tors. *Geogr. J.*, **121**, 470-87.

LINTON, D. L. 1964. The origin of the Pennine tors—an essay in analysis. *Z. Geomorph., Sonderheft Mortensen*, **8**, 5-24.

LOHNES, R. A., and HANDY, R. L. 1968. Slope angles in friable loess. *J. Geol.*, **76**, 247-58.

LOUGHNAN, F. C. 1969. *Chemical weathering of the silicate minerals.* Elsevier, Amsterdam.

LOUIS, H. 1935. Problems der Rumpfflächen und Rumpftreppen. *Verh. u. Wiss. Abh. d.* 25 *Deutsch. Geographentages zu Bad Neuheim* 1934, 118-37.

LOUIS, H. 1957. Rumpfflächenproblem, Erosionzyklus und Klimageomorphologie. *Pett. Georg. Mitt., Erg.* **262**, 9-26.

LOUIS, H. 1961. *Allgemeine Gemorphologie*, Berlin.

LOUIS, H. 1964. Über Rumpfflächen- und Talbildung in den wechselfeuchten Tropen besonders nach Studien in Tanganyika. *Z. Geomorph., Sonderheft Mortensen*, **8**, 43-70.

LOVERING, T. S. 1959. Significance of accumulator plants in rock weathering. *Bull. geol. Soc. Am.*, **70**, 781-800.

LUTZ, H. J. 1960. Movement of rocks by uprooting of forest trees. *Am. J. Sci.*, **258**, 725-6.

LYELL, C. 1841. *Elements of geology.* 2nd edn. Murray, London.

LYELL, C. 1867. *Principles of geology.* 10th edn. 2 vols. Murray, London.

MABBUTT, J. A. 1955. Pediment land forms in Little Namaqualand. *Geogrl J.*, **121**, 77-83.

MABBUTT, J. A. 1966. Mantle-controlled planation of pediments. *Am. J. Sci.*, **264**, 78-91.

MABBUTT, J. A., and SCOTT, R. M. 1966. Periodicity of morphogenesis and soil formation in a savannah landscape near Port Moresby, Papua. *Z. Geomorph.*, **10**, 69-89.

MACAR, P. 1955. Appalachian and Ardennes levels of erosion compared. *J. Geol.*, **63**, 253-67.

MACAR, P. 1963. Études récentes sur les pentes et l'évolution des versants en Belgique. *Slopes Comm. Rep.*, **3**, 71-84.

MACAR, P., and FOURNEAU, R. 1960. Relations entre versants et nature du substratum en Belgique. *Slopes Comm. Rep.*, **2**, 124-8.

MACAR, P., and LAMBERT, J. 1960. Relations entre pentes des couches et pentes des versants dans le Condroz (Belgique). *Slopes Comm. Rep.*, **2**, 129-32.

MACAR, P., and PISSART, A. 1964. Études récents sur l'évolution des versants effectuées à l'Université de Liège. *Slopes Comm. Rep.*, **4**, 74-81.

MACAR, P., and PISSART, A. 1966. Recherches sur l'évolution des vèrsants effectuies à l'Université de Liège. *Tijd. Kon. Nederl. Aardrijkskundig*, **83**, 278-88.

MACGREGOR, D. R. 1957. Some observations on the geographical significance of slopes. *Geography*, **42**, 167-73.

MAIGNEN, R. 1966. Review of research on laterites. *UNESCO Nat. Resour. Res.*, 4.

MALAURIE, J. 1960. Gélifraction, éboulis et ruissellement sur le côte nord-ouest du Groenland. *Slopes Comm. Rep.*, **2**, 59-68.

MAMMERICKX, J. 1964. Quantitative observations on pediments in the Mojave and Sonoran deserts (southwestern United States). *Am. J. Sci.*, **262**, 417-35.

MARR, J. E. 1901. The origin of moels, and their subsequent dissection. *Geogr. J.*, **17**, 63-8.

MAW, G. 1866. Notes on the comparative structure of surfaces produced by subaërial and marine denudation. *Geol. Mag.*, **3**, 439-451.

McDOUGALL, I., and GREEN, R. 1958. The use of magnetic measurements for the study of the structure of talus slopes. *Geol. Mag.*, **95**, 252-60.

McGEE, W. J. 1897. Sheetflood erosion. *Bull. geol. Soc. Am.*, **8**, 87-112.

McKEAGUE, J. A., and CLINE, M. G. 1963. Silica in soils. *Adv. Agron.*, **15**, 339-96.

MEADE, R. H. 1969. Errors in using modern stream-load data to estimate natural rates of erosion. *Bull. geol. Soc. Am.*, **80**, 1265-74.

MECKEL, J. F. M., SAVAGE, J. F., and ZORN, H. C. 1964. Determination of slope. *I.T.C. Pubn*, B, 26.

MEIS, M. R. M. DE, and SILVA, J. X. DA. 1968. Mouvements de mass récents à Rio de Janeiro: une étude de géomorphologie dynamique. *Rev. Géomorph. dyn.*, **18**, 145-52.

MELTON, M. A. 1957. An analysis of the relation among elements of climate, surface properties, and geomorphology. *Tech. Rep. No.* 11, *ONR Project NR* 389-442. New York.

MELTON, M. A. 1960. Intravalley variation in slope angles related to microclimate and erosional environment. *Bull. geol. Soc. Am.*, **71**, 133-44.

MELTON, M. A. 1965. Debris-covered hillslopes of the southern Arizona desert—consideration of their stability and sediment contribution. *J. Geol.*, **73**, 715-29.

MERRIAM, R. 1960. Portuguese Bend landslide, Palos Verdes Hills, California. *J. Geol.*, **68**, 140-53.

METCALFE, G. 1950. The ecology of the Cairngorms. Part II. The mountain Callunetum. *J. Ecol.*, **38**, 46-74.

MEYERHOFF, H. A. 1940. Migration of erosional surfaces. *Ann. Ass. Am. Geogr.*, **30**, 247-54.

MEYNIER, A. 1951. Pieds des vache et terracettes. *Rev. Géomorph. dyn.*, **2**, 81-4.

MEYNIER, A. 1965. Torrents et sheetflood les 1-2 Août 1963 à Allassac (Correze). *Rev. Géomorph. dyn.*, **15**, 61-5.

MICHAUD, J. 1950. Emploi de marques dans l'étude des mouvements du sol. *Rev. Géomorph. dyn.*, **1**, 180-94.

MICHAUD, J., and CAILLEUX, A. 1950. Vitesses des mouvements du sol au Chambeyron (Basses-Alps). *C. R. hebd. Séanc. Acad. Sci. Paris*, **230**, 314-15.

MILLER, O. M., and SUMMERSON, C. H. 1960. Slope-zone maps. *Geogr. Rev.*, **50**, 194-202.

MILLER, R., COMMON, R., and GALLOWAY, R. W. 1954. Stone stripes and other surface features of Tinto Hill. *Geogr. J.*, **120**, 216-19.

MILNE, G. 1947. A soil reconnaissance journey through parts of Tanganyika Territory, December 1935 to February 1936. *J. Ecol.*, **35**, 192-265.

MOLCHANOV, A. K. 1967. (On the study of characteristic and limiting slope angles in the southern regions of the Buryat A.S.S.R.) (In Russian). *Metody Geomorfologicheskikh Issledovannii*, 1, 134-43. Eng. transl., RTS 5175, Nat. Lending Lib., Boston Spa, 1969.

MORARIU, T., and GÂRBACEA, V. 1967. Processus d'évolution des versants en Roumanie. *Slopes Comm. Rep.*, **5**, 175-86.

MORAWETZ, S. 1932. Eine Art von Abtragungsvorgang. *Pett. Geogr. Mitt.*, **78**, 231-3.

MORAWETZ, S. 1937. Das problem der Taldichte und Hangzerschneidung. *Pett. Geogr. Mitt.*, **83**, 346-50.

MORAWETZ, S. 1944. Die Eckbildung, eine Frage der Hangzerschneidung. *Pett. Geogr. Mitt.*, **80**, 188-90.

MORAWETZ, S. 1950. Zur frage der Rinnenbildung. *Mitt. Geogr. Ges. Wien.*, **92**, 101-3.

MORAWETZ, S. 1962. Beobachtungen an Rinnen, Racheln und Tobeln. *Z. Geomorph.*, **6**, 260-78.

MORTENSEN, H. 1956. Über Wandverwitterung und Hangabtragung in semiariden und vollariden Gebieten. *Slopes Comm. Rep.*, **1**, 96-104.

MORTENSEN, H. 1959. Warum ist die rezente Formungsintensität in Neuseeland stärker als in Europa? *Z. Geomorph.*, **3**, 98-9.

MORTENSEN, H. 1960. Zur Theorie der Formenentwicklung freier Felswande. *Slopes Comm. Rep.*, **2**, 103-13.

MORTENSEN, H. 1960b. Neues über den Bergrutsch südlich der Mackenröder Spitze und über die Holozäne Hangformung an Schichtstufen im mitteleuropäischen Klimabereich. *Slopes Comm. Rep.*, **2**, 114-23.

MORTENSEN, H., and HÖVERMANN, J. 1956. Der Bergrutsch an der Mackenröder Spitze bei Göttingen. *Slopes Comm. Rep.*, **1**, 149-55.

MOSELEY, H. 1855. The descent of glaciers. *Proc. Roy. Soc. Lond.*, **7**, 333-42.

MOSELEY, H. 1869. On the descent of a solid body on an inclined plane when subjected to alternations of temperature. *London, Edinburgh, and Dublin Phil. Mag. and J. Sci., 4th Ser.*, **38**, 99-118.

MOSS, R. P. 1965. Slope development and soil morphology in a part of south-west Nigeria. *J. soil Sci.*, **16**, 192-209.

MULCAHY, M. J. 1961. Soil distribution in relation to landscape development. *Z. Geomorph.*, **5**, 211-25.

NEEF, E. 1955. Zur Genese des Formenbildes der Rumpfgebirge. *Pett. Geogr. Mitt.*, **99**, 183-92.

NIKIFOROFF, C. C. 1949. Weathering and soil evolution. *Soil Sci.*, **67**, 219-30.

NORTON, E. A., and SMITH, R. S. 1930. Influence of topography on soil profile character. *J. Am. Soc. Agron.*, **22**, 251-62.

NYE, P. H. 1954-5. Some soil forming processes in the humid tropics. I-IV. *J. soil Sci.*, **5**, 7-21, and **6**, 51-83.

ØDUM, H. 1922. On the nature of so-called sheep-tracks. *Medd. Dansk. Geol. Fören.*, **6**, 1-29.

OERTEL, A. C. 1968. Some observations incompatible with clay illuviation. *Trans. Int. Congr. soil Sci.*, **4**, 481-8.

OLLIER, C. D. 1959. A two-cycle theory of tropical pedology. *J. soil Sci.*, **10**, 137-48.

OLLIER, C. D. 1960. The inselbergs of Uganda. *Z. Geomorph.*, **4**, 43-52.

OLLIER, C. D. 1967. Landform description without stage names. *Austral. geogr. Stud.*, **5**, 73-80.

OLLIER, C. D. 1969. *Weathering*. Oliver and Boyd, Edinburgh.

OLLIER, C. D., and THOMASSON, A. J. 1957. Asymmetrical valleys of the Chiltern Hills. *Geogr. J.*, **123**, 71-80.

OLLIER, C. D., and TUDDENHAM, W. G. 1962. Slope development at Coober Pedy, South Australia. *J. geol. Soc. Austral.*, **9**, 91-105.

ORME, A. R. 1962. Abandoned and composite seacliffs in Britain and Ireland. *Ir. Geogr.*, **4**, 279-91.

OTTMANN, L., and TRICART, J. 1964. Application de la cartographie géomorphologique détaillée à l'étude des versants. *Slopes Comm. Rep.*, **4**, 1-16.

OWENS, I. F. 1969. Causes and rates of soil creep in the Chilton Valley, Cass, New Zealand. *Arctic alpine Res.*, **1**, 213-20.

PACHUR, H.-J. 1970. Zur Hangformung im Tibetsigebirge. *Erde, Berl.*, **101**, 41-54.

PACKER, R. W. 1964. Stability slopes in an area of glacial deposition. *Canad. Geogr.*, **8**, 147-54.

PAGET, E. 1961. Value, valuation and the use of land in the West Indies. *Geogr. J.*, **127**, 493-8.

PALLISTER, J. W. 1956. Slope development in Buganda. *Geogr. J.*, **122**, 80-7.

PALLISTER, J. W. 1956b. Slope form and erosion surfaces in Uganda. *Geol. Mag.*, **93**, 465-72.

PALMER, J. 1956. Tor formation at the Bridestones in north-east Yorkshire and its significance in relation to problems of valley-side development and regional glaciation. *Trans. Inst. Br. Geogr.*, **22**, 55-71.

PALMER, J., and RADLEY, J. 1961. Gritstone tors of the English Pennines. *Z. Geomorph.*, **5**, 37-52.

PARIZEK, E. J., and WOODRUFF, J. F. 1956. Apparent absence of soil creep in the East Georgia Piedmont. *Bull. geol. Soc Am.*, **67**, 1111-16.

PARIZEK, E. J., and WOODRUFF, J. F. 1957. Description and origin of stone layers in soils of the southeastern states. *J. Geol.*, **65**, 24-34.

PARIZEK, E. J., and WOODRUFF, J. F. 1957b. A clarification of the definition and classification of soil creep. *J. Geol.*, **65**, 653-6.

PARKER, G. G. 1963. Piping, a geomorphic agent in landform development of the drylands. *Int. Ass. Scient. Hydrol.*, **65**, 103-13.

PATON, J. R. 1964. The origin of the limestone hills of Malaya. *J. trop. Geogr.*, **18**, 134-47.

PÉCSI, M. 1967. The dynamics of Quaternary slope evolution and its geomorphological representation. *Slopes Comm. Rep.*, **5**, 187-99.

PEEL, R. F. 1941. Denudational landforms of the central Libyan Desert. *J. Geomorph.*, **4**, 3-23.

PEEL, R. F. 1960. Some aspects of desert geomorphology. *Geography*, **45**, 241-62.

PEEL, R. F. 1966. The landscape in aridity. *Trans. Inst. Br. Geogr.*, **38**, 1-23.

PEEL, R. F., and PALMER, J. 1956. The formation of the concave (waning) slope at Saltersgate, N.E. Yorkshire. *Slopes Comm. Rep.*, **1**, 131.

PELTIER, L. C. 1950. The geographic cycle in periglacial regions as it is related to climatic morphology. *Ann. Ass. Am. Geogr.*, **40**, 214-36.

PENCK, W. 1924. *Die morphologische Analyse. Ein Kapitel der physikalischen Geologie.* Engelhorns, Stuttgart. English translation, with summaries, by H. Czech and K. C. Boswell, *Morphological analysis of land forms.* Macmillan, London, 1953. Page citations refer to the English translation.

PENCK, W. 1925. Die Piedmontflächen des südlichen Schwartzwaldes. *Z. Gesell. Erdk. Berlin*, 83-108.

PÉWÉ, T. L. (ed.). 1969. *The periglacial environment: past and present.* McGill-Queen's Univ. Press, Montreal.

PIERZCHAŁKO, L. 1954. Zagadniene dolin asymetrycznych na tle rozwoju geomorfologii klimatycznej. (The problem of asymmetric valleys and the development of climatic geomorphology.) *Czas. Geogr.*, **25**, 359-72.

PIPPAN, T. 1963. Beiträge sur Frage der jungen Hangformung und Hangabtragung in den Salzburger Alpen. *Slopes Comm. Rep.*, **3**, 163-83.

PIPPAN, T. 1964. Hangstudien im Fuschertal in den mittleren Hohen Tauern in Salzburg unter besonderer Berücksichtigung der tektonischen und petrographischen Einflüsse auf die Hangbildung. *Slopes Comm. Rep.*, **4**, 136-66.

PISSART, A. 1962. Les versants des vallées de la Meuse et de la Semois à la traversée de l'Ardenne—Classification des formes et essai d'interpretation. *Ann. Soc. géol. Belg.*, **85**, 113-21.

PISSART, A. 1964. Vitesses des mouvements du sol au Chambeyron (Basses Alpes). *Biul. Peryglac.*, **14**, 303-9.

PISSART, A. 1966. Étude de quelques pentes de l'Ile Prince Patrick. *Ann. Soc. géol. Belg.*, **89**, 377-402.

PISSART, A. 1966b. Le rôle géomorphologique du vent dans la région de Mould Bay (Ile Prince Patrick—N. W. T.—Canada). *Z. Geomorph.*, **10**, 226-36.

PISSART, A. 1967. Quelques résultats de l'étude des versants de l'Ile Prince Patrick. *Slopes Comm., Rep.*, **5**, 215-27.

PITTY, A. F. 1966. Some problems in the location and delimitation of slope profiles. *Z. Geomorph.*, **10**, 454-61.

PITTY, A. F. 1967. Some problems in selecting a ground-surface length for slope-angle measurement. *Rev. Géomorph. dyn.*, **17**, 66-71.

PITTY, A. F. 1968. A simple device for the field measurement of hillslopes. *J. Geol.*, **76**, 717-20.

PITTY, A. F. 1968b. Some comments on the scope of slope analysis based on frequency distributions. *Z. Geomorph.*, **12**, 350-5.

PITTY, A. F. 1969. A scheme for hillslope analysis. I. Initial considerations and calculations. *Univ. Hull Occas. Pubn Geogr.*, 9.

PIWOWAR, A. 1903. Ueber Maximalböschungen trockener Schuttkegel und Schutthalden. *Vierteljahrsschrift Naturf. Gesell. Zürich*, 335-9.

PLAYFAIR, J. 1802. *Illustrations of the Huttonian theory of the earth.* Edinburgh.

POSER, H. 1936. Talstudien aus Westspitz-bergen und Östgrönland. *Z. Gletscherk.*, **24**, 43-98.

POTTS, A. S. 1970. Frost action in rocks: some experimental data. *Trans. Inst. Br. Geogr.*, **49**, 109-24.

PRESTON, H. Creep. *Trans. Leeds geol. Ass.*, **18**, 7-9.

PRIOR, D. B., *et al.* 1968. Composite mudflows on the Antrim coast of north-east Ireland. *Geogr. Annlr*, **50*A***, 65-78.

P.T.R.C. 1967. *Digital terrain models. Proceedings of a P.T.R.C. Seminar.* Planning and Transport Research and Computation Co., London.

PUGH, J. C. 1956. Fringing pediments and marginal depressions in the inselberg landscape of Nigeria. *Trans. Inst. Br. Geogr.*, **22**, 15-31.

PUGH, J. C., and BRUNSDEN, D. 1965. Geographical applications of the Ewing stadialtimeter. *Trans. Inst. Br. Geogr.*, **37**, 157-67.

PUTNAM, T. M. 1917. Mathematical forms of certain eroded mountain sides. *Am. math. Mon.*, **24**, 451-3.

RAGG, J. M., and BIBBEY, J. S. 1966. Frost weathering and solifluction products in southern Scotland. *Geogr. Annlr*, **48A**, 12-23.

RAHM, D. A. 1962. The terracette problem. *Northwest Science*, **36**, 65-80.

RAHN, P. H. 1966. Inselbergs and nickpoints in southwestern Arizona. *Z. Geomorph.*, **10**, 217-25.

RAHN, P. H. 1967. Sheetfloods, streamfloods and the formation of pediments. *Ann. Ass. Am. Geogr.*, **57**, 593-604.

RAISZ, E., and HENRY, J. 1937. An average slope map of southern New England. *Geogr. Rev.*, **27**, 467-72.

RAPP, A. 1957. Studien über Schutthalden in Lappland und auf Spitzbergen. *Z. Geomorph.*, **1**, 179-200.

RAPP, A. 1959. Avalanche boulder tongues in Lappland. *Geogr. Annlr*, **41**, 34-48.

RAPP, A. 1960. Recent development of mountain slopes in Kärkevagge and surroundings, northern Scandinavia. *Geogr. Annlr*, **42**, 65-200.

RAPP, A. 1960b. Talus slopes and mountain walls at Tempelfjorden, Spitzbergen. *Skrifter Norsk Polarinst.*, 119.

RAPP, A. 1963. The debris slides at Ulvådal, western Norway. An example of catastrophic slope processes in Scandinavia. *Slopes Comm. Rep.*, **3**, 195-210.

RAPP, A. 1966. Solifluction and avalanches in the Scandinavian mountains. *Proc. Permafrost Int. Conf.* **1963**. *Wash. D.C.*, 150-4.

RAPP, A. 1967. Pleistocene activity and Holocene stability of hillslopes, with examples from Scandinavia and Pennsylvania. *Slopes Comm. Rep.*, **5**, 229-44.

REICHE, P. 1950. A survey of weathering processes and products. *New Mexico Univ. Publ. Geol.*, **3**.

REVUE DE GÉOMORPHOLOGIE DYNAMIQUE. 1967. Field methods for the study of slope and fluvial processes. *Rev. Géomorph. dyn.*, **17**, 145-88.

RICE, R. M., CORBETT, E. S., and BAILEY, R. G. 1969. Soil slips related to vegetation, topography, and soil in southern California. *Water Resour. Res.*, **5**, 647-59.

RICH, J. L. 1938. Recognition and significance of multiple erosion surfaces. *Bull. geol. Soc. Am.*, **49**, 1695-1722.

RICHTER, E. 1901. Geomorphologische Untersuchungen in den Hochalpen. *Pet. Mitt. Erg.-Hefte*, **132**.

ROBERTS, M. 1903. Note on the action of frost on soil. *J. Geol.*, **11**, 314-17.

ROBINSON, G. 1966. Some residual hillslopes in the Great Fish River Basin, South Africa. *Geogr. J.*, **132**, 386-90.

ROHDENBURG, H. and MEYER, B. 1963. Rezente Mikroformung in Kalkgebieten durch inneren Abtrag und die Rolle der periglazialen Gesteinsverwitterung. *Z. Geomorph.*, **7**, 120-46.

ROSE, C. W. 1960. Soil detachment caused by rainfall. *Soil Sci.*, **89**, 28-35.

ROUGERIE, G. 1956. Études des modes d'érosion et du façonnement des versants en Côte d'Ivoire equatoriale. *Slopes Comm. Rep.*, **1**, 136-41.

ROUGERIE, G. 1960. Le façonnement actuel des modelés en Côte d'Ivoire forestière. *Mém. I.F.A.N.*, 58.

RUBEY, W. W. 1928. Gullies in the Great Plains formed by sinking of the ground. *Am. J. Sci.*, **15**, 417-22.

RUDBERG, S. 1958. Some observations concerning mass movement on slopes in Sweden. *Förhandlingar Geol. Föreningens Stockholm*, **80**, 114-25.

RUDBERG, S. 1962. A report on some field observations concerning periglacial geomorphology and mass movement on slopes in Sweden. *Biul. Peryglac.*, **11**, 311-23.

RUDBERG, S. 1963. Morphological processes and slope development in Axel Heidberg Island, Northwest Territories, Canada. *Slopes Comm. Rep.*, **3**, 211-28.

RUDBERG, S. 1964. Slow mass movement processes and slope development in the Norra Storfjäll area, southern Swedish Lappland. *Slopes Comm. Rep.*, **4**, 192-203.

RUHE, R. V. 1950. Graphic analysis of drift topographies. *Am. J. Sci.*, **248**, 435-43.

RUHE, R. V. 1959. Stone lines in soils. *Soil Sci.*, **87**, 223-31.

RUHE, R. V. 1960. Elements of the soil landscape. *Trans. 7th int. Cong. soil Sci.*, **4**, 165-70.

RUHE, R. V., and WALKER, P. H. 1968. Hillslope models and soil formation. I. Open systems. II. Closed systems. *Trans. 9th int. Cong. soil Sci.*, **4**, 551-60, and 561-8.

RUXTON, B. P. 1958. Weathering and subsurface erosion in granite at the piedmont angle, Balos, Sudan. *Geol. Mag.*, **95**, 353-77.

Ruxton, B. P. 1967. Slopewash under primary rainforest in northern Papua. In Jennings and Mabutt (1967), *q.v.*, 85-94.

Ruxton, B. P., and Berry, L. 1957. Weathering of granite and associated erosional features in Hong Kong. *Bull. geol. Soc. Am.*, **68**, 1263-92.

Ruxton, B. P., and Berry, L. 1961. Weathering profiles and geomorphic position on granite in two tropical regions. *Rev. Géomorph. dyn.*, **12**, 16-31.

Sapper, K. 1935. Geomorphologie der feuchten Tropen. *Geogr. Schriften*, 7.

Savigear, R. A. G. 1952. Some observations on slope development in South Wales. *Trans. Inst. Br. Geogr.*, **18**, 31-51.

Savigear, R. A. G. 1956. Technique and terminology in the investigation of slope forms. *Slopes Comm. Rep.*, **1**, 66-75.

Savigear, R. A. G. 1960. Slopes and hills in West Africa. *Slopes Comm. Rep.*, **2**, 156-71.

Savigear, R. A. G. 1962. Some observations on slope development in north Devon and north Cornwall. *Trans. Inst. Br. Geogr.*, **31**, 23-42.

Savigear, R. A. G. 1963. The morphological basis of geomorphological mapping. *Deltion tis Ellinikis Geogr. Etaireias*, **4**, 104-11.

Savigear, R. A. G. 1965. A technique of morphological mapping. *Ann. Ass. Am. Geogr.*, **55**, 514-38.

Savigear, R. A. G. 1967. The analysis and classification of slope profile forms. *Slopes Comm. Rep.*, **5**, 271-90.

Scheidegger, A. D. 1960. Analytical theory of slope development by undercutting. *J. Alberta Soc. petroleum Geologists*, **8**, 202-6.

Scheidegger, A. E. 1961. *Theoretical geomorphology*. Springer, Berlin.

Scheidegger, A. E. 1961b. Mathematical models of slope development. *Bull. geol. Soc. Am.*, **72**, 37-50.

Scheidegger, A. E. 1961c. Theory of rock movement on scree slopes. *J. Alberta Soc. petroleum Geologists*, **9**, 131.

Scheidegger, A. E. 1964. Lithologic variations in slope development theory. *U.S. geol. Surv. Circular*, (485).

Schmid, J. 1955. *Der Bodenfrost als morphologischer Faktor*. Huthig, Heidelberg.

Schmitthenner, H. 1925. Die Entstehung der Dellen und ihre morphologische Bedeutung. *Z. Geomorph.*, **1**, 3-28.

Schultz, J. R. 1955. Canons of landscape evolution—a discussion. *Bull. geol. Soc. Am.*, **66**, 1207-12.

Schultze, J. H. 1951. Über das Verhaltnis zwischen Denudation und Bodenerosion. *Erde, Berl.*, **3**, 220-32.

Schumm, S. A. 1956. Evolution of drainage systems and slopes in badlands at Perth Amboy, New Jersey. *Bull. geol. Soc. Am.*, **67**, 597-646.

Schumm, S. A. 1956b. The role of creep and rainwash on the retreat of badland slopes. *Am. J. Sci.*, **254**, 693-706.

Schumm, S. A. 1962. Erosion on miniature pediments in Badlands National Monument, South Dakota. *Bull. geol. Soc. Am.*, **73**, 719-24.

Schumm, S. A. 1963. The disparity between present rates of denudation and orogeny. *U.S.G.S. Prof. Paper*, (454-*H*).

Schumm, S. A. 1964. Seasonal variations of erosion rates and processes on hillslopes in western Colorado. *Slopes Comm. Rep.*, **4**, 215-38.

Schumm, S. A. 1967. Rates of surficial rock creep on hillslopes in western Colorado. *Science*, **155**, 560-1.

Schumm, S. A., and Chorley, R. J. 1964. The fall of threatening rock. *Am. J. Sci.*, **262**, 1041-54.

Schumm, S. A., and Chorley, R. J. 1966. Talus weathering and scarp recession in the Colorado Plateaus. *Z. Geomorph.*, **10**, 11-36.

Schumm, S. A., and Hadley, R. F. 1957. Arroyos and the semi-arid cycle of erosion. *Am. J. Sci.*, **225**, 161-174.

Schumm, S. A., and Lichty, R. W. 1965. Time, space and causality in geomorphology. *Am. J. Sci.*, **263**, 110-19.

Schumm, S. A., and Lusby, G. C. 1963. Seasonal variation of infiltration capacity and runoff on hillslopes in western Colorado. *J. geophys. Res.*, **68**, 3655-66.

Schweinfurth, U. 1966. Über eine besondere Form der Hangabtragung im Neuseeländischen Fjordland. *Z. Geomorph.*, **10**, 144-9.

Scrope, G. P. 1866. On the origin of valleys. *Geol. Mag.*, **3**, 193-9.

Selby, M. J. 1966. Methods of measuring soil creep. *J. Hydrol. (N.Z.)*, **5**, 54-63.

Selby, M. J. 1966b. Some slumps and boulder fields near Whitehall. *ibid.*, pp. 35-44.

Selby, M. J. 1968. Cones for measuring soil creep. *J. Hydrol. (N.Z.)*, **7**, 136-7.

SERET, G. 1963. Essai de classification des pentes en Famenne. *Z. Geomorph.*, **7**, 71-85.

SHARP, R. P. 1957. Geomorphology of the Cima Dome, Mojave Desert, California. *Bull. geol. Soc. Am.*, **68**, 273-90.

SHARPE, C. F. S. 1938. *Landslides and related phenomena.* Columbia Univ. Press, New York. Reprinted 1968.

SHARPE, C. F. S. 1941. Geomorphic aspects of normal and accelerated erosion. *Trans. Am. geophys. Un.*, **22**, 236-40.

SHREVE, R. L. 1968. The Blackhawk landslide. *Geol. Soc. Am., Spec. Paper*, 108.

SIMONETT, D. S. 1967. Landslide distribution and earthquakes in the Bewani and Torricelli mountains, New Guinea. In Jennings and Mabbutt (1967), *q.v.*, 64-84.

SIMONS, M. 1962. The morphological analysis of landforms: a new review of the work of Walther Penck (1888-1923). *Trans. Inst. Br. Geogr.*, **31**, 1-14.

SKEMPTON, A. W. 1948. The rate of softening of stiff, fissured clays. *Proc. 2nd int. Conf. soil Mech. foundation Engng, Rotterdam, 2*, 50-3.

SKEMPTON, A. W. 1953. Soil mechanics in relation to geology. *Proc. Yorks geol. Soc.*, **29**, 33-62.

SKEMPTON, A. W. 1964. Long-term stability of clay slopes. *Géotechnique*, **14**, 77-102.

SKEMPTON, A. W., and DELORY, F. A. 1957. Stability of natural slopes on London Clay. *Proc. 4th int. Conf. soil Mech. foundation Engng*, **2**, 378-81.

SLAYMAKER, H. O. 1967. Patterns of subaerial erosion in instrumented catchments, with particular reference to the upper Wye Valley, mid-Wales. *Ph.D. thesis 6224, Cambridge University.*

SMALL, R. J. 1964. The escarpment dry valleys of the Wiltshire Chalk. *Trans. Inst. Br. Geogr.*, **34**, 33-52.

SMALL, R. J., CLARK, M. J., and LEWIN, J. 1970. The periglacial rock-stream at Clatford Bottom, Marlborough Downs, Wiltshire. *Proc. geol. Ass.*, **81**, 87-98.

SMALLEY, I. J., and TAYLOR, R. L. S. 1970. Loess—the yellow earth. *Science J.*, **6**, 28-33.

SMITH, D. D., and WISCHMEIER, W. H. 1962. Rainfall erosion. *Adv. Agron.*, **14**, 109-48.

SMITH, H. T. U. 1949. Physical effects of Pleistocene climatic changes in non-glaciated areas. *Bull. geol. Soc. Am.*, **60**, 1485-616.

SMITH, K. G. 1958. Erosional processes and landforms in Badlands National Monument, South Dakota. *Bull. geol. Soc. Am.*, **69**, 975-1008.

SMITH, R. M., and STAMEY, W. L. 1965. Determining the range of tolerable erosion. *Soil Sci.*, **100**, 414-24.

SOONS, J. M., and RAYNER, J. N. 1968. Microclimate and erosion processes in the Southern Alps. *Geogr. Annlr*, **50**, 1-15.

SORBY, H. C. 1850. On the excavation of the valleys in the Tabular Hills, as shown by the configuration of Yedmandale, near Scarbro'. *Proc. Yorks geol. Soc.*, **3**, 169-72.

SOUCHEZ, R. 1961. Théorie d'une évolution des versants. *Bull. Soc. R. Belg. géogr.*, **85**, 7-18.

SOUCHEZ, R. 1963. Evolution des versants et théorie de la plasticité. *Rev. Belg. géogr.*, **87**, 10-94.

SOUCHEZ, R. 1964. Viscosité, plasticité et rupture dans l'évolution des versants. *Ciel et Terre*, **80**, 3-24.

SOUCHEZ, R. 1966. Slow mass-movement and slope evolution in coherent and homogeneous rocks. *Bull. Soc. Belg. geol.*, **74**, 189-213.

SOUCHEZ, R. 1967. Gélivation et évolution des versants en bordure de l'inlandsis d'Antarctide orientale. *Slopes Comm. Rep.*, **5**, 291-8.

SPARROW, G. W. A. 1965. Observations on slope formation in the Drakensberg and foothills of Natal and East Griqualand. *J. for Geogr.*, **2**, 23-5.

SPARROW, G. W. A. 1966. Some environmental factors in the formation of slopes. *Geogrl J.*, **132**, 390-5.

SPEIGHT, J. G. 1967. Explanation of land system descriptions. In Scott, R. M. *et al. Lands of Bougainville and Buka Islands, Territory of Papua and New Guinea.* C.S.I.R.O., Melbourne, 174-84.

SPEIGHT, J. G. 1968. Parametric description of land form. In Stewart, G. A. (ed.) *Land evaluation.* Macmillan of Australia, 239-50.

SPEIGHT, J. G. 1971. Log-normality of slope distributions. *Z. Geomorph.*, **15**, 290-311.

SPREITZER, H. 1960. Hangformung und Asymmetrie der Bergrücken in den Alpen und im Taurus. *Slopes Comm. Rep.*, **2**, 211-36.

STARKEL, L. 1962. Stan badan nad wspolczesnymi procesami morfogenctycznymi w Karpatach. *Czas. Geogr.*, **33**, 459-73.

STARKEL, L. 1964. The differences in the slope formation of eastern Flysch Carpathians during the Upper Pliocene and the Quaternary. *Slopes Comm. Rep.*, **4**, 107-17.

STEARN, N. H. 1935. Structure and creep. *J. Geol.*, **43**, 323-7.

STEWART, G. A. (ed.). 1968. *Land evaluation.* Macmillan of Australia.

STODDART, D. R. 1969. World erosion and sedimentation. In Chorley (1969), *q.v.*, 43-64.

STODDART, D. R. 1969b. Climatic geomorphology: review and assessment. *Prog. Geogr.*, **1**, 161-222.

STONE, R. O., and DUGUNDJI, J. 1965. A study of microrelief—its mapping, classification and quantification by means of a Fourier analysis. *Eng. Geol.*, **1**, 89-187.

STRAHLER, A. N. 1950. Equilibrium theory of slopes approached by frequency distribution analysis. *Am. J. Sci*, 248, 800-14.

STRAHLER, A. N. 1952. Dynamic basis of geomorphology. *Bull. geol. Soc. Am.*, **63**, 923-38.

STRAHLER, A. N. 1958. Dimensional analysis applied to fluvially eroded landforms. *Bull. geol. Soc. Am.*, **69**, 279-300.

STRATIL-SAUER, G. 1931. Die Tilke. *Z. Geomorph.*, **6**, 255-86.

SWAN, S. B. ST. C. 1970. Relationships between regolith, lithology, and slope in a humid tropical region: Johor, Malaya. *Trans. Inst. Br. Geogr.*, **51**, 189-200.

SWAN, S. B. ST. C. 1970b. Analysis of residual terrain: Johor, Malaya. *Ann. Ass. Am. Geogr.*, **60**, 124-33.

SWAN, S. B. ST. C. 1970c. Landforms of the humid tropics: Johor, Malaya. A quantitative analysis of the landforms of a rainforest area. *Ph.D. thesis, University of Sussex.*

SWEETING, M. M. 1958. The karstlands of Jamaica. *Geogrl J.*, **124**, 184-99.

SWEETING, M. M. 1966. The weathering of limestone. With particular reference to the Carboniferous Limestones of northern England. In Dury, G. H. (ed.) *Essays in geomorphology.* Heinemann, London, 177-210.

SZUPRYCZYŃSKI, J. 1967. Die Entwicklung kleiner Rezenter Erosionstäler an den Stufen des Wda-Sanders (Polen). *Slopes Comm. Rep.*, **5**, 299-303.

TABER, S. 1929. Frost heaving. *J. Geol.*, **37**, 428-61.

TABER, S. 1930. The mechanics of frost heaving. *J. Geol.*, **38**, 303-17.

TALBOT EDWARDS, A. C. 1948. Zomba flood, December, 1946. *Nyasaland J.*, **1**, 53-63.

TATOR, B. A. 1952-3. Pediment characteristics and terminology. *Ann. Ass. Am. Geogr.*, **42**, 295-317, and **43**, 47-53.

TAYLOR, D. W. 1948. *Fundamentals of soil mechanics.* Wiley, New York.

TAYLOR, J. A. 1960. Methods of soil study. *Geography*, **45**, 52-67.

TERZAGHI, K. 1950. Mechanism of landslides. *Geol. Soc. Am., Berkey Vol.*, 83-123.

TERZAGHI, K. 1958. Landforms and subsurface drainage in the Gačka region in Yugoslavia. *Z. Geomorph.*, **2**, 76-100.

TERZAGHI, K. 1962. Stability of steep slopes on hard unweathered rock. *Géotechnique*, **12**, 251-271.

THOMAS, A. S. 1959. Sheep paths. *J. Br. grassland Soc.*, **14**, 157-64.

THOMAS, M. F. 1965. Some aspects of the geomorphology of tors and domes in Nigeria. *Z. Geomorph.*, **9**, 63-81.

THOMAS, M. F. 1966. Some geomorphological implications of deep weathering patterns in crystalline rocks in Nigeria. *Trans. Inst. Br. Geogr.*, **40**, 173-93.

THOMAS, M. F. 1966b. The origin of bornhardts. *Z. Geomorph.*, **10**, 478-80.

THOMAS, M. F. 1967. A bornhardt dome in the plains near Oyo, western Nigeria. *Z. Geomorph.*, **11**, 239-61.

THOMAS, T. M. 1956. Gully erosion in the Brecon Beacons area, South Wales. *Geography*, **41**, 99-107.

THOMSON, C. W. 1877. The movement of the soil cap. *Nature, Lond.*, **15**, 359-60.

THORNBURY, W. D. 1954. *Principles of geomorphology.* Wiley, New York.

THROWER, N. J. W. 1960. Cyprus—a landform study. *Ann. Ass. Am. Geogr., Map Supplement 1.*

TINKLER, K. J. 1966. Slope profiles and scree in the Eglwyseg Valley, North Wales. *Geogr. J.*, **132**, 379-85.

TIVY, J. 1957. Influence des facteurs biologiques sur l'érosion dans les Southern Uplands Écossais. *Rev. Géomorph. dyn.*, **8**, 9-19.

TIVY, J. 1962. An investigation of certain slope deposits in the Lowther Hills, Southern Uplands of Scotland. *Trans. Inst. Br. Geogr.*, **30**, 59-73.

TJIA, H. D. 1969. Slope development in tropical karst. *Z. Geomorph.*, **13**, 260-6.

TOWNSHEND, J. R. G. 1970. Geology, slope form and process and their relation to the occurrence of laterite in the Mato Grosso. *Geogr. J.*, **136**, 392-9.

TRENDALL, A. F. 1962. The formation of 'apparent peneplains' by a process of combined laterisation and surface wash. *Z. Geomorph.*, **6**, 183-97.

TREWARTHA, G. T., and SMITH, G. H. 1941. Surface configuration of the driftless cuestaform hill land. *Ann. Ass. Am. Geogr.*, **31**, 25-45.

TRICART, J. 1957. Mise au point: l'évolution des versants. *L'Information Géogr.*, **21**, 108-16.

TRICART, J. 1970. *Geomorphology of cold environments*. Eng. translation by E. Watson. Macmillan, London.

TRICART, J., and CAILLEUX, A. 1965. *Le modelé des régions chaudes, forêts et savanes*. Soc. d'Édition d'Enseignment Supérieur, Paris.

TRICART, J. *et al.* 1961. Mécanismes normaux et phénomènes catastrophiques dans l'évolution des versants du bassin du Guil. *Z. Geomorph.*, **5**, 277-301.

TROEH, F. R. 1964. Landform parameters connected to soil drainage. *Proc. soil Sci. Soc. Am.*, **28**, 808-12.

TROEH, F. R. 1965. Landform equations fitted to contour maps. *Am. J. Sci.*, **263**, 616-27.

TROLL, C. 1943. Die Frostwechselhaufigkeit in den Luft- und Bodenklimaten der Erde. *Met. Z.*, **60**, 161-71.

TROLL, C. 1944. Strukturboden, solifluktion und Frostklimate der Erde. *Geol. Rundschau*, **34**, 545-694.

TROLL, C. 1947. Die Formen der Solifluktion und die periglaziale Bodenabtragung. *Erdkunde*, **1**, 162-75.

TROLL, C. 1948. Der subnivale oder periglaziale Zyklus der Denudation. *Erdkunde*, **2**, 1-21.

TWIDALE, C. R. 1956. Vallons de gélivation dans le centre du Labrador. *Rev. Géomorph. dyn.*, **7**, 17-23.

TWIDALE, C. R. 1962. Steepened margins of inselbergs from north-western Eyre Peninsula, South Australia. *Z. Geomorph.*, **6**, 51-69.

TWIDALE, C. R. 1964. A contribution to the general theory of domed inselbergs. *Trans. Inst. Br. Geogr.*, **34**, 91-113.

TWIDALE, C. R. 1967. Origin of the piedmont angle as evidenced in South Australia. *J. Geol.*, **75**, 393-411.

TWIDALE, C. R. 1967b. Hillslopes and pediments in the Flinders Ranges, South Australia. In Jennings and Mabbutt (1967), *q.v.*, 95-117.

TYLOR, A. 1875. Action of denuding agencies. *Geol. Mag.*, **22**, 433-73.

VAN BURKALOW, A. 1945. Angle of repose and angle of sliding friction: an experimental study. *Bull. geol. Soc. Am.*, **56**, 669-707.

VAN DIJK, W., and LE HEUX, J. W. N. 1952. Theory of parallel rectilinear slope recession. I and II. *K. Nederl. Akad. Wetens. Proc.*, *B*, **55**, 115-22, and 123-9.

VAN LOPIK, J. R., and KOLB, C. R. 1959. A technique for preparing desert terrain analogs. *Tech. Rep. U.S. Army Waterways Expt. Stn.*, Vicksburg, 3-506.

VARNES, D. J. 1958. Landslide types and processes. In Eckel, E. B. (ed.) Landslides and engineering practice. *Highway Res. Board, Spec. Rep.*, 29, 20-47. Washington, D.C.

VERSTAPPEN, H. T., and VAN ZUIDAM, R. A. 1968. *ITC textbook of photo-interpretation. Vol. VII. Use of aerial photographs in geomorphology. Chapter VII.2: ITC system of geomorphological survey.* Delft.

VITA-FINZI, C. 1964. Slope downwearing by discontinuous sheetwash in Jordan. *Israel J. earth Sci.*, **13**, 88-91.

VON BERTALANFFY, L. 1950. An outline of general system theory. *Br. J. phil. Sci.*, **1**, 134-65.

VON ENGELN, O. D. 1942. *Geomorphology*. Macmillan, New York.

VON ENGELN, O. D. *et al.* 1940. Symposium: Walter Penck's contribution to geomorphology. *Ann. Ass. Am. Geogr.*, **30**, 219-84.

VON FREYBURG, B. 1923. Die Tertiaren Landoberflächen in Thuringien. *Fortsch. Geol. Paläont.*, **6**, 30.

VON FREYBURG, B. 1932. Geomorphologische Probleme des Thuringer Waldes. *Z. Geomorph.*, **7**, 197.

VON ŁOZINSKI, W. 1909. Über die mechanische Verwitterung der Sandsteine im gemässigten Klima. *Bull. Acad. Sci. Cracovie, Klasse Sci. Math. Nat.*, 1-25.

WAHRHAFTIG, C. 1965. Stepped topography of the southern Sierra Nevada, California. *Bull. geol. Soc. Am.*, **76**, 1165-90.

WALKER, P. G., HALL, G. F., and PROTZ, R. 1968. Relation between landform parameters and soil properties. *Proc. soil Sci. Am.*, **32**, 101-4.

WALKER, P. H. 1962. Soil layers on hillslopes: a study at Nowra, N.S.W., Australia. *J. soil Sci.*, **13**, 167-77.

WALTHER, J. 1915. Über den Laterit in Westaustralien. *Z. Deutsch. Geol. Ges.*, **67***B*, 113-40.

WARD, W. H. 1945. The stability of natural slopes. *Geogr. J.*, **105**, 170-97.

WASHBURN, A. L. 1956. Classification of patterned ground and review of suggested origins. *Bull. geol. Soc. Am.*, **67**, 823-65.

WASHBURN, A. L. 1967. Instrumental observations of mass-wasting in the Mesters Vig District, northeast Greenland. *Medd. om Grønland*, 166 (4).

WASHBURN, A. L., and GOLDTHWAIT, R. P. 1958. Slushflows. *Bull. geol. Soc. Am.*, **69**, 1657-8.

WATERS, R. S. 1958. Morphological mapping. *Geography*, **43**, 10-17.

WATSON, E. 1965. Periglacial structures in the Aberystwyth region of central Wales. *Proc. geol. Ass.*, **76**, 443-62.

WATSON, E., and WATSON, S. 1970. The coastal periglacial slope deposits of the Cotentin Peninsula. *Trans. Inst. Br. Geogr.*, **49**, 125-44.

WATT, A. S., and JONES, E. W. 1948. The ecology of the Cairngorms. Part I. The environment and the altitudinal zonation of the vegetation. *J. Ecol.*, **36**, 283-303.

WEBSTER, R., and BECKETT, P. H. T. 1968. Quality and usefulness of soil maps. *Nature, Lond.*, **219**, 680-2.

WEBSTER, R., and WONG, I. F. T. 1969. A numerical procedure for testing soil boundaries interpreted from air photographs. *Photogrammetria*, **24**, 59-72.

WEISCHET, W. 1968. Zur Geomorphologie des Glatthang-Reliefs in der ariden Subtropenzone des Kleinen Nordens von Chile. *Z. Geomorph.*, **13**, 1-21.

WENTWORTH, C. E. 1943. Soil avalanches in Oahu, Hawaii. *Bull. geol. Soc. Am.*, **54**, 53-64.

WENTWORTH, C. K. 1930. A simplified method of determining the average slope of land surfaces. *Am. J. Sci., Ser. 5*, **20**, 184-94.

WHIPKEY, R. Z. 1965. Subsurface storm flow from forested slopes. *Bull. int. Ass. scient. Hydrol.*, **10**, 74-85.

WHITAKER, W. 1867. On subaërial denudation, and on cliffs and escarpments of the Chalk and Lower Tertiary beds. *Geol. Mag.*, **4**, 447-54 and 483-93.

WHITE, S. E. 1949-50. Processes of erosion on steep slopes of Oahu, Hawaii. *Am. J. Sci.*, **247**, 168-86, and **248**, 508-15.

WHITTEN, E. H. T. 1964. Process-response models in geology. *Bull. geol. Soc. Am.*, **75**, 455-64.

WILHELMY, H. 1958. *Klimamorphologie der Massengesteine.* Westermann, Braunschweig.

WILLIAMS, M. A. J. 1968. Termites and soil development near Brocks Creek, Northern Territory. *Austr. J. Sci.*, **31**, 153-4.

WILLIAMS, M. A. J. 1969. Rates of slopewash and soil creep in northern and southeastern Australia. *Ph.D. thesis, Austr. Nat. Univ.*

WILLIAMS, M. A. J. 1969b. Prediction of rainsplash erosion in the seasonally wet tropics. *Nature, Lond.*, **222**, 763-5.

WILLIAMS, M. A. J., and HALL, D. N. 1965. Recent expeditions to Libya from the Royal Military Academy, Sandhurst. *Geogr. J.*, **131**, 482-501.

WILLIAMS, P. J. 1957. The direct recording of solifluction movements. *Am. J. Sci.*, **255**, 705-715.

WILLIAMS, P. J. 1959. An investigation into processes occurring in solifluction. *Am. J. Sci.*, **257**, 481-90.

WILLIAMS, P. J. 1961. Climatic factors controlling the distribution of certain frozen ground phenomena. *Geogr. Annlr*, **43**, 339-47.

WILLIAMS, P. J. 1962. Quantitative investigations of soil movement in frozen ground phenomena. *Biul. Peryglac.*, **11**, 353-60.

WILLIAMS, P. J. 1966. Downslope soil movement at a sub-arctic location with regard to variations with depth. *Canad. Geotech. J.*, **3**, 191-203.

WILLIAMS, R. B. G. 1968. Some estimates of periglacial erosion in southern and eastern England. *Biul. Peryglac.*, **17**, 311-35.

WILLIAMS, W. W. 1956. An east coast survey: some recent changes in the coast of East Anglia. *Geogr. J.*, **122**, 317-34.

WILSON, G. 1952. The influence of rock structures on coast-line and cliff development around Tintagel, north Cornwall. *Proc. geol. Ass.*, **63**, 20-48.

WIMAN, S. 1963. A preliminary study of experimental frost weathering. *Geogr. Annlr*, **45**, 113-21.

WOLFE, P. E. 1943. Soils and subsequent topography. *J. Geol.*, **51**, 204-11.

WOLMAN, M. G., and MILLER, J. P. 1960. Magnitude and frequency of forces in geomorphic processes. *J. Geol.*, **68**, 54-74.

WOOD, A. 1942. The development of hillside slopes. *Proc. geol. Ass.*, **53**, 128-40.

WOOLNOUGH, W. G. 1918. The physiographic significance of laterite in Western Australia. *Geol. Mag.*, **65**, 385-93.

WOOLNOUGH, W. G. 1930. The influence of climate and topography on the formation and distribution of products of weathering. *Geol. Mag.*, **67**, 123-32.

WURM, A. 1936. Morphologische Analyse und Experiment. I. Schichtstufenlandschaft. II. Hangentwicklung, Einebnung, Piedmonttreppen. *Z. Geomorph.*, **9**, 1-24, and 57-87.

WYNNE, A. B. 1867. On denudation with reference to the configuration of the ground. *Geol. Mag.*, **4**, 3-11.

YOUNG, A. 1956. Scree profiles in west Norway. *Slopes Comm. Rep.*, **1**, 125.

YOUNG, A. 1958. A record of the rate of erosion on Millstone Grit. *Proc. Yorks geol. Soc.*, **31**, 149-56.

YOUNG, A. 1958b. Some considerations of slope form and development, regolith, and denudational processes. *Ph.D. thesis, Sheffield University.*

YOUNG, A. 1960. Soil movement by denudational processes on slopes. *Nature, Lond.*, **188**, 120-2.

YOUNG, A. 1961. Characteristic and limiting slope angles. *Z. Geomorph.*, **5**, 126-31.

YOUNG, A. 1963. Some field observations of slope form and regolith and their relation to slope development. *Trans. Inst. Br. Geogr.*, **32**, 1-29.

YOUNG, A. 1963b. Soil movement on slopes. *Nature, Lond.*, **200**, 129-30.

YOUNG, A. 1963c. Deductive models of slope evolution. *Slopes Comm. Rep.*, **3**, 45-66.

YOUNG, A. 1964. Slope profile analysis. *Slopes Comm. Rep.*, **4**, 17-27.

YOUNG, A. 1964b. In discussion on: Slope profiles: a symposium. *Geogr. J.*, **130**, 80-2.

YOUNG, A. 1968. Slope form and the soil catena in savanna and rainforest environments. *Br. geomorph. Res. Group Occas. Paper*, (5), 3-12.

YOUNG, A. 1969. Present rate of land erosion. *Nature, Lond.*, **224**, 851-2.

YOUNG, A. 1969b. The accumulation zone on slopes. *Z. Geomorph.*, **13**, 231-3.

YOUNG, A. 1970. Concepts of equilibrium, grade and uniformity as applied to slopes. *Geogr. J.*, **136**, 585-92.

YOUNG, A. 1970b. Slope form in part of the Mato Grosso, Brazil. *Geogr. J.*, **136**, 383-92.

YOUNG, A. 1971. Slope profile analysis: the system of best units. *Inst. Br. Geogr. Spec. Pubn*, 3, 1-13.

ZAKRZEWSKA, B. 1967. Trends and methods in land form geography. *Ann. Ass. Am. Geogr.*, **57**, 128-65.

ZEITLINGER, J. 1959. Beobachtungen über unterirdische erosion in Verwitterungslehm. *Mitt. Österreich Geogr. Gesell.*, **101**, 94-5.

INDEX

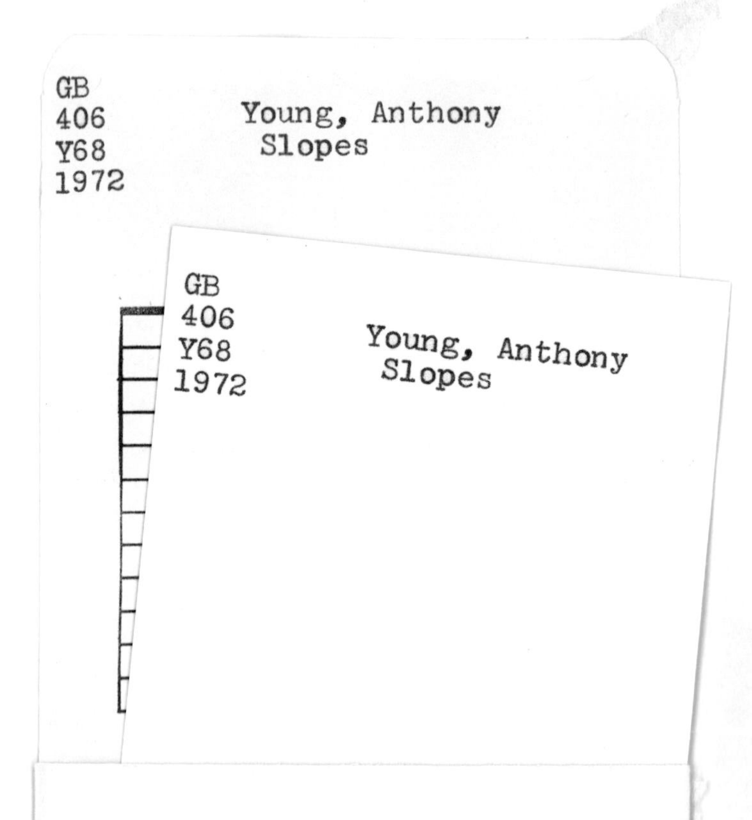
GB
406
Y68
1972
Young, Anthony
Slopes
GB
406
Y68
1972
Young, Anthony
Slopes
COLLINS LIBRARY
UNIVERSITY OF PUGET SOUND
Tacoma, Washington
DEMCO